Contemporary Topics in
Analytical and Clinical Chemistry

Volume 1

Contemporary Topics in
Analytical and
Clinical Chemistry

Volume 1

Edited by
David M. Hercules
University of Pittsburgh
Pittsburgh, Pennsylvania

Gary M. Hieftje
Indiana University
Bloomington, Indiana

Lloyd R. Snyder
Technicon Instruments Corporation
Tarrytown, New York

and
Merle A. Evenson
University of Wisconsin
Madison, Wisconsin

PLENUM PRESS • NEW YORK AND LONDON

Library of Congress Cataloging in Publication Data

Main entry under title:

Contemporary topics in analytical and clinical chemistry.

Includes bibliographical references and index.
I. Hercules, David M.
RB40.C66   616.07'56   77-8099
ISBN 0-306-33521-2 (v. 1)

©1977 Plenum Press, New York
A Division of Plenum Publishing Corporation
227 West 17th Street, New York, N.Y. 10011

Printed in the United States of America

Contributors

S. G. Chang Energy and Environment Division, Lawrence Berkeley Laboratory, University of California, Berkeley, California 94720

Shung-Ho Chang Department of Biochemistry, Purdue University, West Lafayette, Indiana 47907

Edward G. Codding Department of Chemistry, University of Calgary, Calgary, Alberta, Canada T2N 1N4

R. L. Dod Energy and Environment Division, Lawrence Berkeley Laboratory, University of California, Berkeley, California 94720

Karen M. Gooding Department of Biochemistry, Purdue University, West Lafayette, Indiana 47907

Robert L. Grob Department of Chemistry, Villanova University, Villanova, Pennsylvania 19085

Gary Horlick Department of Chemistry, University of Alberta, Edmonton, Alberta, Canada T6E 2E1

Michael P. Neary Department of Chemistry, University of Georgia, Athens, Georgia 30602

T. Novakov Energy and Environment Division, Lawrence Berkeley Laboratory, University of California, Berkeley, California 94720

Fred E. Regnier Department of Biochemistry, Purdue University, West Lafayette, Indiana 47907

W. Rudolf Seitz Department of Chemistry, University of Georgia, Athens, Georgia 30602

Bernard E. Statland   Department of Hospital Laboratories, North
                              Carolina Memorial Hospital, Chapel Hill,
                              North Carolina 27514

Per Winkel   Department of Hospital Laboratories, North
                              Carolina Memorial Hospital, Chapel Hill,
                              North Carolina 27514

Preface

Any addition to the ever-expanding list of scientific publications requires careful consideration and justification. There are already numerous journals in analytical and clinical chemistry adequate for the publication of research results. There does remain a need for a series focused on analytical and clinical chemistry, to provide an overview of instrumental developments relevant to the needs of analytical and clinical chemists. This is the role intended for the present series.

Although the title specifically indicates that the series will deal with analytical and clinical chemistry, our intention is that it will deal with analytical chemistry as related to other areas, such as air and water pollution, oceanography, earth sciences, and various aspects of biomedical science and technology.

It seems appropriate to publish two types of articles in the series. First, we will provide a forum for authoritative, critical reviews for the expert, to enable him to cope with the ever-growing problem of keeping abreast of rapid developments in his own and immediately related fields. In this way we hope the series will stimulate new ideas for research by being at the cutting edge of science. Second, we will publish articles written by experts in the fields being covered but primarily intended for the nonexpert, thereby providing him with some overview of the area. Thus we hope to cross-fertilize research areas, while at the same time serving an educational function. In this sense the series will include articles not only on currently fashionable analytical techniques but on state-of-the-art advances in other areas certain to assume analytical importance. Here special emphasis will be placed on making the topics useful to the practicing experimentalist and indicating why projected analytical impact is predicted.

We will seek to have an international authorship so that the series will accurately reflect the global status of analytical chemistry.

The general format of articles will include a review of the fundamentals of a topic, a description of instrumentation, critical presentation of some interesting applications, and educated speculation as to where and how future improvements of the technique or amplification of its applications may develop. In a series such as this dealing with analytical and clinical chemistry there is always the danger that it may become too narrowly directed or too diffuse. We believe that a balance between these two extremes will best serve the analytical chemistry community. The value of the series will be directly reflected in how close we have come to achieving this goal.

The Editors

Contents

4. Photodiode Arrays for Spectrochemical Measurements   195

Gary Horlick and Edward G. Codding

5. Application of ESCA to the Analysis of Atmospheric Particulates   249

T. Novakov, S. G. Chang, and R. L. Dod

6. Using the Subject as His Own Referent in Assessing Day-to-Day Changes of Laboratory Test Results   287

Per Winkel and Bernard E. Statland

High-Speed Liquid Chromatography of Proteins

Fred E. Regnier, Karen M. Gooding, and Shung-Ho Chang

1. Introduction

1.1. Classical Chromatography of Proteins

A discussion of protein separation should be prefaced with a review of our general knowledge of protein structure. Proteins are biopolymers composed of an ordered sequence of acidic, basic, neutral, and hydrophobic amino acids coupled via peptide bonds. The bending and folding of this primary chain results in a secondary structure that is maintained primarily by internal hydrogen bonding. Further stabilization of the protein occurs by hydrogen, ionic, hydrophobic, covalent, and van der Waals linkages between amino acid side chains. The net result of these secondary and tertiary modifications along the polypeptide chain is a three-dimensional matrix with some amino acids buried in the interior of the protein and others exposed at the surface. As a result, proteins have specific shapes and sizes; in addition, they have areas that may be anionic, cationic, and/or hydrophobic depending on the amino acid sequence. These differences in properties make possible the chromatographic resolution of proteins.

Classical separations of proteins have been achieved on supports of the gel type that are usually composed of carbohydrates or polyacrylamide with a stationary phase immobilized in the gel matrix.[1] It is important that the support matrix be hydrophilic, neutral, and have sufficient porosity to allow the penetration of macromolecules. Carbohydrates, because they are generally neutral and imbibe large quantities of water,

Fred E. Regnier, Karen M. Gooding, and Shung-Ho Chang • Department of Biochemistry, Purdue University, West Lafayette, Indiana 47907

have been used extensively as a support matrix for the chromatography of proteins. These matrices are both hydrophilic and macroporous. The ease with which carbohydrates can be substituted with different ion exchange groups or ligands, in addition to their chemical stability under chromatographic conditions, has also contributed to their popularity for protein separations.

When a hydrophilic support material is cross-linked into a macroporous matrix of controlled porosity without further chemical substitution, a support suitable for gel or exclusion chromatography is generated. Both the polydextran-based Sephadex and polyacrylamide-based Bio-Gel resins are examples of such size-separation supports that can be used in the fractionation of compounds with molecular weights 10^2–10^6. The polar functional groups in either of these support materials cause it to adsorb large quantities of water and swell to many times its dry volume. The volume of this swollen matrix is unfortunately sensitive to pressure and changes in pH, ionic strength, and solvent compositions, which severely limits the use of the material in practice. Thus the high pressures necessary for fast and efficient separations are precluded, and the range of useful solvent compositions is limited.

Substitution of the above gel permeation materials or cellulose with an ionic species such as diethylaminoethanol (DEAE), diethylmethylaminoethanol (QAE), glycolic acid (CM), or propanesulfonic acid (SP) yields ion exchange supports which have been used extensively in the resolution of proteins. However, volume sensitivity is an even greater limitation for ion exchange packings, where, for example, gradient elution is restricted to a narrow range of conditions.

1.2. Inorganic Supports

The need for a rigid, nongel type of support was obvious. The invention by Haller[2,3] of controlled-porosity glass (CPG) supports with pore diameters ranging from 100 to 1500 Å provided a permeable support that answered this need. These supports could be sterilized and cleaned with strong acid without deleterious effects and their pore diameters corresponded closely to the molecular diameters of viruses, proteins, and polysaccharides. It appeared that CPG would be a valuable matrix for gel chromatography. In fact, the purification of large quantities of plant[4] and animal viruses[5] and proteins[6] has been achieved with these supports.

Unfortunately, the glass and silica supports have some severe limitations in the chromatographic separation of biopolymers. As more researchers began using the CPG supports, it was found that their polar surfaces were responsible for the adsorption and denaturation of some sensitive biological compounds. Polio virus, adenoviruses, vesicular exanthema virus, yellow fever, rabies, and the viruses of vaccina were a few of the animal viruses adsorbed,[7] while red clover mottle virus, tobacco mosaic virus, alfalfa mosaic virus, and white clover mosaic virus were some plant viruses[8] that adsorbed under certain conditions. The adsorption and denaturation of enzymes have been observed in numerous cases; the adsorption of polyethylene glycol and polystyrene[7] have also been observed. Surface adsorption was observed in enough cases to seriously diminish the general utility of glass and silica supports in the gel chromatography of proteins.

Adsorption of polyethylene glycol polymers on CPG has been reported as a method for partially controlling the adsorption of both viruses and proteins.[7] Apparently the polyethylene glycol polymers hydrogen bond to surface silanols and shield sample molecules from the surface. Unfortunately, this is a reversible process, and polyethylene glycol elutes from columns under continuous use.

Eltekov et al.[9] were the first to report the ion exchange chromatography of proteins on macroporous inorganic supports. Aminopropylsilyl-bonded silica supports were used in the anion exchange chromatography of a series of proteins. Irreversible adsorption of proteins was reported to be eliminated by this bonded phase. More recently Busby et al.[10] have attempted the high-speed liquid chromatography of creatine phosphokinase isoenzymes from serum on the pellicular–quaternary anion exchanger, Vydac. Vydac is a chromatography support having an impermeable inner 37–74-μm core with a thin (2-μm) outer silica layer. The ion exchanger is covalently coupled to the surface of the porous outer layer. Both the ion exchange capacity and porosity of pellicular supports are considerably lower than the totally porous silica and CPG supports. It should be noted that the MB isoenzyme of creatine phosphokinase was not recovered from the support, indicating its irreversible adsorption or denaturation.

The successful use of inorganic bonded phase supports for affinity chromatography of proteins has been discussed extensively.[11] Application of these supports to high-speed affinity chromatography will depend very much on the rate of formation and dissociation of the protein–ligand complex. In those cases where rates are rapid, high mobile-phase velocities and rapid elution will be possible; slower rates of complex formation will preclude high mobile-phase velocities. Antigen–antibody complex formations requiring hours are frequently encountered cases of slow complex formation.

Literature on the high-speed liquid chromatography of proteins is sparse. During the past two years our laboratory has been heavily involved in the development of new supports for high-speed liquid chromatography of proteins. The remainder of this article will refer mainly to this work.

2. Gel Permeation Chromatography (GPC)

2.1. Rationale

In the introduction we endeavored to establish criteria for a high-speed support based on the properties of proteins and on the classical protein chromatography literature. It may be concluded that (1) the support must be macroporous to allow the penetration and partitioning of large molecules, (2) its surface must be neutral and hydrophilic so that the imbibed water layer will protect the protein from denaturation, (3) it must be easily derivatized to provide the various classes of chromatography supports, and (4) it must be capable of withstanding high mobile-phase velocities. The most severe limitation on the choice of basic support material is the requirement of mechanical stability under high flow rates and pressure. This criterion virtually forces the use of macroporous inorganic supports.

2.2. Coating Chemistry

Our initial efforts in the preparation of high-speed supports for proteins were directed toward gel permeation materials.[12] During the penetration of inorganic support pores, a protein comes in contact with large areas of surface ranging from 10 to 150 m² per cubic centimeter of support. Since these surfaces are polar and often charged, they are primarily responsible for the deleterious effects of inorganic supports on proteins. We concluded that by bonding a thin layer of carbohydrate to the support surface we could (1) deactivate the surface, (2) provide a neutral organic layer that excludes sensitive compounds from contact with the surface, and (3) provide an organic layer that imbibes water and creates a hydrated surface layer in the pore similar to carbohydrate gel filtration supports.

Studies with a series of mono-, oligo-, and polysaccharide bonded phases indicated that a glycerylropylsilyl coating was the most effective in controlling protein adsorption and/or denaturation. Bonding of organosilanes to the surface of silica and glass through an Si–O–Si–C linkage is well documented[13] and used in the preparation of most commercial bonded supports. By the reaction outlined in Figure 1, glycidoxypropyltrimethoxy silane (I) was bonded to controlled-porosity glass (CPG). This compound (I) readily hydrolyzes in water to yield the trihydroxy intermediate (II).

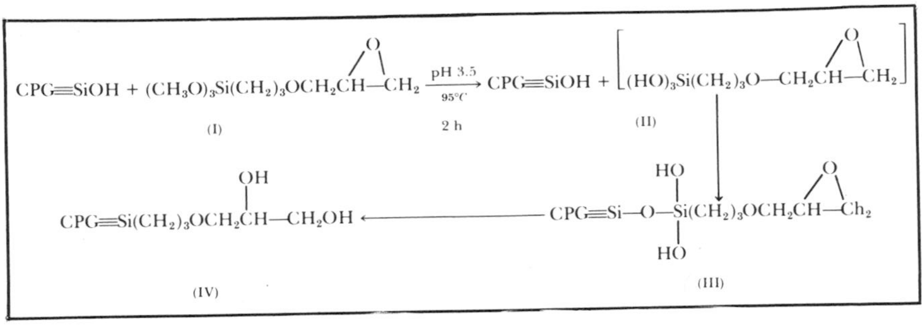

Figure 1. Bonding chemistry of Glycophase G/CPG.

Initial bonding of (II) to a silanol on the surface of CPG yields intermediate (III). Further heating at pH 3.5 completes the bonding to additional silanols and hydrolyzes the oxirane to yield glycerylpropylsilyl-CPG (IV). These bonded carbohydrate stationary phases have been given the trivial name "glycophases." Glycerylpropylsilyl-CPG (IV) is designated "Glycophase G/CPG" because the carbohydrate residue is glycerol.

Adsorption and/or denaturation of proteins on CPG were significantly decreased by the "glycerol" coating. When a series of enzymes was exposed to 507-Å pore diameter CPG with 74 m²/g surface area, large quantities of enzyme activity were lost with half of the enzymes tested, as indicated in Table 1. Addition of the glycerylpropylsilyl coating (IV) to CPG increased the recovery of enzyme activity to 90% or better in most cases. Recovery of activity from phosphodiesterase and guanidinobutyraldehyde dehydrogenase were, however, still only 70 and 78%, respectively. This may be attributed to a slight hydrophobicity of the organic coating;

Table 1. Relative Recovery of Enzymes (%)

Enzymes	Glycophase-G CPG	Uncoated CPG
Chymotrypsin	96	16
Lactic acid dehydrogenase	100	23
Lipoxygenase	95	8
Alphaketoarginine decarboxylase	100	100
Guanidinobutyaldehyde dehydrogenase	78	100
Anthranilate synthetase	91	49
DAHP synthetase	96	2
Pyrophosphatase (corn leaf)	100	93
Phosphodiesterase (calf intestine)	70	95
Threonine deaminase	79	34
Dihydroxy acid dehydratase	90	51
Acetohydroxy acid synthetase	95	95
Acetohydroxy acid isomeroreductase	100	62

we have found phosphodiesterase to be an intensely hydrophobic protein. It has also been noted that protein recoveries decrease for some proteins at basic pH and buffer concentrations less than 0.1 M.

2.3. Properties of Glycophage G/CPG

Collins and Haller[14] have shown that the distribution coefficient (K_D) of a sodium-dodecylsulfate–protein complex is related to the pore diameter of CPG. Therefore, any organic coating inside the pores of CPG that decreases its pore diameter would decrease the elution volume and exclusion limits of proteins. Using a series of native CPG supports, distribution coefficients (K_D) were determined with chymotrypsin, as summarized in Figure 2. When the 507-Å support was coated with glycerylpropylsilane (IV), the pore diameter was decreased to 470 Å as indicated by the K_D of chymotrypsin. This 37-Å decrease in pore diameter shows that the surface layer is, on the average, 19-Å thick. The theoretical value for a glycerylpropyl monolayer is 14 Å.

Quantitative analysis by periodate oxidation indicated that this 507-Å Glycophase G/CPG support had 80 μmol of vicinal diol/g. Since the surface area of this support is 74 m²/g, it may be calculated that a glycerylpropyl residue occupies 154 Å² of surface and the distance between organic residues is 12.5 Å.

Molecular exclusion limits taken from commercial literature[15] for a series of CPG supports are seen in Figure 3.

The speed with which an analysis may be performed is closely related to the linear velocity of the mobile phase (u) and to column efficiency.

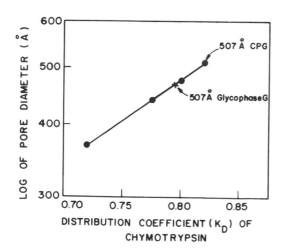

Figure 2. K_D vs. pore diameter of CPG.

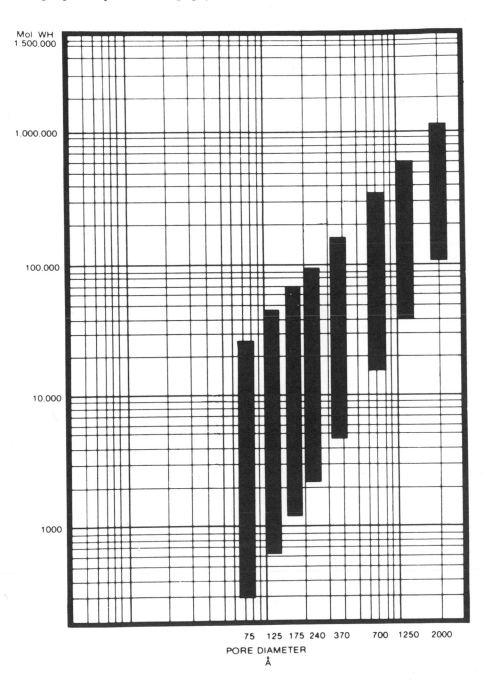

Figure 3. Exclusion limits and operate range of the Corning CPG-10 series for dextran polymers in distilled water.[15]

Column efficiency is commonly measured by either the theoretical plate number N or the height equivalent to a theoretical plate H.

$$N = 16(V_e/\Delta V)^2 \tag{1}$$

where V_e is the elution volume of the solute and ΔV is the solute bandwidth (see Figure 4). The plate height (H) may then be found by the formula $H = L/N$ where L is the column length. Good analytical liquid chromatography (LC) columns have H values of 0.1–1 mm for small-molecule separations.[17–19] A plot of H versus u for the protein ovalbumin on various Glycophase G/CPG supports is shown in Figure 5. Reviews of the factors affecting H versus u are presented in a series of papers.[16,17,20,21] It is seen in these discussions that the particle diameter (d_p), the solute diffusion coefficient in the mobile phase (D_m) and u play an important role in band spreading.

The comparison of efficiencies of different LC columns has been facilitated by formulation of the reduced parameters h and v,[16,22] where $h = H/d_p$ and $v = ud_p/D_m$. Using these reduced parameters it is possible to equate systems with different d_p and D_m values; plots of h versus v or log h versus log v would be identical for well-packed columns. Such a plot in Figure 6 at similar linear velocities illustrates the inequality of reduced

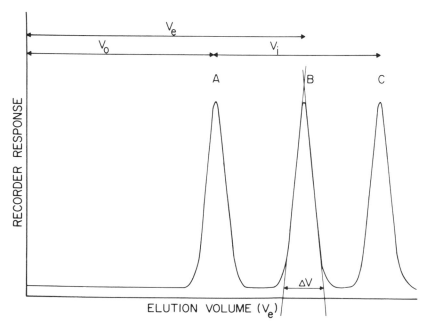

Figure 4. A model chromatogram for a gel permeation chromatographic separation.

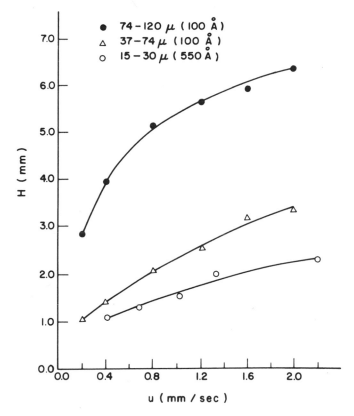

Figure 5. Effect of mobile phase velocity on efficiency for ovalbumin on Glycophase G columns of various particle sizes. Reprinted from Ref. 12 by courtesy of the publisher of *J. Chromatogr. Sci.*

velocities of ovalbumin and phenylalanine. If their reduced velocities are set equal, the relationship between the resulting linear velocities can be determined. Thus

$$\left(\frac{u_p d_p}{D_{mp}}\right)_{\text{protein}} = \left(\frac{u_s d_p}{D_{ms}}\right)_{\text{small solute}}$$

where u_p and u_s are the mobile-phase velocities of the protein and small molecule, respectively, and D_{mp} and D_{ms} are the diffusion coefficients of the protein and small molecule, respectively. The diffusion coefficient of ovalbumin is 8×10^{-7} cm²/s and that of phenylalanine is 10^{-5} cm²/s. The u_p/u_s ratio from the above equation indicates that to achieve similar reduced velocities, the linear velocity of ovalbumin must be 8% of that of phenylalanine. In view of this, it is not surprising that the column

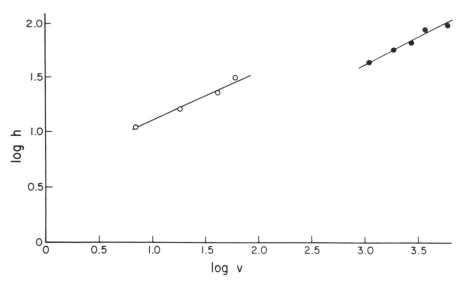

Figure 6. Logarithmic plots of *h* vs. *v* for ovalbumin (●—●) and phenylalanine (○—○) on Glycophase G to illustrate the difference in reduced velocities (Ref. 23).

efficiency for ovalbumin is normally lower than phenylalanine. This is not suggesting that flow rates must be slowed down by a factor of ten. Such a rate would defeat the purpose of high-speed analysis. Rather, it is suggested that the lower efficiency for high-molecular-weight solutes be acknowledged and compensated for by using high-efficiency columns (small particles and good packing techniques) to minimize band spreading.

2.4. Applications

A typical gel permeation separation of serum proteins on a 170-Å-pore-diameter Glycophase G/CPG support is shown in Figure 7. The 37–74-μm Glycophase G support in a 4.2 mm × 100 cm stainless steel column was eluted with 0.05 M sodium phosphate buffer (pH 7.0) at a flow rate of 1.6 mm/s and a head pressure of 100 psi. Under these conditions the protein eluting in the void volume (V_0) at 100 min consisted of immuno-globins (IgG, IgE, IgM) and other high-molecular-weight proteins while the peak at 13 min was predominantly albumin. The peaks eluting between 15 and 25 min are unknowns. These compounds are probably a complex mixture of compounds under 50,000 daltons. A substantially higher recovery of albumin is obtained from this support than native CPG.[6]

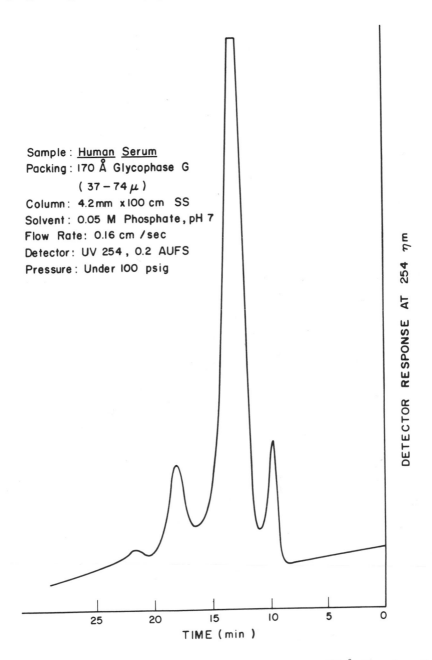

Figure 7. A gel permeation separation of serum proteins on 170-Å Glycophase G. Reprinted from Ref. 12 by courtesy of the publisher of *J. Chromatogr. Sci.*

Separations of nucleic acids and polysaccharides are shown in Figures 8 and 9, respectively. The first peak to elute from the nucleic acid mixture in 18 min was salmon sperm DNA followed by a mixture of tRNA species and mononucleotides. The polysaccharide mixture consisted of blue dextran with an average molecular weight of 2×10^6 (peak A), an unresolved mixture of dextrans with molecular weights 1×10^4 to 5×10^4 (peak B), and glucose (peak C).

The GPC separation of albumin, cytochrome c, and glycylphenylalanine on 5–10-μm Glycophase G/CPG with 100-Å pore diameter is shown in Figure

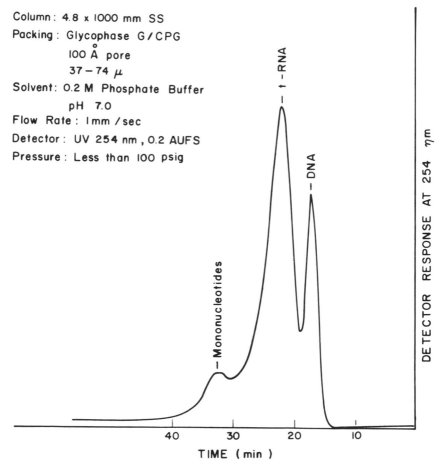

Figure 8. A gel permeation separation of nucleic acids on a Glycophase G column. Reprinted from Ref. 12 by courtesy of the publisher of *J. Chromatogr. Sci.*

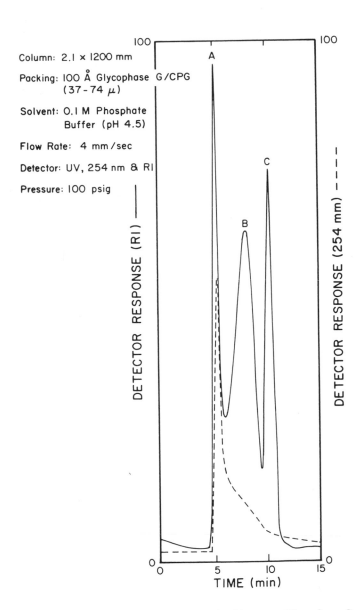

Figure 9. A gel permeation separation of polysaccharides on a Glycophase G column. Reprinted from Ref. 12 by courtesy of the publisher of *J. Chromatogr. Sci.*

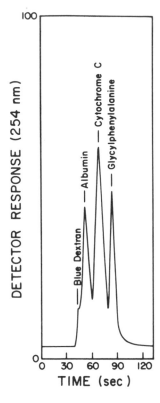

SAMPLE: Proteins

PACKING: Glycophase G/CPG
             100 Å Pore, 5-10µ

COLUMN: 4.1 × 300 mm SS

SOLVENT: 0.1 M Phosphate
             Buffer, pH 6.0

FLOW RATE: 7 mm/sec

PRESSURE: 2700 psi

DETECTOR: 254 nm

Figure 10. Steric exclusion chromatography of proteins in 90 s. Reprinted from *J. Chromatogr.*, *125*, 103–114 (1976) by courtesy of Elsevier Scientific Publishing Co.

10. This analysis required only 90 s and could be used for the rough fractionation of high-, intermediate-, and low-molecular-weight compounds.

In Figure 11 the digestion of calf thymus DNA by bovine spleen deoxyribonuclease II is monitored by GPC on 10-µm Glycophase G/Lichrospher Si 100. Over the course of 261 min, the totally excluded DNA (2,000,000 daltons) disappears and degradation products of lower molecular weights appear.

2.5. Column Preparation

There are basically two procedures for packing HPLC columns—dry packing and slurry packing. Selection of a method depends on the size and shape of support material. Irregular particles down to 40–50 µm may be dry-packed well.[20,22] For small particles, particularly the 5–10-µm size, slurry packing is necessary. If columns of less than "infinite diameter" are used, they should be either glass or precision-bore stainless steel tubing. Halász has suggested that the precision-bore tubing be drilled to make the inner surface uniform and smooth.[25] Stainless steel systems must be used for proteins since copper or brass produce deleterious effects on proteins. Since glass columns have a low pressure limit, stainless steel is usually used, although the glass is more inert. Columns should be scrubbed with a detergent and rinsed with water and acetone before use.[20,26]

The best dry-packing procedure is the "tap-fill" method.[20,26] A frit is placed on the outlet of the column and enough support added to the vertical column to fill 3–5 mm. The column is vertically tapped on the floor or a hard surface 80–100 times while rotating slowly and the side is rapped gently at the level of packing. This rapping loosens the particles from the sides of the tube. Rapping is then discontinued and the column

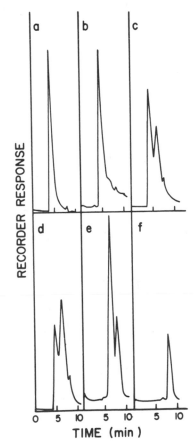

Figure 11. A time study of the digestion of calf thymus DNA by bovine spleen deoxyribonuclease II (DNase). Analysis is on a 300-×-4.1-mm column of Glycophase G/Lichrospher Si 100. The mobile phase is 0.05 M KH_2PO_4, pH 6 at 0.35 mliter/min and 670 psi. DNase was added at room temperature and the solution monitored. (a) Before DNase addition, (b) after 2 min digestion with DNase, (c) after 38 min, (d) after 61 min, (e) after 170 min, (f) after 261 min. Reprinted from *J. Chromatogr. 125*, 103–114 (1976) by courtesy of Elsevier Scientific Publishing Co.

gently tapped vertically another 15–20 s. Another increment of packing is added and the procedure repeated until the column is filled. It should then be gently tapped for 5 min more and a frit or porous plug placed in the inlet end of the column.

The balanced-density method of slurry packing is the most widely accepted.[20,25,27,28,29] This method is not very effective for columns longer than 50 cm; it is better to put several short columns in series if a column longer than 50 cm is desired.[18] The viscosity method is an alternative method of slurry packing[25]; balanced density solvents are required with solvents of viscosities of 40–60 cP.

Another slurry-packing technique uses the Micromeritics column packer. Isopropanol is stirred in the cylindrical base of the packer with a magnetic stirring bar. About 2 g of support (for a column 4.1 × 300 mm) is added gradually, and the bar can be adjusted to keep the particles in suspension. The cover of the packer, which has the solvent inlet and column attached, is then bolted on. Isopropanol is pumped at a constant flow of 2 ml/min until the pressure stabilizes (about 30 min). A 5-μm column terminator is then attached. This method gives reproducible H values of about 0.2 mm for glycylphenylalanine in sodium phosphate buffer.

2.6. Resolution

The separating power of an inorganic gel support for proteins is often less than that of a comparable soft gel support. These differences in relative resolution strongly influence the column length and number of theoretical plates required for a given separation. In an effort to explain this phenomenon we will examine the resolution of various supports.

The total liquid volume (V_T) in a GPC column is expressed by the equation

$$V_T = V_0 + V_i \tag{2}$$

where V_0 is the void volume or column volume outside the support particles and V_i is the internal volume or pore volume of the support. Figure 4 is a typical elution profile from a gel permeation column. Peak A is a solute whose molecular dimensions exceed the pore diameter of the support and cause it to be eluted in the void volume (V_0) while peak C is a small molecule that totally penetrates the support and has an elution volume (V_T). From equation (2) it is seen that $V_i = V_T - V_0$. The elution volume (V_e) of a solute, such as peak B in Figure 4, is expressed by the equation

$$V_e = V_0 + K_D V_i \tag{3}$$

where K_D is the distribution coefficient. K_D ranges from zero to one in pure GPC. If resolution of two compounds is to be achieved, it is apparent that their distribution coefficients must be different. The relative difference in distribution coefficients is referred to as the relative retention (α) and is expressed as

$$\alpha = K_D''/K_D' \tag{4}$$

The ratio of the internal volume of a support to its void volume (V_i/V_0) has an important effect on resolution. This ratio shows the permeability (p) of a support and will be defined as $p = V_i/V_0$. Inorganic support permeabilities range from 0.8 to 1.2 while those of gel-type supports vary from 1 to 3. A general resolution equation for GPC that

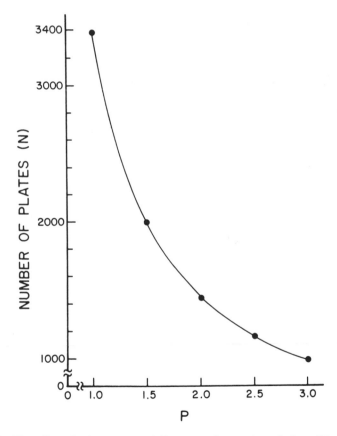

Figure 12. The effect of column permeability (p) on the number of plates (N) required to give a resolution of 1.0 at an α value of 1.25 at $K_D = 0.5$ in gel permeation chromatography. Reprinted from *J. Chromatogr. 125*, 103–114 (1976) by courtesy of Elsevier Scientific Publishing Co.

relates support permeabilities (p) and distribution coefficients (K_D) to the average number of plates (N) required for a given resolution (R_S) is

$$R_S = \frac{\sqrt{N}}{2} \frac{pK_D''(\alpha - 1)}{2\alpha + pK_D''(\alpha + 1)} \tag{5}$$

This equation is derived in the Appendix (Section 6).

By assuming fixed values of 1 and 1.25 for R_S and α, respectively, it is possible to evaluate the influence of p on N at a K_D value of 0.5. Figure 12 shows that as the permeability of a support decreases, the number of plates required to obtain the same resolution increases. Using 550-Å CPG ($p = 1$) and Sephadex G-200 ($p = 2.2$) as a practical comparison, it is seen that the CPG support requires 2.6 times as many plates to achieve comparable resolution. Assuming that plate height (H) is equal on the two supports, a column 2.6 times as long is required with the inorganic support.

The difference in resolving power for a soft gel vs. the inorganic support cited above is as large as will be found. As pore diameter is decreased in inorganic supports, p remains approximately 1, while in soft gel supports p decreases with cross-linking to $p = 1.1$ in Sephadex G-10.

3. Ion Exchange Chromatography on Carbohydrate-Coated Supports

3.1. Rationale

Two reported cases[9,10] of ion exchange chromatography of proteins on inorganic supports were not totally successful, owing to the nature of the organic coatings and the resulting poor recoveries of sample. Also, the types of stationary phases chosen for these studies were not those commonly used in the resolution of proteins. An enormous body of literature is available describing the resolution of a myriad of proteins on four basic types of ion exchangers: DEAE, QAE, CM, and SP. Since elution protocols and properties of so many proteins are recorded on these stationary phases, it is highly desirable that any new high-speed support be related to these classical supports.

Having introduced criteria for high-speed supports and demonstrated the utility of carbohydrate coatings in the GPC resolution of proteins, we now discuss the extension of these findings to the preparation of carbohydrate-bonded ion exchange supports.[30,31]

3.2. Bonded Phases

3.2.1. Non-Cross-Linked Supports

Non-cross-linked supports were prepared by grafting a stationary phase (P_S) to the carbohydrate-bonded supports previously described for gel permeation chromatography.[30] In Figure 1, it will be noted that γ-glycidoxypropylsilyl/CPG (III) is obtained in the synthesis of Glycophase G/CPG (IV). Due to slightly acidic reaction conditions, the glycidoxy derivative (III) is rapidly hydrolyzed to a diol (IV). By adjusting the reaction conditions to pH 5.8, oxirane hydrolysis decreases and the glycidoxy derivative (III) accumulates. Nucleophilic stationary phases (P_S) were used to open the oxirane ring and complete the attachment of P_S to the support according to Figure 13.

3.2.2. Cross-Linked Supports

Under basic conditions and high ionic strength some of the non-cross-linked stationary phases are stripped from CPG. Stable bonded phases are most easily obtained by forming on the surface of CPG a thin glycerol polymer layer that is attached to the support through an organosilane coupling agent and has the stationary phase bonded to the polymer matrix.[31] This polymer has the general formula

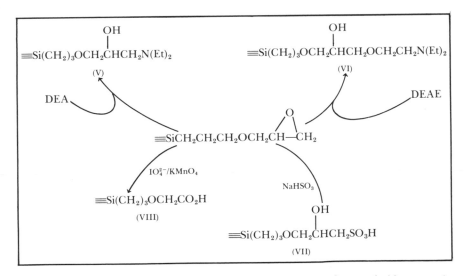

Figure 13. Synthesis of non-cross-linked ion exchange supports from γ-glycidoxypropyl-silyl/CPG.

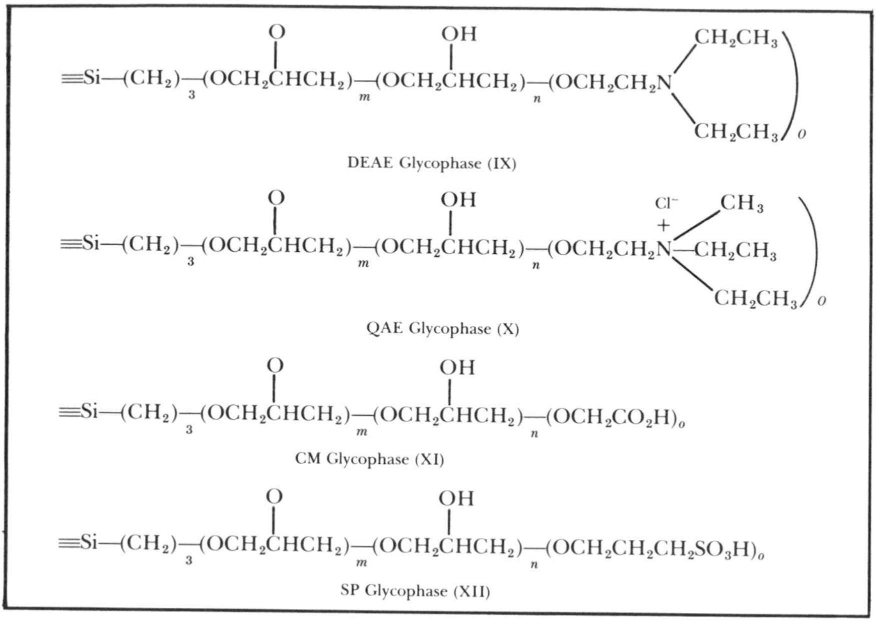

Figure 14. Structures of cross-linked ion exchange supports.

$\equiv$SiCH$_2$CH$_2$CH$_2$R$_1$P$_S$ where R$_1$ is a glycerol polymer and P$_S$ is the stationary phase. The structural formulas for four polymeric supports, DEAE Glycophase (IX), QAE Glycophase (X), CM Glycophase (XI), and SP Glycophase (XII), are shown in Figure 14. Since the exact ratios of monomers ($m/n/o$) are not known, they are not specified in the figure.

The glycerol polymers used on these supports are hydrophilic, neutral, and stable so they will not partition proteins. The hydroxy groups of the glycerylpropylsilyl-bonded CPG (Glycophase G) are incorporated into the polymer

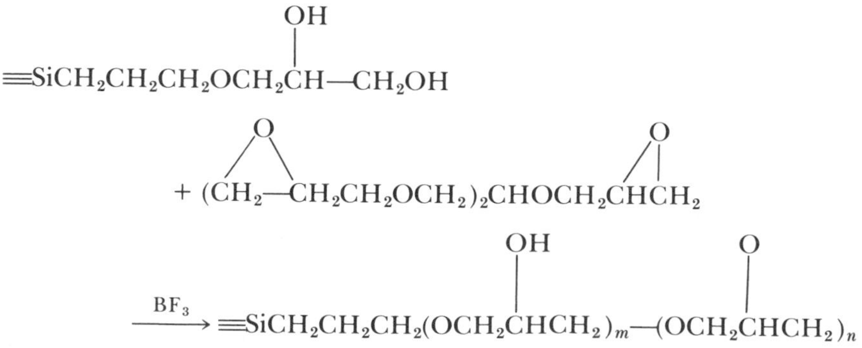

The multifunctional oxirane monomer in this example is triglycidylglycerol although any of the monomers in Table 2 can be used. To ensure a thin uniform layer, the monomer is coated by either a slurry[32] or filtration[33] method followed by drying in a fluidized bed. The BF_3 polymerization is achieved by sweeping BF_3 in a stream of nitrogen through the fluidized bed; this tumbling during polymerization prevents aggregation of the particles.

Various stationary phases are incorporated into the polymer matrix by either copolymerization or coupling during polymerization. In the former, a monomer containing the stationary phase (P_S) or a group to be converted to P_S is coated with the multifunctional monomer onto the support and copolymerized in a fluidized bed. The reaction is as follows:

$$\equiv SiCH_2CH_2CH_2OCH_2\overset{\overset{\displaystyle OH}{|}}{C}HCH_2OH$$

$$+ (\overset{\overset{\displaystyle O}{\diagup\!\!\diagdown}}{CH_2}CHCH_2OCH_2)_2CHOCH_2\overset{\overset{\displaystyle O}{\diagup\!\!\diagdown}}{C}HCH_2 + \overset{\overset{\displaystyle O}{\diagup\!\!\diagdown}}{CH_2}CHCH_2OP_S$$

$$\xrightarrow{BF_3} \equiv SiCH_2CH_2CH_2(OCH_2\overset{\overset{\displaystyle O}{|}}{C}HCH_2)_m(OCH_2\overset{\overset{\displaystyle OH}{|}}{C}HCH_2)_n(OP_S)_o$$

Stationary phases with nucleophilic substituents (X) may also be coupled to the oxirane monomer during polymerization:

$$\equiv SiCH_2CH_2CH_2OCH_2\overset{\overset{\displaystyle OH}{|}}{C}HCH_2OH$$

$$+ (\overset{\overset{\displaystyle O}{\diagup\!\!\diagdown}}{CH_2}CHCH_2OCH_2)_2CHOCH_2\overset{\overset{\displaystyle O}{\diagup\!\!\diagdown}}{C}HCH_2 + XP_S$$

$$\xrightarrow{catalyst} \equiv SiCH_2CH_2CH_2(OCH_2\overset{\overset{\displaystyle O}{|}}{C}HCH_2)_m(OCH\overset{\overset{\displaystyle OH}{|}}{C}HCH_2)_n(XP_S)_o$$

The choice of catalyst is dependent on the nature of P_S.[31]

3.3. Ion Exchange Properties

It is becoming common practice to report the ion exchange capacities of supports for proteins in terms of the hemoglobin bound per unit

Table 2. Oxirane Monomers Used in Support Preparation

Monomer	Oxiranes/molecule	Structure	Postpolymerization function
Triglycidyl glycerol	3	$CH_2\text{-}OCH_2CHCH_2$ (epoxide) $CH\text{-}OCH_2CHCH_2$ (epoxide) $CH_2\text{-}OCH_2CHCH_2$ (epoxide)	Hydrophilic support layer
Diglycidyl[a] glycerol	2	$CH_2OCH_2CHCH_2$ (epoxide) $CHOH$ $CH_2OCH_2CHCH_2$ (epoxide)	Hydrophilic support layer

Diglycidyl ethylene glycol	2	$CH_2-OCH_2CHCH_2$ / $CH_2-OCH_2CHCH_2$ (epoxide)	Hydrophilic support layer
Butadiene diepoxide	2	$CH_2CHCHCH_2$ (diepoxide)	Hydrophilic support layer
Allyl glycidyl ether	1	$CH{=}CH_2CH_2OCH_2CHCH_2$ (epoxide)	Intermediate in preparation of cation exchanger

a This compound is probably a mixture of α, α' and α, β diglyceryl ethers with the former predominating.

Table 3. Ion Exchange Capacities of Supports

Text reference No.	Stationary phase	Inorganic support	Pore diameter (Å)	Surface area (m^2/g)	Ion exchange capacity (mg/ mliter)[c]
Non-cross-linked[a]					
VIII	CM	CPG	550	70	21
VIII	CM	CPG	250	100	38
VIII	CM	CPG	100	170	17
VII	SP	CPG	550	70	18
V	DEA	CPG	250	100	54
VI	DEAE	CPG	250	100	55
Cross-linked[b]					
XI	CM	CPG	550	70	20
XI	CM	Silica	500	—	19
XI	CM	CPG	250	100	40
XII	SP	CPG	550	70	14
IX	DEAE	CPG	550	70	36
IX	DEAE	CPG	250	100	71
X	QAE	CPG	250	100	62

[a] Reference 30.
[b] Reference 31.
[c] Hemoglobin was used to determine protein ion exchange capacity.

volume of support. In this way, the ion exchange capacities of a variety of supports can be compared. The hemoglobin ion exchange capacities for a series of supports are seen in Table 3. It is interesting to note that in the CM supports (VIII), ion exchange capacity is maximum with the 250-Å-pore-diameter material. Ion exchange capacity is probably a function of *available* surface area. Since the hemoglobin molecule is not able to totally penetrate the 100-Å support, its ion exchange capacity is lower than that of the 250-Å and 550-Å supports even though the 100-Å material has a larger surface area. For this reason, supports with smaller than 150-Å pores will be of less utility for the resolution of high-molecular-weight proteins than those of larger pore sizes.

There appears to be little difference between the ion exchange capacities of cross-linked and non-cross-linked supports. Thus, it may be concluded that the construction of a polymer layer in the pores of CPG has not significantly changed the surface area of the support. It may be calculated from Table 3 that 10^{-6} mol of hemoglobin are bound per 30 m^2 of surface. This is equal to a μmol per 3.0×10^{21} $Å^2$. Since a hemoglobin molecule occupies approximately 4000 $Å^2$ of surface, it may be calculated that a μmol of hemoglobin occupies 2.4×10^{21} $Å^2$ of surface. Thus 80% of the surface of the DEAE support is covered with hemoglobin at maximum loading. Since more than 99% of the surface area of a 250-Å support is inside the pores of the support, the bulk of the protein is immobilized on the pore walls. The amount of protein bound on the exterior surface of the ion exchange particle is negligible even in large-pore-diameter supports.

It may be concluded that the ion exchange capacity of an inorganic support for a particular macromolecule is a function of two factors: (1) the extent to which the molecule can penetrate the pores of the support and (2) the surface area in the pores.

3.4. Applications

The way in which an ion exchange support can be used is dependent on the particular mixture to be resolved. When the difference in the partition coefficients of individual components is large, the support may be used in a batch mode. When the difference in partition coefficients is small, more efficient separations are required, and the packed column mode is used.

3.4.1. Batch Separations

Three creatine phosphokinase isoenzymes (MM, MB, and BB) have been isolated from human tissue and plasma. Plasma from normal subjects contains primarily the MM isoenzyme. The MB form accounts for less than 5% of the total enzyme activity and the amount of BB form is very low.[34,35] After myocardial infarction, the MB isoenzyme increases significantly and is diagnostic of heart damage.[34] The column[36] and electrophoretic[34] methods used to determine creatine phosphokinase activities require more time and attention than is desirable for routine clinical use.

Henry et al.[37] have developed a method for the rapid separation of the two plasma creatine phosphokinase isoenzymes by batch adsorption on DEA/CPG (V). By judicious control of pH and ionic strength, the MB isoenzyme is quantitatively adsorbed onto the anion exchange support while the MM form remains in solution. Sedimentation of the ion exchanger followed by filtration and washing with the initial buffer completes the resolution of the isoenzymes. Desorption of the MB isoenzyme with high-ionic-strength buffer is followed by spectrophotometric or kinetic fluorescence assay. When the BB isoenzyme is present, as in cases of malignant hyperthermia or infarction of an organ rich in BB, this method is not effective. The BB isoenzyme is adsorbed with the MB so that the two are indistinguishable in this assay.

Although other DEAE anion exchange supports could function for this batch separation, the rapid attainment of ion exchange equilibrium,

Fred E. Regnier *et al.*

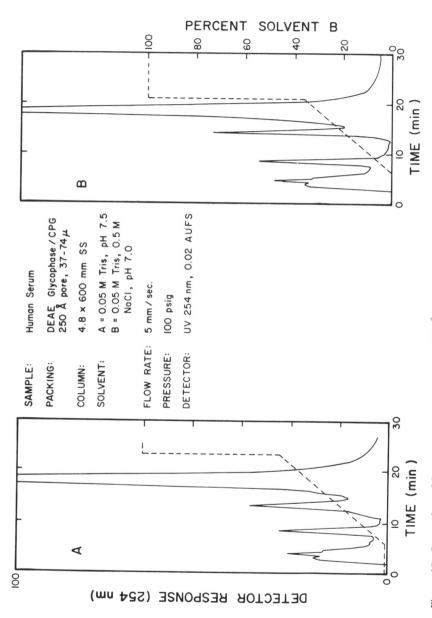

Figure 15. Separation of human serum proteins on 250-Å DEAE Glycophase/CPG chromatographed 48 h apart to illustrate column reproducibility. Reprinted with permission from *Anal. Chem.*, *48* (1976). Copyright by the American Chemical Society.

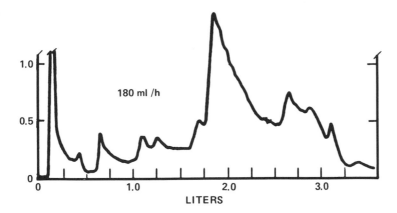

Figure 16. Separation of human serum proteins on DEAE cellulose. Reprinted from *Federation Proceedings 17,* 1116–1126 (1958) by courtesy of the Federation of American Societies for Experimental Biology.

rapid gravity sedimentation, and the obviation of the need to preswell supports significantly contribute to the utility of the inorganic (CPG) support.

3.4.2. Column Separations

The packing and operation of CPG ion exchange columns is similar to that described for GPC supports. Since support volume is insensitive to pressure, pH, and ionic strength changes, columns may be reproducibly operated through many gradient elution cycles without repacking. Elution of specific proteins from a CPG ion exchange support has been found to correspond closely with that on soft gel supports using the protocol described in both cases. The technique of using strong base for recycling must, however, be modified since glass degrades much more rapidly in strong base than carbohydrate gels. The useful life of CPG ion exchange supports is decreased by repeated or prolonged use of strong base. Reversed gradient cycling after column elution appears to be a good recycling technique.

Separation of human serum proteins on a 37–74-μm 250-Å-pore-diameter DEAE Glycophase/CPG (IX) is shown in Figure 15.[31] The first peak to elute consists primarily of immunoglobulins while the major component is albumin. This serum sample was run twice with a 48-h time interval between analyses to demonstrate reproducibility. Although a detailed comparison of proteins eluted from the DEAE Glycophase/CPG

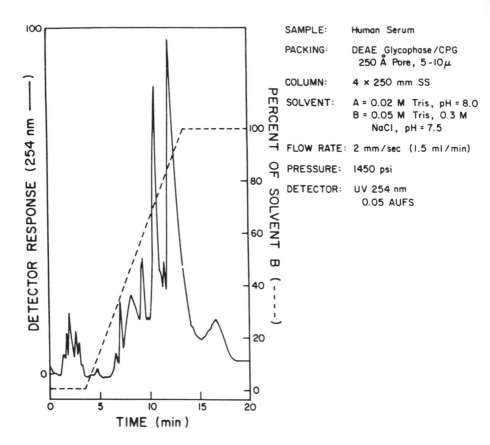

Figure 17. Separation of human serum proteins on DEAE Glycophase/CPG. Reprinted from *J. Chromatogr. 125*, 103–114 (1976) by courtesy of Elsevier Scientific Publishing Co.

and DEAE cellulose has not been made, the elution profiles appear similar. A separation on DEAE cellulose is given in Figure 16.[38] The improved resolution of human serum proteins on 5–10-μm 250-Å-pore-diameter DEAE Glycophase/CPG is obvious in Figure 17.[24] The elution time of this latter analysis was 1/1000th of that on the cellulose column of Figure 16.

Use of the 37–74-μm DEAE Glycophase/CPG column in the resolution of the proteins in a crude rat liver homogenate is illustrated in Figure 18.[31] Again, the column is able to resolve a number of protein components in a short time at 100 psi.

Resolution of the three creatine phosphokinase isoenzymes from bovine tissue on a 37–74-μm DEAE Glycophase/CPG column is shown in Figure 19.[31] The MM isoenzyme elutes first followed by the MB and BB isoenzymes, respectively. Separation of these isoenzymes on DEAE cellulose is shown in Figure 20.[39] The DEAE Glycophase column was developed in one-sixth the time.

The separation of components of a hemoglobin control mixture on a 5–10-μm DEAE Glycophase/CPG column is shown in Figure 21.[24] Hemoglobin A_1 and A_2 are usually found in normal adult blood. Hemoglobin F is fetal hemoglobin and Hemoglobin S is found in persons with sickle cell trait or disease.

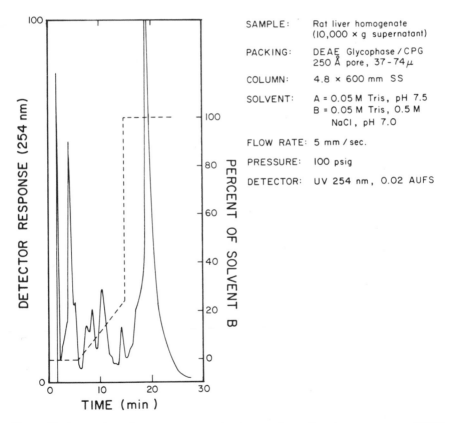

Figure 18. Separation of component proteins in a crude rat liver homogenate on DEAE Glycophase/CPG. Reprinted with permission from *Anal. Chem., 48* (1976). Copyright by the American Chemical Society.

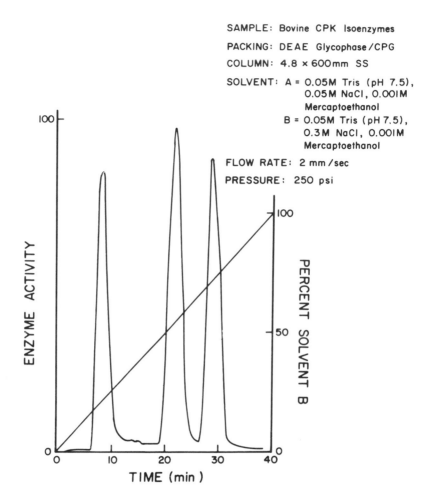

Figure 19. Separation of creatine phosphokinase isoenzymes on 250-Å DEAE Glyco-phase/CPG. Reprinted with permission from *Anal. Chem., 48* (1976). Copyright by the American Chemical Society.

A separation of several proteins on 74–128-μm QAE Glycophase/CPG is seen in Figure 22.[31] This strong anion exchanger is not as versatile as DEAE since it can denature very labile compounds, but it can be used effectively for compounds with high pI values.

Purification of soybean trypsin inhibitor on 250-Å-pore-diameter CM Glycophase/CPG is illustrated in Figure 23.[31] Soybean trypsin inhibitor is the major component eluting from the column at 20 min. The elution of

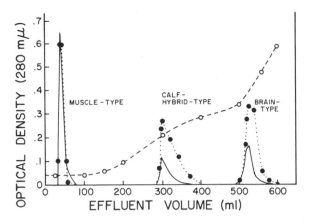

Figure 20. Distribution of tissue CK isoenzymes MM, MB, and BB on DEAE-cellulose columns. Reprinted from *Arch. Biochem. Biophys., 150,* 648 (1972) by courtesy of Academic Press.

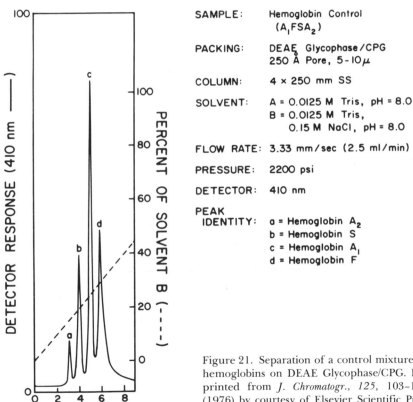

SAMPLE: Hemoglobin Control
            (A_1FSA_2)

PACKING: DEAE Glycophase/CPG
            250 Å Pore, 5-10 μ

COLUMN: 4 × 250 mm SS

SOLVENT: A = 0.0125 M Tris, pH = 8.0
            B = 0.0125 M Tris,
                0.15 M NaCl, pH = 8.0

FLOW RATE: 3.33 mm/sec (2.5 ml/min)

PRESSURE: 2200 psi

DETECTOR: 410 nm

PEAK
IDENTITY: a = Hemoglobin A_2
            b = Hemoglobin S
            c = Hemoglobin A_1
            d = Hemoglobin F

Figure 21. Separation of a control mixture of hemoglobins on DEAE Glycophase/CPG. Reprinted from *J. Chromatogr., 125,* 103–114 (1976) by courtesy of Elsevier Scientific Publishing Co.

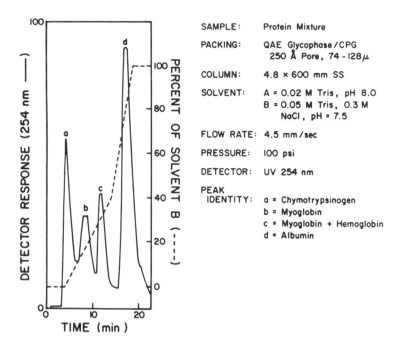

Figure 22. Resolution of proteins on a QAE Glycophase/CPG column. Reprinted with permission from *Anal. Chem.*, *48* (1976). Copyright by the American Chemical Society.

soybean trypsin inhibitor from a CM cellulose column in 10 h is shown in Figure 24.[40]

Resolution of components in the rat liver homogenate on CM Glycophase/CPG is illustrated in Figure 25.

Purification of commercial bovine pancreas α-chymotrypsin on 74–128-μm SP Glycophase/CPG is shown in Figure 26.[31] Enzyme assays of collected fractions with *p*-nitrophenylacetate show the peak at 13 min to be chymotrypsin.

3.5. Resolution

During the analysis of a variety of proteins with inorganic ion exchange supports, it was noted that both ion exchange capacity and pore diameter influence resolution. Resolution (R_S) in partition chromatogra-

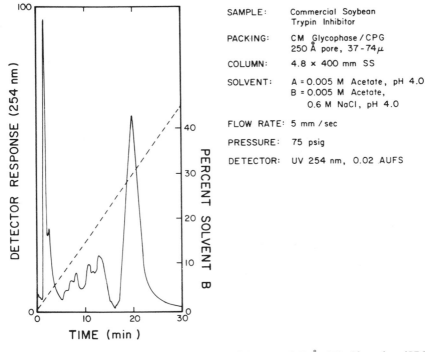

Figure 23. Purification of soybean trypsin inhibitor on 250-Å CM Glycophase/CPG. Reprinted with permission from *Anal. Chem.*, *48* (1976). Copyright by the American Chemical Society.

phy is expressed by the equation

$$R_S = \left(\frac{\sqrt{N}}{4}\right)\left(\frac{\alpha - 1}{\alpha}\right)\left(\frac{k'}{1 + k'}\right) \qquad (6)$$

The capacity factor (k') in ion exchange chromatography is generally defined by the equation

$$k' = K_D\left(\frac{V_S}{V_m}\right) \qquad (7)$$

where V_S is the volume of the stationary phase, V_m is the volume of the mobile phase, and K_D is the partition coefficient. The partition coefficient is defined as

$$K_D = C_S/C_m \qquad (8)$$

where C_S and C_m are the concentrations of solute in the stationary and mobile phases, respectively. Substituting equation (7) in equation (6) produces a resolution equation specific for ion exchange chromatography:

$$R_S = \left(\frac{\sqrt{N}}{4}\right)\left(\frac{\alpha - 1}{\alpha}\right)\left(\frac{K_D V_S}{V_m + K_D V_S}\right) \tag{9}$$

It has been noted above that in the case of inorganic ion exchange supports, ion exchange equilibrium occurs on the support surface. C_S may therefore be expressed in moles per square meter and the ion exchange surface area substituted for V_S.[41] K_D then has units of liters per square meter. Therefore, V_S must be expressed in square meters and V_m in liters. The influence of V_S on R_S will be noted in equation (9). Since the molecular dimensions of a biopolymer and the pore diameter determine the extent to which a solute penetrates a support, it is obvious that V_S may

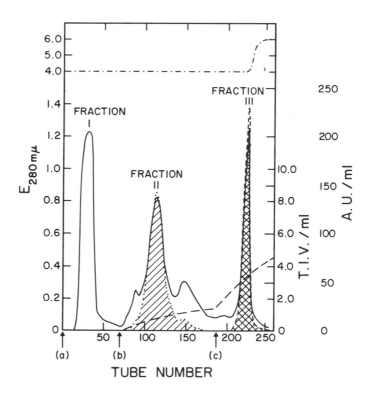

Figure 24. Purification of soybean trypsin inhibitor on a CM cellulose column. Reprinted from Ref. 40 by courtesy of the Biochemistry Society.

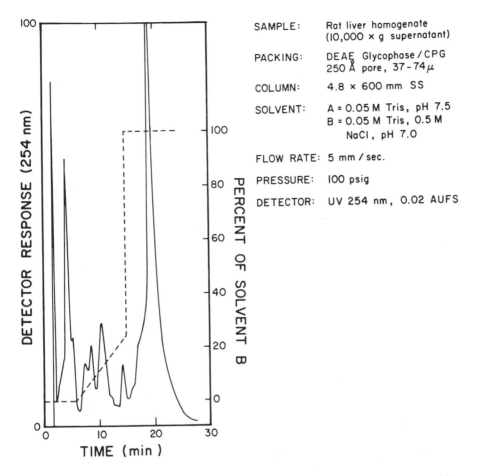

SAMPLE: Rat liver homogenate (10,000 × g supernatant)

PACKING: DEAE Glycophase / CPG 250 Å pore, 37-74 μ

COLUMN: 4.8 × 600 mm SS

SOLVENT: A = 0.05 M Tris, pH 7.5
B = 0.05 M Tris, 0.5 M NaCl, pH 7.0

FLOW RATE: 5 mm / sec.

PRESSURE: 100 psig

DETECTOR: UV 254 nm, 0.02 AUFS

Figure 25. Separation of component proteins in a crude rat liver homogenate on CM Glycophase/CPG.

not be the same for all large molecules. V_S is determined by available surface area and is proportional to ion exchange capacity (I_c). The surface area of a milliliter of support may be calculated by the equation

$$V_S = I_c S \qquad (10)$$

where I_c is the ion exchange capacity of the support in moles per milliliter and S the surface area (in square meters) occupied by a mole of protein. Using Table 3 it may be calculated that with the 250-A DEAE Glycophase G support $I_c = 10^{-6}$ mol/mliter. It may also be calculated from the above

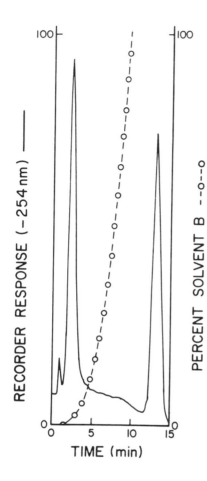

Figure 26. The purification of commercial α-chymotrypsin on a 74–128-μm, 250-Å pore diameter SP Glycophase/CPG. A 4-×-300-mm stainless steel column was gradient eluted at a linear velocity of 1 mm/s with 50 psi of column inlet pressure. Solvent A was 10 mM sodium phosphate buffer (pH 8). Solvent B was 10 mM sodium phosphate buffer (pH 8) with 0.5 M sodium chloride. Reprinted with permission from *Anal. Chem.*, **48** (1976). Copyright by the American Chemical Society.

discussion that $S = 24 \times 10^5$ m²/mol for a hemoglobin monolayer. V_m is 0.66 mliter for a series of CPG supports ranging from 100 Å to 550 Å. When a protein has a $k' = 3$ it may be calculated from equation (7) that $K_D = 8.25 \times 10^{-5}$ liter/m². Combining equations (9) and (10) we get

$$R_S = \left(\frac{\sqrt{N}}{4} \right) \left(\frac{\alpha - 1}{\alpha} \right) \left(\frac{K_D I_c S}{V_m + K_D I_c S} \right) \tag{11}$$

Using the above values for K_D, S, and V_m and assuming that R_S and α are constant, it is possible to show the relationship between I_c and N for an ion exchange support. Doubling the ion exchange capacity of a support for a given protein reduces the plate requirements for resolution by 23% while trebling the capacity reduces the requirements by 31%.

4. Enzyme Detectors

4.1. Rationale

When dealing with crude mixtures of proteins such as those in serum, specific enzymes must be quantitated by enzyme assay of fractions rather than spectrophotometric monitoring of the effluent. Often the component of interest is in very low concentration and an enzyme assay provides sensitivity besides selectivity. This is especially important in clinical applications such as the assay for isoenzymes; isoenzyme profiles can be used to diagnose diseases such as myocardial infarctions,[34] pulmonary infarctions, and liver disease.

Once protein separations within 10 min or less were made possible, the enzymatic analysis of the components became the limiting step; i.e., the enzyme assays of the fractions required much more time than the chromatography. For example, the five isoenzymes of lactic dehydrogenase (LDH) can be resolved in 6 min and the three isoenzymes of creatine phosphokinase (CPK) in 4 min; however, the enzyme assay of all the generated fractions required hours. It was obvious that a continuous-flow detector was needed.

4.2. Enzyme Kinetics

The activity of an enzyme can be measured by incubating an enzyme (E) with an appropriate substrate (S) and monitoring the formation of product (P) after a fixed reaction time. The Michaelis–Menten equation describes the relationship between the enzyme reaction rate, substrate concentration, and enzyme concentration:

$$v = \frac{V_{\max}[S]}{K_m + [S]} \tag{12}$$

In this equation, v is the initial rate of the reaction, S is the substrate concentration, K_m is the Michaelis–Menten constant for the enzyme, and $V_{\max}$ is the maximum velocity where $V_{\max} = k_3[E]$. E is the enzyme concentration. If the substrate concentration is much larger than K_m, then the velocity is proportional to the enzyme concentration:

$$v = V_{\max} = k_3[E] \tag{13}$$

Since the reaction rate is independent of substrate concentration, the reaction is zero-order with respect to substrate. This results in a linear accumulation of product with time. As substrate is depleted, the reaction becomes first-order with respect to substrate and may not be used to determine enzyme concentration. An enzyme detector must maintain zero-order reaction kinetics with respect to substrate during enzyme detection.

4.3. Detector Design

There are two major considerations in the design and construction of a chromatographic flow-through detector: (1) the minimization of chromatographic profile distortion by band spreading in the detector and (2) the prevention of enzyme–substrate demixing during passage through the system. A basic enzyme-monitoring system is illustrated in Figure 27. The effluent from an analytical column is mixed with a substrate solution and allowed to react for a fixed time in a reaction bed. The amount of product formed is then determined by a detector and recorded.

The Technicon AutoAnalyzer has overcome the problems of band spreading and demixing by segmenting the liquid reactant stream with air bubbles. The alternating gas/liquid segmentation prevents intersegmental mixing while still allowing intrasegmental mixing of reactants. Long reaction times and fast separations are possible. Proper choice of segmentation rate, flow rate, and tube diameter can keep band broadening in the detector at a minimum.[42]

Theoretically the problems of band spreading and demixing may also be overcome in a packed column flow-through detector. Through the efforts of Giddings,[16] Kirkland,[20] Snyder,[20,41] Knox,[22] Karger,[21,41] and Grushka,[17] band spreading in packed columns is well understood. Parameters such as particle size, mobile phase velocity, solute diffusion coefficients, and stationary-phase properties all influence solute band spreading in a separation system. It must be noted however that the packed chromatography column and the packed flow-through enzyme detector are functionally different. The chromatography column is intended to separate compounds while the flow-through reactor must

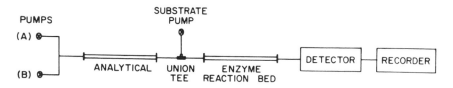

Figure 27. A diagrammatic representation of an enzyme monitoring system. Reprinted from *J. Chromatogr.*, *125*, 103–114 (1976) by courtesy of Elsevier Scientific Publishing Co.

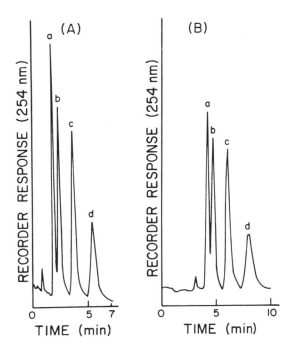

Figure 28. Band spreading of a nucleotide mixture in a post column packed with nonporous support. In A, a 5–10-μm DEAE analytical column is connected directly to a UV 254 detector. In B, a 4.8-×-600-mm post column packed with 40-μm nonporous sodium silicate glass was inserted between the analytical column and the detector. The sample was eluted isocratically with 0.015 M KH_2PO_4 buffer (pH = 3.35) at a mobile phase velocity of 3.33 mm/s. Peak identity is a, CMP; b, AMP; c, UMP; d, GMP. Reprinted from *J. Chromatogr., 125,* 103–114 (1976) by courtesy of Elsevier Scientific Publishing Co.

prevent any separation of the components of a mixture. To achieve this, it is necessary to eliminate all of those features in the flow-reactor support that could cause any separation of the components in a mixture. The packing material must be completely inert to all compounds; i.e., no solute partitioning may occur. Pores must be eliminated to prevent size separation of large protein molecules from their small substrates. Differential rates of diffusion in stagnant mobile-phase pools within supports would also necessitate using nonporous materials. These requirements are fulfilled with an inert nonporous spherical support, i.e., sodium silicate glass beads.

The post column reactor is a 5-×-600-mm stainless steel column that

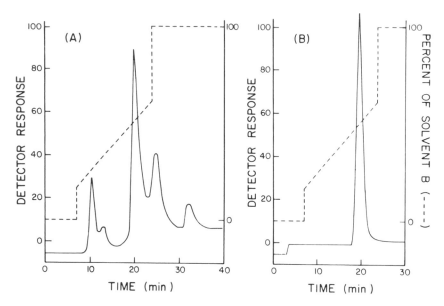

Figure 29. Fractionation of a commercial calf intestinal alkaline phosphatase sample with and without the enzyme detector. Reprinted from *J. Chromatogr., 125,* 103–114 (1976) by courtesy of Elsevier Scientific Publishing Co. Operational parameters are as follows: Sample: (A) 8 mg/mliter calf intestinal alkaline phosphatase, (B) 0.4 mg/mliter calf intestinal alkaline phosphatase. Buffer: initial buffer—0.05 M Tris (pH = 8), final buffer—0.05 M Tris (pH 8), 0.3 M NaCl. Analytical column: 0.05 × 50 cm, 37–74-μm particle size, 250-Å pore diameter, DEAE Glycophase/CPG. Post column: 0.5 × 60 cm 40-μm particle size, Glycophase G/nonporous sodium silicate. Substrate: 8 mM *p*-nitrophenylphosphate in initial buffer. Flow rate: (a) 1.37 mliter/min (analytical pump), (b) 0.35 mliter/min (substrate pump). Detector: (A) 280 nm, (B) 410 nm. Temperature: room temperature.

is packed with 40-μm sodium silicate glass beads. Initially these beads were deactivated with glycerylpropylsilane (Glycophase G), but they were later found to be inert without the coating. The analytical column exit is connected by a union tee to the packed post column reactor as illustrated in Figure 27.[24] The tee also serves as the inlet for the substrate pump. The substrate solution may include other enzymes which must be coupled to produce a spectrophotometrically measurable end product. The band spreading produced by this system is minimal as shown in Figure 28 for a nucleotide mixture chromatographed with and without the detector on a 5–10-μm DEAE Glycophase/CPG column.[24]

4.4. Applications

Figure 29A shows a commercial sample of calf intestinal alkaline phosphatase detected by a 254-nm UV detector.[24] Figure 29B is the same sample using the enzyme detector with a p-nitrophenylphosphate substrate. This shows that the enzyme activity is present in only one peak, illustrating the selectivity of the detector. Besides being specific, this detection system was also more than 20 times as sensitive. Figure 30 shows the detector to have a hundredfold linear range.[24] Since the sensitivity of a detector is also important, alkaline phosphatase was run with an 8-×-250-mm post column detector at 60°C. The wider column increased the incubation time, and the higher temperature increased the reaction rate, so that small amounts of alkaline phosphatase could be detected. The

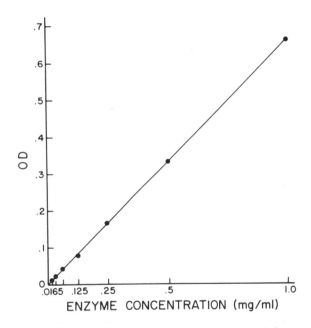

Figure 30. Linearity of flow-through enzyme detector. Reprinted from *J. Chromatogr.*, *125*, 103–114 (1976) by courtesy of Elsevier Scientific Publishing Co. Operational parameters are as follows: Sample: hog intestinal alkaline phosphatase. Buffer: 0.005 M borate buffer (pH = 8) with 2 mM $MgCl_2$ and 0.03 M NaCl. Analytical column: 0.48 × 60 cm, 37–74-μm particle size 250-Å pore diameter, DEAE Glycophage/CPG. Post column: 0.48 × 40 cm, 40-μm Glycophase G/nonporous sodium silicate. Substrate: 4 mM p-nitrophenylphosphate in above buffer. Flow rate: (a) 1.37 mliter/min (analytical pump), (b) 0.70 mliter/min (substrate pump). Detector: 410 nm. Temperature: room temperature.

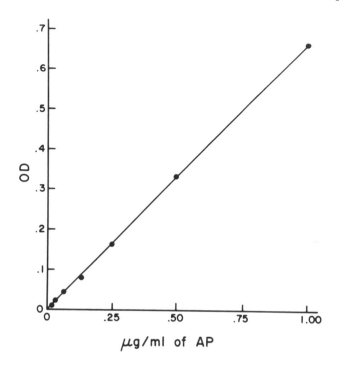

Figure 31. Sensitivity of flow-through enzyme detector. Operational parameters are as follows: Sample: hog intestinal alkaline phosphate (275 μliter). Buffer: 0.05 M Tris, pH = 8.0, 2 mM MgCl₂. Analytical column: None. Post column: 8 × 250 mm SS with 40-μm Glycophase G/nonporous sodium silicate. Substrate: 4 mM *p*-nitrophenylphosphate in 0.05 M Tris (pH 8.0) with 0.3 M NaCl. Flow rate: (a) 1.37 mliter/min (analytical pump), (b) 0.70 mliter/min (substrate pump). Detector: 410 nm. Temperature: 60°C water bath.

results of this sensitivity study are shown in Figure 31. The lower detection limit of alkaline phosphatase was 25 ng/mliter.

Separation of the three creatine phosphokinase (CPK) isoenzymes on 5–10-μm DEAE Glycophase/CPG is shown in Figure 32.[24] The post column reactor was used for detection with a substrate solution of 20 mM phosphocreatine, 1.5 mM ADP, 3 mM glucose, 0.8 mM NADP, 2.7 mM AMP, 2.5 mM glutathione, and 50 units/deciliter of hexokinase and glucose-6-phosphate dehydrogenase in 0.1 M Tris-HCl buffer (pH 6.8) with 0.03 M MgSO₄ and 0.1% albumin.

The five isoenzymes of lactic dehydrogenase (LDH) were separated in 6 min on DEAE Glycophase/CPG as seen in Figure 33.[24] The post column reactor was used for detection with a substrate solution of 0.279 M lactate and 1 mM NAD⁺ in 0.05 M Tris-HCl buffer (pH 8.8) with 0.1% albumin.

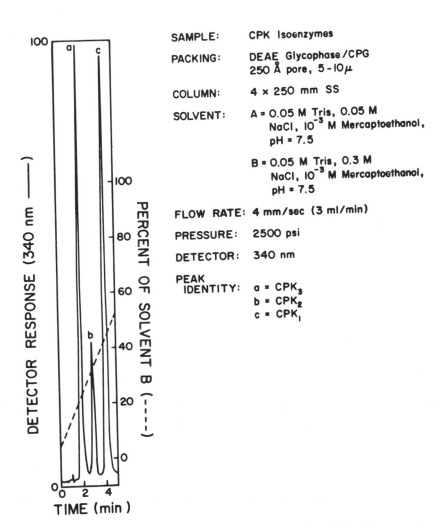

SAMPLE: CPK Isoenzymes

PACKING: DEAE Glycophase/CPG
            250 Å pore, 5-10μ

COLUMN: 4 × 250 mm SS

SOLVENT: A = 0.05 M Tris, 0.05 M
            NaCl, 10^{-3} M Mercaptoethanol,
            pH = 7.5

            B = 0.05 M Tris, 0.3 M
            NaCl, 10^{-3} M Mercaptoethanol,
            pH = 7.5

FLOW RATE: 4 mm/sec (3 ml/min)

PRESSURE: 2500 psi

DETECTOR: 340 nm

PEAK
  IDENTITY: a = CPK_3
             b = CPK_2
             c = CPK_1

Figure 32. Separation of CPK isoenzymes on DEAE Glycophase/CPG with a post column enzyme detector. Reprinted from *J. Chromatogr.*, *125*, 103–144 (1976) by courtesy of Elsevier Scientific Publishing Co.

Fred E. Regnier *et al.*

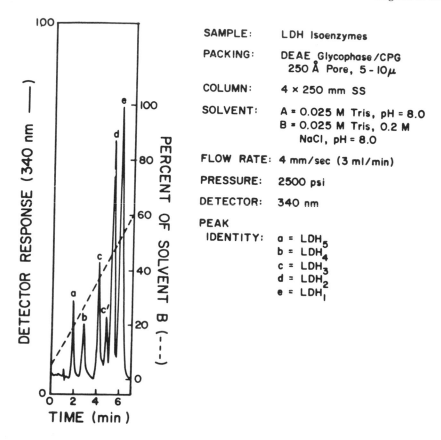

SAMPLE: LDH Isoenzymes

PACKING: DEAE Glycophase/CPG
250 Å Pore, 5 - 10μ

COLUMN: 4 × 250 mm SS

SOLVENT: A = 0.025 M Tris, pH = 8.0
B = 0.025 M Tris, 0.2 M
NaCl, pH = 8.0

FLOW RATE: 4 mm/sec (3 ml/min)

PRESSURE: 2500 psi

DETECTOR: 340 nm

PEAK
IDENTITY: a = LDH$_5$
b = LDH$_4$
c = LDH$_3$
d = LDH$_2$
e = LDH$_1$

Figure 33. Separation of LDH isoenzymes on DEAE Glycophase/CPG with a post column enzyme detector. Reprinted from *J. Chromatogr., 125,* 103–114 (1976) by courtesy of Elsevier Scientific Publishing Co.

5. Summary

It has recently been demonstrated that proteins can be separated at 10–1000 times the rate possible on classical gel supports using bonded stationary phases on an inorganic support. A steric exclusion chromatography support has been synthesized, in addition to four ion-exchangers which duplicate those traditionally used: DEAE, a weak anion-exchanger; QAE, a strong anion-exchanger; CM, a weak cation-exchanger; and SP, a strong cation-exchanger. All of these supports are capable of withstanding 5000 psi and a variety of chromatography solvents without deleterious effects. The smaller diffusion coefficients of proteins have a major effect on column efficiency and separation speed, but fast separations are still possible on microparticulate (e.g., 5–10-μm particle) columns. A flow-through detector makes it possible to detect enzymes with specificity and enhanced sensitivity. These systems should be particularly useful in clinical situations where rapid identification and quantitation of enzymes and isoenzymes are of diagnostic importance.

6. Appendix

In gel chromatography, peak widths usually remain constant as the elution volume increases.[20] This implies that efficiency increases with elution volume, which is different from other forms of chromatography. When deriving equations that express the resolution of GPC systems, this fact is commonly ignored. We will define resolution (R_S) of two solutes as

$$R_S = \frac{V''_e - V'_e}{\Delta V} \tag{A1}$$

where V'_e and V''_e are the elution volumes of the first and second solute, respectively, and ΔV is their common peak width. Recalling that $N = 16$ $(V_e/\Delta V)^2$, it is seen that

$$\Delta V = (4/\sqrt{N})V_e \tag{A2}$$

Since the column efficiencies at V'_e and V''_e are different, the average elution volume of the two solutes, $(V''_e + V'_e)/2$ will be used in the derivation of a resolution equation. Substitution of the average elution volume in equation (A2) produces the formula

$$\Delta V = \frac{2}{\sqrt{N}} (V''_e + V'_e) \tag{A3}$$

where N is the number of plates at the average elution volume. Combining equations (A1) and (A3) the resolution at the average elution volume is

$$R_S = \frac{\sqrt{N}}{2} \frac{V''_e - V'_e}{V''_e + V'_e} \tag{A4}$$

Recalling that $V'_e = V_0 + K'_D V_i$ and $V''_e = V_0 + K''_D V_i$ and substituting these values in equation (A4) gives

$$R_S = \frac{\sqrt{N}V_i}{2} \frac{(K''_D - K'_D)}{2V_0 + V_i(K''_D + K'_D)} \tag{A5}$$

Substituting $\alpha = K''_D/K'_D$ and $p = V_i/V_0$, from Section 2.6 of the text, produces a general resolution equation for GPC:

$$R_S = \frac{\sqrt{N}}{2} \frac{pK_D''(\alpha - 1)}{2\alpha + pK_D''(\alpha + 1)} \tag{A6}$$

This equation employs dimensions encountered in GPC such as $p = V_i/V_0$ and K_D. α is easily calculated from the relative K_D values and N is purely a function of the system.

7. References

1. H. Fasold, G. Gundlach, and F. Turba, in: *Chromatography* (E. Heftmann, ed.), pp. 378–427, Reinhold, New York (1961).
2. W. Haller, Material and Method for Performing Steric Separations, U. S. Patent 3,549,524.
3. W. Haller, Chromatography on glass of controlled pore size, *Nature, 206,* 693–696 (1965).
4. W. Haller, Virus isolation with glass of controlled pore size, MS-2 Bacteriophage and Kilham Virus, *Virology, 33,* 740–743 (1967).
5. H. H. Gschwender, W. Haller, and P. H. Hofschneider, Large-scale preparation of viruses by steric chromatography on columns of controlled-pore glass: 0X174-, M13-, M12-, QB- and T4-bacteriophages, *Biochim. Biophys. Acta, 190,* 460–469 (1969).
6. W. Haller, K. D. Tympner, and K. Hannig, Preparation of immunoglobulin concentrates from human serum by chromatography on controlled pore glass, *Anal. Biochem., 35,* 23–31 (1970).
7. C. W. Hiatt, A. Shelokov, E. J. Rosenthal and J. M. Galimore, Treatment of controlled pore glass with poly(ethylene-oxide) to prevent adsorption of rabies virus, *J. Chromatogr., 56,* 362–364 (1971).
8. K. Marcinka, Application of permeation chromatography on controlled-pore glass in the purification of plant viruses, *Acta Virol., 16,* 52–62 (1972).
9. Y. A. Eltekov, A. V. Kiselev, T. D. Khokhlova, and Y. S. Nikitin, Adsorption and chromatography of proteins on chemically modified macroporous silica-aminosilochrom, *Chromatographia, 6,* 187–189 (1973).
10. P. J. Kudirka, M. G. Busby, R. N. Carey, and E. C. Toren, Separation of creatine kinase isoenzymes by high-pressure liquid chromatography, *Clin. Chem., 21,* 450–452 (1975).
11. C. R. Lowe and P. D. G. Dean, *Affinity Chromatography,* John Wiley and Sons, Inc., New York (1974).
12. F. E. Regnier and R. Noel, Glycerolpropylsilane bonded phases in the steric exclusion chromatography of biological macromolecules, *J. Chromatogr. Sci., 14,* 316–32 (1976).
13. D. C. Locke, Chemically bonded stationary phases for liquid chromatography, *J. Chromatogr. Sci., 11,* 120–128 (1973).
14. R. C. Collins and W. Haller, Protein–sodium dodecyl sulfate complexes: Determination of molecular weight, size and shape by controlled pore glass chromatography, *Anal. Biochem., 54,* 47–53 (1973).
15. Corning Controlled Pore Glass CPG-10 Series Column Packing, Corning Chromatography Products Product Information. Data Sheet 104 (1969).

16. J. C. Giddings, *Dynamics of Chromatography Part I*, Marcel Dekker, Inc., New York (1965).
17. E. Grusha, L. R. Snyder, and J. H. Knox, Advances in band spreading theories, *J. Chromatogr. Sci.*, *13*, 25–37 (1975).
18. R. E. Majors, Effect of particle size on column efficiency in liquid–solid chromatography, *J. Chromatogr. Sci.*, *11*, 88–95 (1973).
19. J. Vermont, M. Deleuil, A. J. deVries, and C. L. Guillemin, Modern liquid chromatography on Spherosil, *Anal. Chem.*, *47*, 1329–1337 (1975).
20. L. R. Snyder and J. J. Kirkland, *Introduction to Modern Liquid Chromatography*, John Wiley and Sons, Inc., New York (1974).
21. B. L. Karger in: *Modern Practice of Liquid Chromatography* (J. J. Kirkland, ed.), pp. 3–53, John Wiley and Sons, Inc., New York (1971).
22. G. J. Kennedy and J. H. Knox, The performance of packings in high performance liquid chromatography (HPLC). 1. Porous and surface layered supports, *J. Chromatogr. Sci.*, *10*, 549–556 (1972).
23. F. E. Regnier and K. M. Gooding, unpublished results.
24. S. H. Chang, K. M. Gooding, and F. E. Regnier, High speed liquid chromatography of proteins, *J. Chromatogr.*, *125*, 103–114 (1976).
25. J. Asshauer and I. Halasz, Reproducibility and efficiency of columns packed with 10 μ silica in liquid chromatography, *J. Chromatogr. Sci.*, *12*, 139–147 (1974).
26. J. J. Kirkland, Performance of zipax controlled surface porosity support in high speed liquid chromatography, *J. Chromatogr. Sci.*, *10*, 129–137 (1972).
27. R. E. Majors, High performance liquid chromatography on small particle silica gel, *Anal. Chem.*, *44*, 1722–1726 (1972).
28. R. M. Cassidy, D. S. LeGay, and R. W. Frei, Study of packing techniques for small-particle silica gels in high-speed liquid chromatography, *Anal. Chem.*, *46*, 340–344 (1974).
29. J. J. Kirkland, High speed liquid-partition chromatography with chemically bonded organic stationary phases, *J. Chromatogr. Sci.*, *9*, 206–214 (1971).
30. S. H. Chang, K. M. Gooding, and F. E. Regnier, The use of oxiranes in the preparation of bonded phase supports, *J. Chromatogr.*, *120*, 321–333 (1976).
31. S. H. Chang, R. Noel, and F. E. Regnier, High speed ion exchange chromatography of proteins, *Anal. Chem.*, *48*, 1839–1845 (1976).
32. H. Purnell, *Gas Chromatography*, John Wiley and Sons, Inc., New York, 1962.
33. E. C. Horning, E. A. Moscatelli, and C. C. Sweeley, Polyester liquid phases in gas–liquid chromatography, *Chem. Ind.* (London) *1959*, 751–752.
34. R. Roberts, P. D. Henry, S. A. G. J. Witteveen, and B. A. Sobel, Quantification of serum creatine phosphokinase isoenzyme activity, *Am. J. Cardiol.*, *33*, 650–654 (1974).
35. V. Anido, R. B. Conn, H. F. Mengoli, and G. Anido, Diagnostic efficacy of myocardial creatine phosphokinase using polyacrylamide disk gel electrophoresis, *Am. J. Clin. Pathol.*, *61*, 599–605 (1974).
36. D. Mercer, Separation of tissue and serum creatine kinase isoenzymes by ion-exchange column chromatography, *Clin. Chem.*, *20*, 36–40 (1974).
37. P. D. Henry, R. Roberts, and B. E. Sobel, Rapid separation of plasma creatine kinase isoenzymes by batch adsorption on glass beads, *Clin. Chem.*, *21*, 844–849 (1975).
38. H. A. Sober and E. A. Peterson, Protein chromatography on ion exchange cellulose, *Fed. Proc. Fed. Am. Soc. Exp. Biol.*, *17*, 1116–1126 (1958).
39. H. J. Keutel, K. Okabe, H. K. Jacobs, F. Ziter, L. Maland, and S. A. Kuby, Studies and adenosine triphosphate transphosphorylase, *Arch. Biochem. Biophys.*, *150*, 648–678 (1972).

40. Y. Birk, A. Gertler, and S. Khalef, A pure trypsin inhibitor from soya beans, *Biochem. J., 87,* 281–284 (1963).
41. B. L. Karger, L. R. Snyder, and C. Horvath, *An Introduction to Separation Science,* John Wiley and Sons, Inc., New York (1973).
42. L. R. Snyder, Reaction colorimeters as detectors in high performance liquid chromatography. Extra-column band broadening with segmented flow through the reaction coil, Address at the Second International Symposium on Column Liquid Chromatography, Wilmington, Delaware (1976).

2

Chemiluminescence and Bioluminescence Analysis

W. Rudolf Seitz and Michael P. Neary

1. Introduction

Certain chemical reactions lead to the formation of products in excited electronic states that return to the ground state by emitting a photon. Because excitation is chemical rather than photolytic, this light is known as chemiluminescence (CL) even though the emission spectra match the photoluminescence (fluorescence or phosphorescence) spectra of the excited-state products. If an excited-state product does not itself luminesce, addition of a luminescent energy acceptor can lead to sensitized chemiluminescence.

Three conditions are necessary for a reaction to lead to chemiluminescence:

1. The free energy of the reaction must be sufficient to form product in an excited electronic state.
2. The reaction pathway must favor the formation of excited state product. The efficiency of excited-state formation, ϕ_R, can be defined:

$$\phi_R = \frac{\text{the number (or rate) of molecules forming an excited state}}{\text{the number (or rate) of molecules reacting}}$$

W. Rudolf Seitz and Michael P. Neary • Department of Chemistry, University of Georgia, Athens, Georgia 30602. Present address of W. R. S.: Department of Chemistry, University of New Hampshire, Durham, New Hampshire 03824. Present address of M. P. N.: Los Alamos Scientific Laboratories WX-3, M. S. 938, Los Alamos, New Mexico 87545.

3. The excited state must luminesce. The efficiency of luminescence, ϕ_L, is defined:

$$\phi_L = \frac{\text{the number (or rate) of molecules luminescing}}{\text{the number (or rate) of molecules forming an excited state}}$$

ϕ_L is equivalent to the fluorescence of phosphorescence efficiency. The chemiluminescence efficiency, ϕ_{CL}, is defined:

$$\phi_{CL} = \frac{\text{the number (or rate) of molecules luminescing}}{\text{the number (or rate) of molecules reacting}}$$

As can be seen from the definitions:

$$\phi_{CL} = \phi_R \phi_L \tag{1}$$

Although the number of reactions satisfying the conditions for CL is relatively small, many different types of reactions can lead to CL. Table 1 lists some representative CL reactions. Energetic gas-phase reactions often lead to CL, particularly at low pressures, since alternate mechanisms of energy dissipation require a third body which is not readily available in the gas phase. CL is observed both from homogeneous gas phase reactions,

Table 1. Some Representative Reactions Leading to Chemiluminescence

		Reference
1.	$O_3 + NO \longrightarrow NO_2^* + O_2$	(1)
	$NO_2^* \longrightarrow NO_2 + h\nu \ (\lambda_{max} = 1200 \text{ nm})$	
2.	$H + PO \longrightarrow POH^*$	(2)
	$POH^* \longrightarrow POH + h\nu \ (\lambda_{max} = 525 \text{ nm})$	

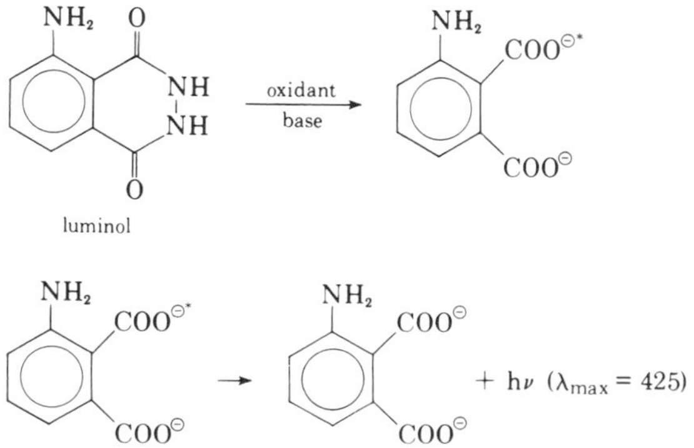

luminol

(3)

Table 1. (*Continued*)

	Reference

4. $H_2O_2 + OCl^- \longrightarrow O_2^*$ (4)

$$O_2 + h\nu \; (\lambda_{max} = 703 \text{ nm})$$

5.

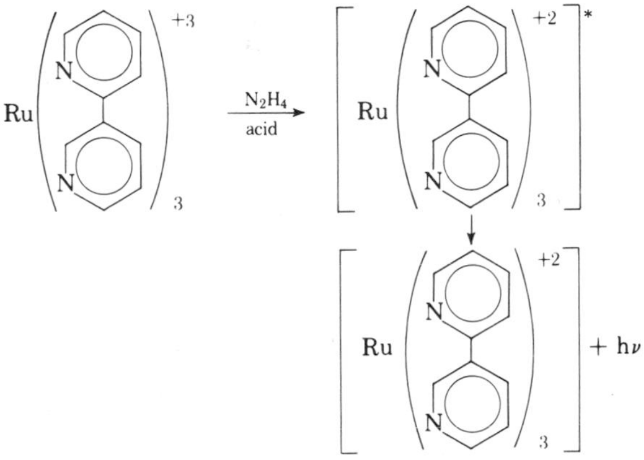

$$(\lambda_{max} = 595 \text{ nm})$$

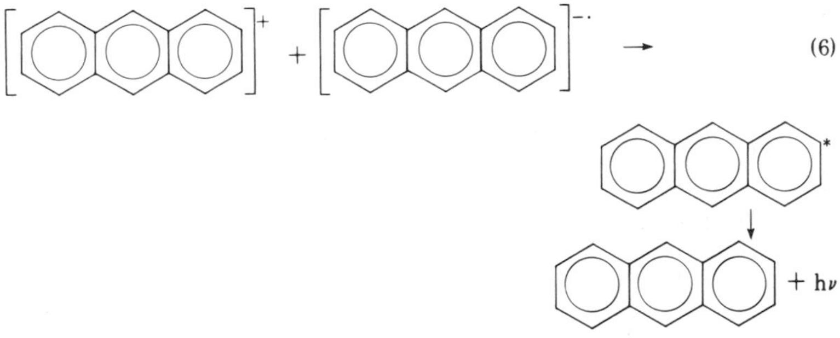

       (5)

6.

$$(\lambda_{max} = 440 \text{ nm})$$

       (6)

7.[a] $\overset{\displaystyle O \quad O}{\underset{\displaystyle \parallel \quad \parallel}{Cl-C-C-Cl}} + H_2O_2 + flr \longrightarrow flr^*$

$$flr + h\nu$$

       (7)

8.[b] $LH_2 + E + ATP + Mg^{+2} + O_2 \longrightarrow [Oxyluciferin]^*$

$$[Oxyluciferin] + h\nu$$

$$(\lambda_{max} = 562 \text{ nm})$$

       (8)

---

[a] Flr, a fluorescing molecule.
[b] LH$_2$, luciferin; E, enzyme; ATP, adenosine triphosphate.

such as the ozone–nitric-oxide reaction (reaction 1, Table 1), and from flame reactions such as the reaction between hydrogen atoms and PO molecules when phosphorus compounds are burned in a hydrogen–air flame (reaction 2, Table 1). Liquid-phase CL has been observed for the oxidation of organic compounds such as luminol (reaction 3) and inorganic compounds (reaction 4), as well as from electron transfer reactions of metal chelates (reaction 5), and anion-radical–cation-radical annihilations (reaction 6). Sensitized CL has been observed for energetic liquid-phase reactions that do not directly lead to luminescent products (reaction 7).

An important subclass of CL reactions occurs in biological systems and involves an enzyme–substrate interaction. Light generated in this fashion is known as bioluminescence (BL). Although the firefly reaction (reaction 8, Table 1) is the best known BL system, BL is most frequently observed in species living deep in the ocean, where the absence of sunlight has favored the evolution of creatures capable of producing their own light.

Chemiluminescence efficiencies vary widely. Bioluminescence reactions are very efficient, with ϕ_{CL} often equal or close to unity. For nonbiological systems ϕ_{CL} rarely exceeds 0.01.

For more details on CL and BL the reader is referred to other work. A listing of some of the more useful references is given in Table 2.

1.1. Chemical Analysis Using Chemiluminescence

Chemiluminescence reactions are well suited for chemical analysis. The progress of a reaction can be followed by measuring CL intensity as a function of time. Even with simple instrumentation CL intensities can be measured very sensitively. With photon counting, individual CL events can be directly observed. Bioluminescence is even better for chemical analysis since the inherent specificity of enzyme reactions plus the high efficiency of BL is added to the other advantages of CL. The applications of CL and BL to chemical analysis have been recently reviewed.[9,10]

If an analyte is reacted with an excess of CL-generating reagents, the CL intensity is related to analyte concentration by

$$I_{CL}(t) = \phi_{CL} \frac{dC}{dt} \qquad (2)$$

where $I_{CL}(t)$ is CL intensity (in photons per second) at time t, ϕ_{CL} is the CL efficiency based on the analyte (photons/molecules reacting), and dC/dt is the rate at which analyte is reacting (molecules reacting per second). It does not matter whether the electronically excited product is derived from the analyte or from a reagent as long as ϕ_{CL} is based on the analyte.

Table 2. Books Dealing with Chemiluminescence and Bioluminescence

Chemiluminescence and Bioluminescence, M. J. Cormier, D. M. Hercules, and J. Lee, eds., Plenum Press, New York, (1973).

Chemilumineszenz Organischer Verbindungen, K. D. Gundermann, Springer-Verlag, New York, (1969).

Light and Life, W. D. McElroy and B. Glass, eds., Johns Hopkins Press, Baltimore, Md., (1961).

Bioluminescence in Progress, F. H. Johnston and Y. Haneda, eds., Princeton University Press, Princeton, N. J., (1966).

Bioluminescence, E. N. Harvey, Academic Press, New York, (1952).

Photophysiology, Vols. II, IV, V, A. C. Gese, Academic Press, New York, (1970).

If the kinetics of the CL reaction are first order with respect to the analyte, then the rate of change in I_{CL} as analyte is consumed can be directly related to concentration. In practice, however, CL is usually integrated for a known time period:

$$\int I_{CL}dt = \phi_{CL} \int \frac{dC}{dt} dt \tag{3}$$

Integrated CL intensity is directly proportional to analyte concentration. Such measurement can be done by stopped-flow mixing of analyte and reagents in front of a light detector and integrating the resulting light-vs.-time measurement. Alternatively, analyte and reagents can be continuously mixed in front of a light detector and steady-state CL measured. This, in effect, integrates the CL from the time of mixing for the average length of time the sample remains in front of the detector. While the stopped-flow measurement provides information about the kinetics of CL that cannot be obtained from the steady-state measurement, the steady-state measurement makes it possible to do continuous analysis using CL.

Chemical analysis using CL requires a reagent that reacts with the analyte to generate CL. Because the total number of known CL reactions is relatively small, the number of possible direct CL analyses is limited. For existing CL reactions, three possible situations are important analytically:

1. In the ideal case, the CL reaction is specific for the species of interest. No resolution, either optical or chemical, is required. An example is the firefly reaction which is specific for adenosine triphosphate (ATP).

2. Several different excited-state products are formed in the reaction between the sample and the CL-generating reagent. In this case, optical resolution can be used to selectively measure the CL of interest, provided that it occurs at a unique frequency. In practice,

a filter is usually satisfactory for optical resolutions. An example of this situation is the use of the O_3–NO reaction to monitor atmospheric NO. Since O_3 can react with several other possible components of the atmosphere (e.g., olefins) to produce CL, a cutoff filter is necessary to insure selectivity for NO.

3. The CL-generating reagent itself luminesces upon reacting with a variety of possible analytes. Since several species all produce the same luminescence, optical resolution is not possible. Instead, resolution must be based on differences in the chemistry of the CL-generating species. For example, several metal ions catalyze the oxidation of luminol by hydrogen peroxide to produce CL. Selectivity must be based on metal ion chemistry. One way of achieving selectivity is to do an ion exchange separation prior to reacting the metals with hydrogen peroxide and luminol.

Indirect analysis using CL and BL offers the greatest possibility for future development. In this case, an analyte is reacted to a product that participates in a CL or BL reaction. For example, enzymatic oxidation of glucose by glucose oxidase yields hydrogen peroxide, which can be sensitively and conveniently determined using the luminol reaction.[11,12] Another instance of indirect analysis is the conversion of creatine phosphate to ATP, which is then measured using the firefly reaction.[13] The specificity of these assays depends on the chemistry involved. For this reason, this approach is particularly well suited for work with enzyme reactions that are by nature specific.

1.2. Sensitivity

The detection limits for analytical methods based on CL are frequently determined by factors other than the capability of modern instrumentation to measure low light levels. This is because many CL methods are so sensitive that contamination and other "low concentration effects" become significant before the limits of light-measuring capabilities are approached.

Nevertheless, it is instructive to calculate the minimum detectable concentrations for CL analysis assuming that they are determined by light-measuring capability. The theoretical minimum reaction rate detectable by CL can be defined as the rate where the measured photon flux from the reaction equals three times the square root of the background flux:

$$I_{CL\ measured} = 3\sqrt{N(t)} \tag{4}$$

where $N(t)$ equals the number of background counts per second and I_{CL} is in photons per second. The measured photon flux is given by

$$I_{CL\ measured} = \phi_{det}\phi_{geo}\phi_{CL} \frac{dC(t)}{dt} \tag{5}$$

where

ϕ_{det} = quantum efficiency of the detector for the wavelength distribution of the CL spectra (photons measured/photons impinging on the detector).

ϕ_{geo} = the fraction of emitted photons impinging on the detector (photons impinging on the detector/photons emitted).

ϕ_{CL} depends on the reaction being used analytically. ϕ_{geo} can be maximized by proper choice of geometry. ϕ_{det} depends both on the quantum efficiency of the detector as a function of wavelength and the CL spectra. Since equations (4) and (5) define a minimum detectable reaction rate, the minimum detectable amount will depend on the kinetics of the analytical reaction and the time span of the measurement. If the kinetics are first order in the species being determined,

$$\frac{dC}{dt} = KC(t) \tag{6}$$

If CL is measured from t_1 to t_2, the minimum detectable concentration, C_{min}, can be calculated from equations (4), (5), and (6):

$$C_{min}(e^{-Kt_1} - e^{-Kt_2}) = \frac{3\sqrt{N_{t_1-t_2}}}{\phi_{det}\phi_{geo}\phi_{CL}} \tag{7}$$

where $N_{t_1-t_2}$ equals the number of background counts integrated from t_1 and t_2. The fraction of the total analyte concentration reacted between t_1 and t_2 is given by $(e^{-Kt_1} - e^{-Kt_2})$.

The values of these parameters depend on the actual experimental conditions. ϕ_{geo} can easily be made as high as 0.5 using reflectors to direct luminescence on active detector surface. ϕ_{det} can also approach 0.5 for sensitive photocathodes at their optimum wavelengths. Background counting rates may be on the order of 100 counts per second. This can be reduced if necessary by cooling the photomultiplier tube. ϕ_{CL} is close to unity for most BL reactions such as the firefly reaction but rarely exceeds 0.01 for reactions not derived from living systems. Kinetics of course are highly variable from reaction to reaction.

For illustrative purposes the detection limit will be calculated for an arbitrary set of conditions $\phi_{det} = 0.2$, $\phi_{geo} = 0.2$, $\phi_{CL} = 0.01$, background count rate = 100 counts per second, intensity integrated from 0 to 60 seconds after mixing, and reaction complete after 60 s. Plugging these values into equation (7), the minimum detectable amount is calculated to be 4.5×10^7 molecules or 7.5×10^{-17} mol.

Theoretical detection limits are rarely realized in practice but they do illustrate some important points:

1. CL methods are very sensitive. Because of this sensitivity, response is often linear over several orders of magnitude, since effects leading to nonlinearity occur at relatively high concentrations.
2. When working with efficient CL reactions, lowest detection limits will often be achieved by taking maximum precautions to avoid contamination rather than using the most sensitive light-measuring instrumentation.
3. A very important conclusion is that ϕ_{CL} need not be large for a CL reaction to be sensitive for chemical analysis. Most of the research on CL has been directed toward achieving maximum efficiency in order to find a reaction that could serve as light source. There has been little work done to find reactions with lower ϕ_{CL} that could be useful analytically.

2. Gas-Phase Chemiluminescence

Chemiluminescence used analytically has had its greatest impact in the field of air pollution. In the last few years a new generation of CL techniques has replaced earlier methods used for air pollutants that lacked specificity and which required bubbling the sample through a scrubbing solution for an extended period of time. The new methods involve both gas-phase CL, considered in this section, and flame CL, discussed in the next section. There are several recent reviews in this area by researchers active in the field.[14-18]

2.1. Ozone

Ozone has been determined by CL in several ways. In the earliest methods, ozone was determined by its reaction with luminol (5-amino-2,3-dihydrophthalazine-1,4-dione).[19,20] A better method involves passing an air sample over rhodamine-B adsorbed on silica gel.[21-24] A diagram of the apparatus is shown in Figure 1. A pump draws the air sample over the rhodamine-B surface, which is positioned in front of a photomultiplier tube. The ozone scrubber is used to generate ozone-free air for setting zero response, while an ozone source is used to calibrate the surface. The observed CL emission corresponds to the dye fluorescence. The dye is not

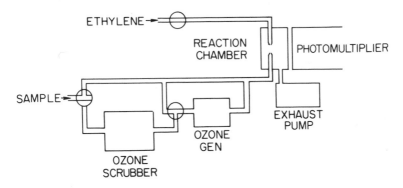

Figure 1. Diagram of apparatus for determining ozone concentrations based on homogeneous gas phase CL from the ethylene–ozone reaction.

consumed in the CL reaction, indicating that the mechanism involves energy transfer from an excited intermediate to the dye. Other dyes fluorescing in the red can be used in place of rhodamine-B.

This method of determining ozone has been used in monitoring instruments that have performed reliably in the field. With a freshly prepared surface the detection limit is 0.2 ppb, and response is linear up to 0.4 ppm.[22] The main problem with the method is a gradual loss in sensitivity with time due to slow oxidation of the rhodamine-B by ozone. The surface response must be frequently recalibrated using the internal ozone source. The lifetime of a surface is 2 to 3 months of continuous monitoring with the sample diluted by a factor of 10. The sample dilution is possible because the sensitivity of the method is more than sufficient to monitor ambient ozone levels.

Ozone has also been determined by passing an air sample through a solution of rhodamine-B and gallic acid in ethanol.[25] The function of the gallic acid is to protect the rhodamine-B from oxidation. The ozone reacts preferentially with the gallic acid to form an excited intermediate which then energy transfers to the rhodamine-B.

An alternative method for ozone is based on the gas phase reaction between ozone and ethylene at atmospheric pressure.[26,27] The emission is a continuum from 300–600 nm (λ_{max} = 435 nm). The emitting species formed in the reaction include formaldehyde and OH.[28] The mechanism and kinetics of the reaction between ozone and olefins have been studied because of their importance in the formation of photochemical smog.[28,29]

The apparatus used is similar to that used for the ozone–rhodamine-B monitors, except that the active surface is replaced by a source of excess ethylene that reacts with ozone in a chamber in front of a photomultiplier. Typically, the flow rate of sample is 975 cm³/min, while the ethylene flow rate is 25 cm^3/min. This method has a detection limit of 1 ppb and responds linearly up to 30 ppm ozone. It is now the reference method for monitoring ozone in the atmosphere.

The reaction of O_3 with excess NO to generate red CL (see below) can also be used to determine O_3.[30] The O_3–ethylene reaction is preferred, however, since the O_3–NO reaction requires an expensive red-sensitive photomultiplier tube.

2.2. Analysis for Oxides of Nitrogen Using the O₃–NO Reaction

Several CL reactions can be used to monitor for oxides of nitrogen. Of these, the reaction most widely applied is between ozone and nitric oxide.[30,31] The mechanism for this reaction is[31]

$$NO + O_3 \longrightarrow NO_2{}^* + O_2 \tag{8}$$

$$NO + O_3 \longrightarrow NO_2 \text{ (gnd state)} + O_2 \tag{9}$$

$$NO_2{}^* \longrightarrow NO_2 + h\nu \tag{10}$$

$$NO_2{}^* + M \longrightarrow NO_2 + M \tag{11}$$

The rate constants for these reactions have been measured.[32] The rate constant for quenching (reaction (11) varies for different quenching gases. In the atmosphere, the quenchers N_2 and O_2 are present at constant concentration so this is not a problem for monitoring. Because of third-body quenching, greatest analytical sensitivity is achieved at reduced pressures.

The emission from $NO_2{}^*$ extends from 600–3000 nm ($\lambda_{max} = 1200$ nm). A cooled, red-sensitive photomultiplier is used to increase sensitivity of detection.

Figure 2 shows a diagram of the instrument used for monitoring oxides of nitrogen in the atmosphere. The sample is reacted with oxygen containing approximately 2.5% ozone in a chamber positioned in front of the photomultiplier. A filter absorbing emission at wavelengths shorter than 600 nm blocks out CL from ozone reactions with other possible constituents of the atmosphere such as olefins.

In the most sensitive monitors the pressure is maintained at 1–5 torr by a small vacuum pump. The flow rates at STP for both sample and reactant gases are in the range from 50–100 cm^3/min. The detection limit for this instrument is 1 ppb with linear response up to 10,000 ppm, a linear dynamic range of 10^7. Higher pressures can be used with a reduction in sensitivity. The quenching effects at high pressure are partially compensated for by using higher sample flow rates. The advan-

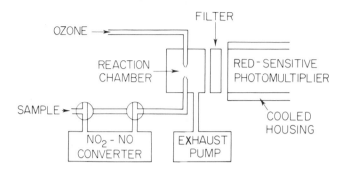

Figure 2. Diagram of apparatus for determining oxides of nitrogen concentrations based on homogeneous gas phase CL from the ozone–NO reaction.

tage of working at higher pressures is that a mechanical vacuum pump is not required.

The CL reaction measures only NO and not NO_2. To measure total oxides of nitrogen, NO_2 is reduced to NO before entering the reaction chamber. Most monitors use some form of carbon at high temperatures for the reduction. The concentration of NO_2 in a sample can be determined from the difference in CL intensity when the sample passes through the NO_2-to-NO converter and when it bypasses the converter.

2.3. Ammonia

Ammonia can be oxidized to NO at high temperatures and detected by CL.[17] It is a potential positive interference in the CL method for oxides of nitrogen. The NH_3 interference can be removed by passing an air sample through an acidic scrubber or, preferably, by reducing NO_2 to NO at conditions that will not cause NH_3 oxidation.

Ammonia can be determined by using conditions where it is oxidized to NO. The difference between the CL intensity with and without an acidic scrubber is proportional to NH_3 concentration.[33] NH_3 has also been measured in a CL instrument containing two converters, one that converts both NO_2 and NH_3 to NO while the other converts only NO_2 to NO.[34]

Total atmospheric acid can be determined by reacting an air sample with excess ammonia at a known concentration.[35] The resulting ammonium salt particles are removed by filtration. The remaining NH_3 is then converted to NO and determined by CL. The reduction in NH_3 concentration caused by the neutralization reaction is proportional to the acid concentration.

2.4. Reactions of Atomic Oxygen

Atomic oxygen is a potential reagent for determining several air pollutants by CL. It can be used to determine total oxides of nitrogen using the reactions[36]

$$O + NO_2 \longrightarrow NO + O_2 \qquad (12)$$

$$O + NO \longrightarrow NO_2* \qquad (13)$$

$$NO_2* \longrightarrow NO_2 + h\nu \qquad (14)$$

Because O atoms reduce NO_2 to NO, the determination measures total oxides of nitrogen. The NO_2 emission is shifted to shorter wavelengths (400–1400 nm) because atomic oxygen is more energetic than ozone. This is an advantage for analysis of oxides of nitrogen since it does not require as expensive a photomultiplier as the ozone–NO reaction.

Oxygen atoms can be generated using a microwave or electrical discharge through oxygen or oxygen/argon mixtures. The problem is that the discharge produces NO from traces of nitrogen present in even the purest gases, producing a fluctuating background signal.[17] This difficulty has been overcome by using high temperature to dissociate ozone into oxygen molecules and oxygen atoms.[37] The reported instrument measures total oxides of nitrogen when the thermal converter is on, generating O atoms and NO only when the thermal converter is off.

The O–NO reaction has been used to determine NO_2 in air.[38] The sample is first photolyzed at reduced pressure dissociating NO_2 to NO and O. The O atoms are then reacted with excess NO in front of a photomultiplier. The concentration of O atoms is proportional to NO_2 concentration. This method has a detection limit of 1 ppb and is linear up to 1 ppm. Chemiluminescence from the O_3–NO reaction is a positive interference. It can be compensated for by turning off the photolysis source so that no O atoms are produced. The remaining signal from the O_3–NO reaction can be subtracted from the total signal. By turning the photolysis source on and off alternately, the same system can be used to determine NO_2 and O_3.

Atomic oxygen reacts with SO_2 producing CL[39]:

$$O + O + SO_2 \longrightarrow O_2 + SO_2* \qquad (15)$$

$$SO_2* \longrightarrow SO_2 + h\nu(\lambda_{max} = 280 \text{ nm}) \qquad (16)$$

SO_2 can be determined with a detection limit of 1 ppb, using an interference filter to resolve the SO_2 emission from the other emissions generated by atomic O. Problems with O-atom sources have hindered practical use of this reaction.

Oxygen atoms also react with carbon monoxide[15]:

$$O + CO \longrightarrow CO_2^* \tag{17}$$

$$CO_2^* \longrightarrow CO_2 + h\nu \tag{18}$$

The emission extends from 300 to 500 nm. A detection limit of about 1 ppb CO in air has been estimated for this reaction.

The reaction of O atoms with hydrocarbons produces CL from OH, CH, C_2 and HCO.[40,41] Unfortunately the emission spectra are too general for detection of specific compounds.

2.5. Other CL Reactions

Ozone reacts with various sulfur compounds including H_2S and organic sulfides producing electronically excited SO_2 as a product.[42] This could be used for nonspecific detection of atmospheric sulfur. The rate at which CL is generated varies for different compounds. This means that sensitivity will depend on the compound measured when the CL reaction is performed in a flow system.

Peroxyacetyl nitrate (PAN), one of the compounds involved in photochemical smog, reacts with triethyl amine to produce broad-band emission peaking at 665 nm.[43] The detection limit is 6 ppb.

2.6. N-Nitrosoamines

Recently, gas-phase CL has been used for selective detection of N-nitrosoamines.[44,45] Many N-nitrosoamines are carcinogenic, often at rather low concentrations. Since sodium nitrite is used as a preservative in foods, there is concern that it may react with amines in the food to form carcinogens. Because of the complexity of foodstuffs as an analytical matrix, a selective and sensitive method for detecting N-nitroso compounds is needed.

To analyze for N-nitrosoamines, a sample is injected into a catalytic converter at 300°C which cleaves the N–N bond, producing NO. The sample then passes through a cold trap which freezes out organic fragments and enters a CL NO analyzer. This technique is highly selective. Organic nitrites are the only compounds besides N-nitroso compounds that are detected. By combining gas chromatography with this detector, identification of specific N-nitrosoamines is possible.

3. Flame Chemiluminescence

3.1. General Characteristics

Chemical reactions in flames can produce electronically excited atoms and molecules at concentrations greater than would be expected for purely thermal excitation. Often, the excited species is not directly involved in a chemical reaction, but receives some of the energy liberated from highly exothermic reactions between other flame species. For example, the energy from hydrogen atom recombination in a hydrogen–air flame can excite emission from sodium atoms in the flame.

This article will confine itself to molecular emission from nonmetals, C (as C_2 and CH), N (as HNO), P (as POH), S (as S_2), Se (as Se_2) and the halogens (as indium halides) excited in cool hydrogen–air flames. These elements (except Se) are not conveniently determined by atomic absorption because their principal resonance lines lie in the vacuum ultraviolet region of the spectrum.

The molecular emission spectra typically consist of several bands corresponding to transitions between various vibrational levels. Filters transmitting one or more of the emission bands are suitable for wavelength resolution.

The most important analytical applications of flame CL are element-selective gas chromatography detection and determination of air pollutants. General discussions of flame CL can be found in reviews of these two areas.

3.2. Sulfur

Figure 3 shows the CL emission spectrum observed from S_2 molecules when sulfur compounds are burned in a cool hydrogen-rich flame. The intensity is greatest when the level of oxygen entering the flame is just barely enough to sustain combustion, i.e., when the concentration of H atoms in the flame is high.[46,47] This suggests that the excitation energy comes from the reaction

$$H + H + S_2 \longrightarrow H_2 + S_2* \tag{19}$$

$$S_2* \longrightarrow S_2 + h\nu^{(47,48)} \tag{20}$$

Typical burners designed for observing this emission are shown in Figure 4.[47,49] In the burner of 4A hydrogen and air are mixed at the base of the flame. When the flame is shielded to keep out oxygen, emission can be observed several centimeters above the flame. This makes it possible to detect the CL emission without looking directly at the flame.

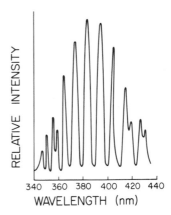

Figure 3. Emission spectrum for S_2 in a hydrogen–air flame.

In the burner in Figure 4B, sulfur emission is observed from the interior of a nitrogen–hydrogen diffusion flame, where the flame is relatively cool and hydrogen is in excess.

The optimum temperature for sulfur emission is 390°C.[49] To keep temperatures in and above the flame from getting too hot, the flame gases need to be diluted by inert gases, or special cooling devices are required.

This emission is widely used in the flame photometric detector (FPD) to selectively detect sulfur compounds separated by gas chromatography.[50–52] A diagram of a commercial detector is shown in Figure 5. The carrier gas used for the chromatographic separation dilutes and cools the flame. An interference filter passes the emission band, peaking at 394 nm. A mirror or lens is used to increase sensitivity by focusing more light on

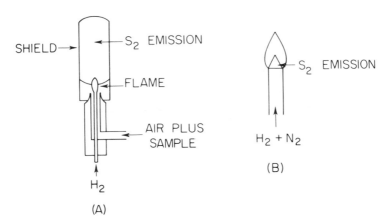

Figure 4. Burners for flame CL: (A) shielded hydrogen–air flame, (B) Hydrogen–nitrogen diffusion flame.

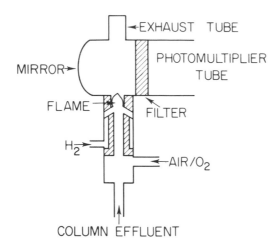

Figure 5. Diagram of flame photometric detector for gas chromatography.

the filter. The same flame can be used for selective determination of phosphorus compounds (see section 3.3) and for flame ionization detection.[52]

Because there are two sulfur atoms in S_2, emission intensity is proportional to the square of the sulfur concentration of the flame. For this reason, the selectivity of the detector for sulfur relative to other elements depends on concentration.[52] At all concentrations, the selectivity for sulfur to carbon and sulfur to phosphorus is greater than 10^4. The minimum detectable concentration is 2×10^{-10} g.

The flame photometric detector has been widely applied to such problems as determining sulfur-containing pesticides, sulfur compounds in cigarette smoke, sulfur compounds in coal, etc. A review of applications is beyond the scope of this article.

Another important application of flame CL is measuring sulfur in the atmosphere. Identification of specific sulfur compounds SO_2, H_2S, and CH_3SH can be done by gas chromatography with flame photometric detection.[53,54] Over 90% of the sulfur in urban atmospheres is in the form of SO_2, as has been shown by comparing total sulfur determined by flame CL with the SO_2 concentration determined by GC with flame photometric detection.[55] In the total sulfur monitor using flame CL, sample is drawn into the flame by an exhaust pump downstream from the burner. The detection limit is 5 ppb sulfur.

Sulfur analysis in liquids is also possible by flame CL but presents several problems. Organic solvents cannot be used because they quench S_2 emission, probably by reducing the concentration of hydrogen atoms in the flame.[49] Sensitivity is reduced for samples introduced to the flame as aqueous aerosols because the cool flame required for CL does not have sufficient energy to efficiently vaporize the water droplets. Greatest sensitivity is achieved by using an extremely fine aerosol or by passing the aerosol through a small furnace to vaporize the solvent before it enters the flame.[56]

The emission intensity for a given amount of sulfur varies substantially for different compounds.[49] The order of emissive response in unbuffered solutions is sulfide > sulfite > thiosulfate ≫ ammonium sulfate ≈ sulfuric acid ≫ alkali metal sulfates. Metal ions depress emission from sulfur-containing anions by reducing the volatility of sulfur in the flame. One approach to analyzing for total sulfur in aqueous samples is to oxidize it all to sulfate and then remove metal ions by treatment with a cation exchange resin. Another approach has been to precipitate sulfate as perimidyl ammonium sulfate.[57] Pyrolysis of this compound produces SO_2 that can be determined by flame CL.

The amount of sample required for an analysis can be decreased by using a filament vaporization system.[58] One microliter of sample is placed on a carbon filament, dried, and heated to convert the sample to a vapor which is then analyzed for sulfur by flame CL. This vaporization technique is subject to the same interferences as observed when the sample is introduced to the flame as an aerosol, i.e., emission intensity per unit sulfur varies according to the volatility of the sulfur compound.

3.3. Selenium and Tellurium

Selenium does not emit when aspirated into a cool hydrogen flame, and tellurium emits only weakly. However, a new sample introduction technique, molecular emission cavity analysis (MECA), leads to emission from these elements.[59-60] A small liquid or solid sample (microliter or milligram amounts) is placed in a cavity at the end of a stainless steel rod. The cavity can be reproducibly positioned in a hydrogen–nitrogen diffusion flame in line with a detector as shown in Figure 6. The flame heats the cavity to volatilize the sample and provides the radicals that react to produce CL.

Selenium compounds produce a blue emission from Se_2 in the cavity while tellurium compounds give both a weak blue emission and a stronger green emission which develops at higher temperatures. Figure 7 shows emission from Se and Te as a function of time after the cavity is introduced into the flame. The detection limits are 50 ng of Se and 0.5 ng Te. The shape and magnitude of the intensity-vs.-time curve varies from compound to compound. The emission from less volatile compounds is reduced in intensity and develops more slowly.

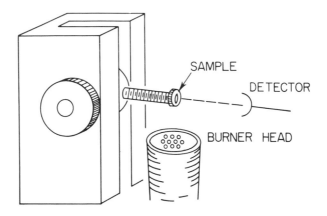

Figure 6. Diagram of apparatus for MECA (taken from Ref. 59).

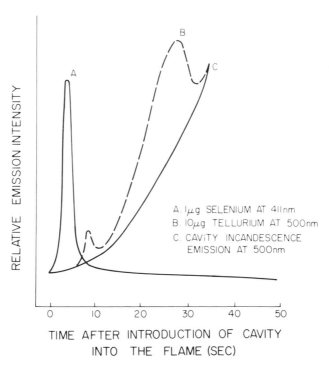

Figure 7. MECA signals for selenium and tellurium (taken from Anacon technical literature).

The MECA technique also applies to the other emissions discussed in this review, S_2, POH, indium halide, etc., as well as to many emissions not covered here.[61] The advantages are small sample size (typically 5 μliter) and increased sensitivity for liquid samples. Concentration can be related to the height of the observed emission peak or to the intensity integrated with respect to time. The principal problem is the dependence of emission intensity on the nature of the compound. In many potential applications, preliminary chemical treatment is required to convert the element of interest into a known form.

3.4. Phosphorus

Phosphorus emits as POH in a hydrogen-rich flame. The emission spectrum is shown in Figure 8. As in the case of sulfur, emission per unit phosphorus is greatest when the level of oxygen is just enough to sustain combustion. The reaction leading to excitation is

$$H + PO + M \longrightarrow POH^* + M \tag{21}$$

$$POH^* \longrightarrow POH + h\nu^{(62)} \tag{22}$$

The most important application of POH emission is selective detection of phosphorus compounds separated by gas chromatography.[50-52] The detector is the same as for sulfur (see above) except for the filter which selectively passes the emission band, peaking at 526 nm. Simultaneous detection of phosphorus and sulfur can be accomplished using the same flame with two-channel detection.[51] The detection limit for phosphorus is 4×10^{-11} g.[52] Phosphorus-to-sulfur selectivity depends on sulfur concentration because of the squared relationship between emission intensity and sulfur concentration. At high S concentrations (100–200 ng), the response to P is only four times greater than the response to sulfur.

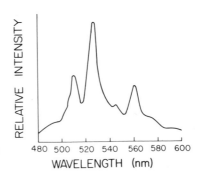

Figure 8. Emission spectrum for HPO in a hydrogen–air flame.

Selective detection of phosphorus compounds has been applied to a wide variety of samples: pesticides, phosphorus compounds in tobacco smoke, etc. Flame CL can also be used for monitoring total phosphorus in air with a detection limit of 1 ppb.

The same problems occur with phosphorus in liquid samples as with sulfur.[47,56,63] Organic solvents quench emission, sensitivity is reduced with aqueous solutions, and analysis must be preceded by treatment with ion exchange resin to remove metal ions. Sensitivity is increased by introducing the sample as a very fine aerosol or by passing the aerosol through a heated chamber to vaporize the solvent prior to entering the flame.[56] With ultrasonic nebulization the detection limit for phosphorus in water is 3 ppb.[64] Flame CL has been used to analyze for phosphorus in detergents, rock, oil, and natural waters (dissolved phosphorus only).[64-67]

The problem of metal ion interference in phosphorus emission can be reduced by vaporizing an aqueous sample in a graphite furnace.[68] The vapor is carried out of the furnace by a stream of nitrogen, which is then mixed with air and drawn into a hydrogen–air flame. Large excesses of calcium and ferric ions added to the sample reduce but do not eliminate the POH emission signal. The detection limit on a 100-μliter sample is close to 1 ppb.

3.5. Nitrogen

The reaction

$$H + NO \longrightarrow HNO* \tag{23}$$

$$HNO* \longrightarrow HNO + h\nu \tag{24}$$

in a hydrogen-rich flame produces the emission spectrum shown in Figure 9.[69] As with phosphorus and sulfur, maximum intensity is observed when the level of oxygen is just enough to sustain combustion.

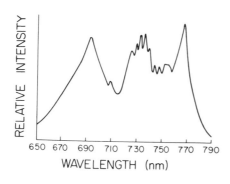

Figure 9. Emission spectrum for HNO in a hydrogen–air flame (taken from Ref. 69).

This emission was investigated as a means of determining atmospheric oxides of nitrogen simultaneously with sulfur determination as S_2 using a two-channel detector. HNO emission was observed through an interference filter with peak transmittance at 690 nm. Response for a given amount of N is the same for NO, NO_2, NH_3 and CH_3NH_2 (N_2 produces no response). Flame CL is inadequate for air monitoring for two reasons. (1) Sulfur interferes: 2 ppm SO_2 produces a signal equivalent to 1 ppm NO. (2) Sensitivity is not high enough for NO levels normally found in the atmosphere: The detection limit is 0.15 ppm NO in air.

Nevertheless, there are several possible applications of flame CL analysis for nitrogen compounds. It can be used to measure NO_x emissions from gasoline engines where NO_x levels are high and SO_2 is absent, and it could be adapted to specific detection of nitrogen compounds separated by gas chromatography.

3.6. Carbon

The CH_2 and C_2 emissions in hydrogen–air flames can be used to determine organic compounds separated by gas chromatography.[70] This application is not important because of the much greater sensitivity of flame ionization detection. However, it has been shown that emission intensity ratios at different wavelengths depend on compound structure.[71] Thus, by measuring the emission intensity at more than one wavelength, it is possible to obtain qualitative structural information in addition to quantitative information.

3.7. Halogens

When halogens burn in a hydrogen–air flame in the presence of indium, intense emission is observed from indium (I) halides.[72-73] The usual method for introducing indium to the flame involves a burner of the type diagramed in Figure 10. An excess of hydrogen burns in air at the base of the inner chimney. The combustion products move up the chimney past an indium surface located below the upper end of the inner chimney. The gas containing excess hydrogen that emerges from the inner chimney burns in a second flame within an outer chimney carrying an independent flow of air sufficient to complete the combustion. When halogens in any form are introduced to the lower flame, they are converted to hydrogen halide. The hydrogen halide is then apparently converted to indium halide as it passes over the indium surface. Indium halide emission is observed from the upper flame. The lower flame can be

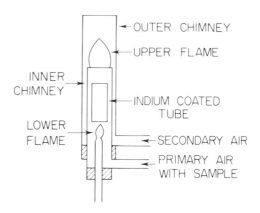

Figure 10. Burner used for observing indium halide emissions (based on Ref. 73).

used for phosphorus and sulfur analysis by flame CL, thus permitting simultaneous analysis for three elements.[74]

A flame ionization detector can be modified for simultaneous halide analysis by surrounding the base of the flame with indium metal as shown in Figure 11.[75] The heat from the flame volatilizes indium which reacts with halide in the flame.

The emission spectra for InCl, InBr, and InI are shown in Figure 12. Emission is also observed from In, InH and In_2 (or possibly InOH). The indium halide emission has been most frequently studied as a selective detector for halogen-containing compounds separated by gas chromatography. Interhalogen selectivity and sensitivity depend on the wavelength and bandwidth of the observed emission.[72,76-78] Using a Beckman DU monochromator with a slit width of 2 mm, the detection limits are 5.2×10^{-10} g chlorine (at 360 nm), 4.0×10^{-9} g iodine (at 410 nm) and 4.3×10^{-10} g bromine (at 373 nm). When the slit is decreased to 0.1 mm, the detection limits are poorer by a factor of about 20. The range of linear response is greater than 10^3 for all three halides. No signal is observed for fluorine compounds. At 360-mm and 0.1-mm slit the chlorine-to-iodine selectivity is greater than 100, and chlorine-to-bromine selectivity is about 20. At 350 nm chlorine-to-bromine selectivity is 170 although there is a loss in sensitivity by about 30%. At 373 nm, the bromine-to-chlorine selectivity is 29 and bromine-to-iodine selectivity is 38. At 410 nm the selectivity for iodine is about 5 relative to both chlorine and bromine. The reason for the poor selectivity for iodine is that all the halides cause emission from atomic indium which is superimposed on the InI spectrum.

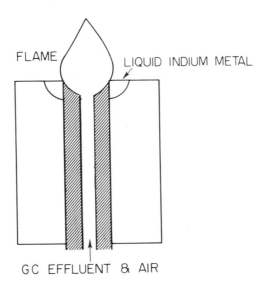

Figure 11. Modified flame ionization detector for observing indium halide emissions (taken from Ref. 75).

Because of the superior sensitivity of the electron-capture detector for halogen compounds, a commercial flame CL detector for halogens has not been developed. Nevertheless, this type of detector has been used for the determining pesticides[79-81] and bromine in urine.[82]

Halides in solution can be determined by adding an excess of indium nitrate and aspirating them into a hydrogen–air flame.[83] The detection limits are 0.7 ppm Cl, 1.6 ppm Br, and 2 ppm I. Emission intensity is effected by other ions present in solution as well as the concentration of indium nitrate.

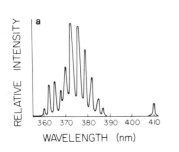
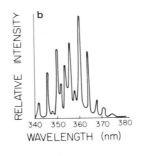
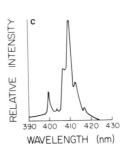

Figure 12. Indium halide emission spectra: (a) InCl, (b) InBr, and (c) InI.

4. Chemiluminescence from Organic Compounds in the Liquid Phase

4.1. Luminol

Of all the organic compounds that chemiluminesce, luminol has attracted by far the most interest as a possible analytical reagent. In basic solution luminol is oxidized to the aminophthalate anion producing intense blue CL. (See Table 1 for the reaction.) The most common oxidant is hydrogen peroxide in the presence of a catalyst or cooxidant. Catalysts include hemin, Cu(II), Cr(III), Fe(II), VO(II), Mn(II), Ni(II) and Co(II). Cooxidants include $Fe(CN)_6^{-3}$ and $S_2O_8^{=}$. Other oxidants that react with luminol to produce CL are the hypohalogenites, and permanganate and oxygen in the presence of a suitable catalyst such as Fe(II), V(II), and Cr(II).

4.1.1. Mechanism

The mechanism of luminol CL in aqueous solution has not been agreed upon, despite numerous studies. Both the radical formed by one-electron oxidation of luminol and the azaquinone formed by two-electron oxidation of luminol have been proposed as intermediates preceding CL.[3] The evidence suggests that both intermediates can occur but with different oxidizing agents. Azaquinones are formed when two-electron oxidizing agents react with luminol. This includes persulfate and the hypohalogenites. In the persulfate–luminol–hydrogen-peroxide system CL comes from the following sequence of reactions.[84]

$$\text{(luminol)} + S_2O_8^{2-} \longrightarrow \text{(azaquinone)} \qquad (25)$$

$$\text{(azaquinone)} + H_2O_2 \longrightarrow \text{(aminophthalate)} + N_2 + h\nu \qquad (26)$$

In this system the rate and efficiency of CL is unaffected by the addition of radical scavengers indicating that no radicals are formed. Hypohalogenite oxidation of luminol similarly proceeds via two consecutive two-electron oxidations.[85,86]

$$OX \quad + \qquad\qquad\qquad \longrightarrow \qquad\qquad\qquad (27)$$

$$OX \quad + \qquad\qquad\qquad \longrightarrow \qquad\qquad\qquad + N_2 + h\nu \quad (28)$$

Oxygen can also react with azaquinone with a low CL efficiency.

However, when ferricyanide is used as a cooxidant with hydrogen peroxide, luminol is oxidized by one electron.[87]

$$Fe(CN)_6^{3-} + \qquad\qquad\qquad \longrightarrow \quad Fe(CN)_6^{4-} +$$

$$(29)$$

The radical then reacts with hydrogen peroxide leading to efficient CL. It can also react with oxygen but the CL efficiency of this reaction is lower.

The mechanism of metal-catalyzed oxidation of luminol by hydrogen peroxide is of particular interest for chemical analysis, since trace metal analysis is one of the important applications of luminol CL. A bidentate–luminol–metal-ion complex has been proposed as an intermediate.[88,89]

However, the fact that high concentrations of luminol reduce the CL intensity from metal-catalyzed peroxide oxidation of luminol, coupled with the observation that metal ions catalyze CL from other hydrazides that cannot form bidentate metal complexes, suggests that a luminol–metal-ion complex is not required for CL. Instead, the properties of the reaction are consistent with the following sequence of reactions[90]:

$$M^{+n} + HO_2^- \longrightarrow M^{+n}\cdot HO_2^- \tag{30}$$

$$M^{+n}\cdot HO_2^- + H_2O + luminol^- \longrightarrow M^{+n+1} + 3OH^- + luminol^{\bar{}} \tag{31}$$

$$luminol^{\bar{}} + HO_2^- \longrightarrow CL \tag{32}$$

Reaction 24 is the rate-determining step. This mechanism explains why all metal ions that catalyze luminol oxidation by peroxide can be oxidized by one electron. The effects of pH, peroxide concentration, luminol concentration, and added chelating agent on CL intensity are all qualitatively consistent with their expected effect on the concentration of metal-ion–peroxide complex while they are difficult to explain in terms of other possible mechanisms.

The mechanisms for other oxidizing systems that generate luminol CL such as MnO_4^- and oxygen-catalyzed Fe(II) have not been studied. They are not as efficient in generating CL or as important analytically as other oxidizing systems.

4.1.2. Analytical Applications

Babko and associates have shown that luminol CL can be proportional to a variety of catalysts and oxidants. A list of some systems studied by this group is given in Table 3. In addition to direct analysis where CL is directly proportional to concentration, Babko et al. have investigated a

Table 3. Metals Determined Using Luminol Chemiluminescence

	Metals	Reference
Catalysts	Co	91
	Cu	92
	Fe	93
	Mn	94
	Hg	95
Inhibitors	V	96
	Zr	97
	Ce	98
	Th	99

number of indirect systems where concentration is related to reduction in CL intensity. Unfortunately, only a few of Babko's publications are available in English.

An interesting feature of Babko's early work is the apparatus used. The reagents are injected into a sealed container surrounded by a piece of film. The reaction is allowed to proceed for a specified length of time. Then the film is developed and the degree of exposure is related to concentration. Despite such simple apparatus, he and his colleagues were able to detect concentrations in the part-per-billion range for metal catalysts of luminol oxidation by peroxide.

Various other analytical applications of luminol have been proposed. Analysis for amino acids, phenols, and other organic compounds based on their quenching of copper-catalyzed CL has been suggested.[100-104] Since CL is proportional to peroxide concentration, the luminol reaction can be used to determine peroxide.[105-108] Luminol has also been proposed as a reagent for detecting blood stains since hemoglobin is an efficient catalyst of CL.[109]

In spite of all these possibilities, luminol CL is not yet used for any practical applications. The problem is the nonselectivity of the luminol reaction. Methods for masking or separating interfering species must be developed for use with the luminol reaction if its analytical potential is to be realized. The remainder of this section will deal with more recent studies where the practical aspects of using the luminol reaction with real systems have been considered.

4.1.3. Trace Metal Analysis

The analytical results reported by Babko et al. can be improved by performing the luminol reaction in a flow system such as diagramed in Figure 13.[110-111] Reactants and background solution are continuously mixed in a cell positioned in front of a photomultiplier that measures CL intensity. Luminol and peroxide are mixed in the flow system since a luminol–peroxide solution slowly decomposes with time. Two- to three-mliter slugs of metal catalyst solution are inserted into the background using a sampling valve. As the metal ion solution passes through the cell, steady-state CL is recorded. The cell can be stirred by bubbling gas through it, in which case the gas can serve as one of the reactants. However, overhead rotational or vibrational stirring reduces fluctuations in mixing. Oxidizing systems where CL kinetics are rapid are very sensitive to small fluctuations in mixing, while systems with slow kinetics give stable steady-state CL for all mixing techniques. Figure 14 shows some typical data using this system.

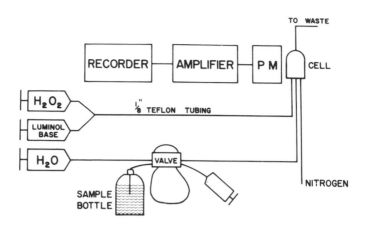

Figure 13. Diagram of flow system for making steady-state CL measurements (taken from Ref. 110).

The performance of this system for various metal catalysts is given in Table 4. The detection limits are imposed by background CL catalyzed by trace metal contaminants in the reagents. This background is over 100 times greater than the dark current. The upper limit of linear response appears to be determined by the solubility of the metal ion at the pH at which the reaction is carried out. The optimum pH is around 11 for most metals.[112] Cu(II) and Mn(II) both show unusual behavior. Steady-state CL-vs.-Cu concentration is greater than first order. The reason for this is unknown. Mn(II) catalyzes CL only in the presence of amine complexing agents. Possibly amines stabilize the Mn(III) oxidation state. For all other metals the presence of organic complexing agents reduces CL intensity, presumably by reducing the concentration of metal-ion–peroxide com-

Table 4. Analytical Performance of Flow System for Determining Metal Catalysts of the Luminol Reaction

Catalyst	Detection limit (M)	Linear range (M)	Remarks
Co(II)	10^{-11}	10^{-11} to 10^{-7}	—
Cu(II)	10^{-9}	—	Nonlinear
Ni(II)	10^{-8}	10^{-8} to 10^{-5}	—
Cr(III)	10^{-9}	10^{-9} to 10^{-6}	—
V(IV)	—	—	Quite sensitive
Mn(II)	10^{-8}	—	Requires amines
Fe(II)	10^{-9}	10^{-9} to 6×10^{-7}	Requires O_2 only

plex. The complexing agents reduce background CL also indicating that the background CL is due to trace metal contamination.

Cr(III) can be determined by adding an excess of EDTA to the sample bottle and background.[110] EDTA complexation drastically reduces CL catalyzed by metal ions. However, at a pH of 4.4 at room temperature Cr(III)–EDTA is kinetically slow to form. Therefore, Cr(III) will retain its efficiency as a catalyst even after EDTA is added. In this way, the luminol reaction can be used to specifically analyze for Cr(III). Residual catalysis from the metal–EDTA complexes can be accounted for by running a blank. The sample loop is filled with the solution of Cr(III) to be analyzed. The loop is then immersed in hot water, 90 to 100°C to accelerate Cr(III)–EDTA formation. After cooling, the sample is reacted with luminol. Since the heating procedure converts all Cr(III) to Cr(III)–EDTA complex, all that remains is the residual catalysis from other metals complexed by EDTA. This is subtracted from sample catalyzed CL. CL has been used to determine Cr in water, orchard leaves, bovine liver and blood,[113] using the method of standard additions. For the biological samples analysis is preceding by wet ashing. The CL results are in good agreement with values for Cr determined by atomic absorption.

Figure 14. Typical data using flow system in Fig. 13. Peaks are for slugs of Cr(III) passing through cell (taken from ref. 110): peak 2 = 2×10^{-8} M Cr(III), peak 4 = 4×10^{-8} M Cr(III), peak 6 = 6×10^{-8} M Cr(III).

5 min

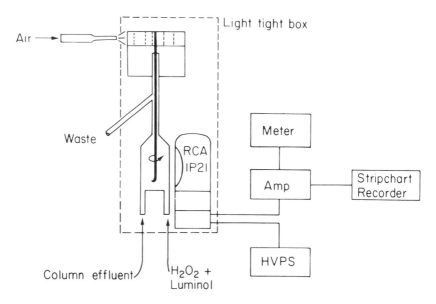

Figure 15. Small-volume cell suitable for CL detection of species separated by liquid chromatography (taken from Ref. 114).

Fe(II) can be determined specifically by reacting it with luminol in the presence of oxygen.[111] The only other metal ions that catalyze luminol CL in the presence of oxygen are V(II) and Cr(II). This method has been applied to river water and orchard leaves using sulfite to reduce ferric iron to ferrous. The results agree with the values determined by atomic absorption and spectrophotometry.

The most promising analytical application is using luminol CL to detect metal ions separated by high-speed ion exchange chromatography.[114–115] In this way several metals can be determined at trace levels in one analysis. Figure 15 shows a detector cell with a small volume to minimize loss of resolution. The cell has been demonstrated to respond to very low metal concentrations. Coupling CL detection to a good separation of metal ions has been a problem because the best methods are done in strong HCl which is incompatible with the basic pH required for luminol CL. A partial separation of cobalt and copper has been achieved using LiCl as an eluant. Since the metal ion concentrations are very low, it is possible to use pellicular ion exchange resins, thereby improving resolution.

The most efficient metal catalyst is cobalt. Because vitamin B-12 is a cobalt complex, luminol CL has been evaluated as a means of determining vitamin B-12.[116] The efficiency of cobalt catalysis in vitamin B-12 is drastically reduced because the cobalt is tied up in a complex. However, by adding EDTA to suppress background CL from trace metal contaminants, it is possible to measure B-12 levels from 10^{-8} to 10^{-6} M. Before reacting with luminol, the B-12 is reduced in a Jones Reductor which is directly inserted in the flow system.

4.1.4. Coupling Enzyme Reactions to CL Detection

Many enzyme reactions generate hydrogen peroxide, which can then be determined by measuring the CL from its reaction with excess luminol and catalyst or cooxidant. Several catalysts/cooxidants are possible; however, a detailed study showed that the best performance in terms of sensitivity and precision was observed with ferricyanide.[11]

CL has been used for glucose analysis after enzymatic oxidation to gluconic acid and peroxide.[11,117] The apparatus used for this study is diagramed in Figure 16. Ferricyanide and buffered luminol are mixed in the flow system before entering the cell. A column of glucose oxidase immobilized on a solid support is inserted in the background flow line

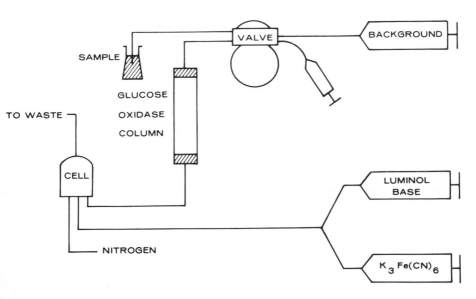

Figure 16. Diagram of flow system for CL determination of glucose (taken from Ref. 117).

between the cell and the sample valve. A glucose sample is introduced into the background flow line using the sample valve. As it passes over the immobilized enzyme, the glucose is completely oxidized to hydrogen peroxide which is then measured in the cell. In the initial study glucose oxidase immobilized on sepharose was used in the column. Sepharose has a tendency to compact, causing it to have a high resistance to flow. In a subsequent study, the glucose oxidase was immobilized on porous glass, leading to a significant reduction in the pressure required to drive solution through the column.[118]

The method has been successfully applied to serum and urine samples. The detection limit for glucose is 2×10^{-8} M. Response is linear up to 10^{-5} M. The detection limit is imposed by background CL from the reaction between ferricyanide, luminol, and oxygen.

Another technique for determining glucose is to incubate the sample with the enzyme under controlled conditions.[12] This solution is then mixed with a buffered solution of ferricyanide and luminol, and the resulting CL is related to concentration. This approach has also been demonstrated to work on serum samples.

Any enzyme reaction that can be coupled to the formation of hydrogen peroxide can be followed by CL. Oxidases generate peroxide directly and, therefore, can be used for CL analysis in the same system already demonstrated for glucose oxidase. Many other enzymes can also be coupled to hydrogen peroxide formation. For example, NADH reduces oxygen to peroxide in the presence of methylene blue, which serves as a mediator.[118] Since most enzymes can be coupled to the formation or consumption of NADH, this greatly increases the number of possible applications of CL detection.

The combination of enzyme reactions with CL detection creates an almost ideal analytical situation. Enzymes are more specific than any chemical reagent while CL is superior to all other detection techniques in sensitivity and simplicity. This is a fertile field for future developments.

Table 5. Metals Determined Using Lucigenin
Chemiluminescence (All Catalysts)

Metal	Reference
Mn	119
Co	120
Ag	121
Tl	122
Pb	123
Bi	124

4.2. Other Reactions

4.2.1. Lucigenin

Lucigenin (bis-N-methyl acridinium nitrate) is oxidized by hydrogen peroxide in basic solution yielding CL:

$$(33)$$

$$(34)$$

The reaction proceeds without catalysts; however, it is accelerated by metal ions leading to an intensification of CL.

The Russians, Babko and his co-workers, primarily Dubovenko, have extensively investigated the possibility of trace metal analysis using lucigenin. Some of their work is summarized in Table 5. In our laboratory the CL response of lucigenin to various metal ions has been surveyed using the flow system of Figure 13.[125] Only Ag^+, Co^{++}, and Pb^{++} were observed to give significant response. The optimum pH was greater than 13, considerably higher than for luminol.

It appears that the mechanism of metal catalysis of lucigenin CL differs from luminol since different metals catalyze lucigenin CL. Lucigenin deserves further study since it could prove to complement luminol as a sensitive detection system for metal ions separated by liquid chromatography.

4.2.2. Ruthenium–Bipyridyl CL

Ruthenium(III)–(bipyridyl)$_3$ complex is reduced in dilute sulfuric acid to electronically excited Ru(II)–(bipyridyl)$_3$ (Reaction 5, Table 1). The analytical potential of this reaction is based on the fact that hydrazine is the only reducing agent leading to CL. It has been shown that substituted hydrazines also generate CL so that this reaction can be used for specific detection of substituted hydrazines in the presence of other reducing compounds. In preliminary studies 10^{-7} M hydralazine has been detected by Cl.[126]

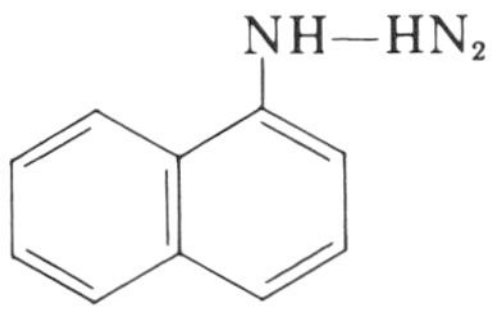

Hydralazine

4.2.3. Sensitized CL

Sensitized CL refers to those reactions where the emitting molecule is not formed directly as the product of a chemical reaction but instead is excited by energy transfer from some other excited species. Examples of sensitized CL already considered in this article include peroxyoxalate reactions (reaction 7, Table 1) and ozone-induced CL of rhodamine-B and other dyes. Another widely studied reaction is the decomposition of dioxetane in the presence of an energy acceptor:

$$(CH_3)_2C \overset{O-O}{\underset{}{-}} C(CH_3)_2 \xrightarrow{\text{heat}} CH_3\text{-}\overset{O}{\overset{\|}{C}}\text{-}CH_3 + CH_3\overset{O}{\overset{\|}{C}}\text{-}CH_3* \tag{35}$$

$$CH_3\text{-}\overset{O}{\overset{\|}{C}}\text{-}CH_3* + \text{flr} \longrightarrow CH_3\text{-}\overset{O}{\overset{\|}{C}}\text{-}CH_3 + \text{flr}* \tag{36}$$

$$\text{flr}* \longrightarrow \text{flr} + h\nu \tag{37}$$

(* = fluorescer)

These reactions do not lead to appreciable CL in the absence of an energy acceptor.

If the other reactants are present in excess, CL intensity is proportional to the concentration of energy acceptor. Thus, it should be possible to use these sensitized CL reactions to analyze for any fluorescent molecule that can serve as the energy acceptor. Ozone-induced CL has been proposed as an alternative to fluorescence analysis and shown to have about the same detection limits as fluorescence.[127] However, this system has two weaknesses: it requires mixing a gaseous reactant with a liquid or solid phase, and its energy is sufficient to transfer only to fluorescent molecules emitting in the red. The dioxetane reaction requires that oxygen be completely removed from the system. Therefore, peroxyoxalate reactions appear to be best suited for analytical reactions. These reactions are thought to involve energy transfer from electronically excited CO_2. They are capable of transferring energy to molecules emitting at wavelengths shorter than 300 nm, although the efficiency of energy transfer is greater in the visible. They produce CL in a variety of solvents including mixed aqueous–organic systems.

Present work is being directed toward evaluating the analytical possibilities of peroxyoxalate CL. The approach is to adapt established fluorescence methods to CL by having the fluorescer serve as the emitter for the peroxyoxalate reaction. This approach should greatly simplify fluorescence analysis, eliminating much of the instrumentation, the source and excitation monochromator, and many of the experimental problems, such as scattering and absorption of the excitation radiation by the sample.

4.2.4. Chemiluminescent Indicators

Indicators that chemiluminesce at the endpoint have been extensively investigated because they can be sensitively detected and can be used in colored or turbid solutions where color changes may be obscured. This area has been recently covered in detail elsewhere and will not be considered here.[10]

5. Chemiluminescence for Evaluation of Material Degradation

Recently, it has been demonstrated that various types of materials give off low levels of CL associated with free-radical oxidative degradation.[128] The reaction sequence leading to CL is summarized in Table 6. Reaction 4 in this Table is the only reaction sufficiently exothermic to produce the observed CL emission, hence it is inferred that this step in the free-radical oxidation scheme must be responsible for electronic excitation. Reactions 6 and 8 assume the presence of a fluorescent energy acceptor. Experimentally, it is observed that addition of a fluorescer such as dibromoanthracene or diphenylanthracene to liquid samples leads to an increase in CL intensity, presumably because the fluorescer has a much

Table 6. Reactions in the Oxidative Degradation of Organic Materials[a]

$$\text{ROOH} \xrightarrow{\text{Catalyst } k_1} 2R \cdot (+ROH + H_2O)$$

$$R \cdot + O_2 \xrightarrow{k_2} RO_2 \cdot$$

$$RO_2 \cdot + RH \xrightarrow{k_3} ROOH + R \cdot$$

$$RO_2 \cdot + RO_2 \cdot \xrightarrow{k_4} \alpha R' \!-\! \overset{\overset{\displaystyle O}{\|}}{C} \!-\! R''^* + ROH + O_2 + (1-\alpha)R' \!-\! \overset{\overset{\displaystyle O}{\|}}{C} \!-\! R''$$

$$R' \!-\! \overset{\overset{\displaystyle O}{\|}}{C} \!-\! R''^* \xrightarrow{k_5} R' \!-\! \overset{\overset{\displaystyle O}{\|}}{C} \!-\! R''$$

$$R' \!-\! \overset{\overset{\displaystyle O}{\|}}{C} \!-\! R''^* + F \xrightarrow{k_6} F^* + R' \!-\! \overset{\overset{\displaystyle O}{\|}}{C} \!-\! R''$$

$$R' \!-\! \overset{\overset{\displaystyle O}{\|}}{C} \!-\! R''^* + O_2 \xrightarrow{k_7} R' \!-\! \overset{\overset{\displaystyle O}{\|}}{C} \!-\! R'' + O_2^*$$

$$F^* \xrightarrow{k_8} F + \phi h\nu$$

[a] R depends on the organic material being studied. F is a fluorescer capable of accepting energy from the electronically excited ketone.

higher luminescence efficiency than the electronically excited ketone produced in the chemical reaction.

Because the CL produced in this type of reaction is very weak, sensitive light detection is required to observe it. The work to date has all been performed at Battelle Memorial Institute using photon counting with a cooled photomultiplier tube. Twenty different filters can be interposed between the sample and detector, making it possible to determine the wavelengths of CL emission. The sample compartment is designed so that the temperature and atmosphere can both be controlled. Using this instrument, it is estimated that reaction rates as low as 10^{-9} mol/year for solid samples and 10^{-14} mol/year for liquid samples can be measured. (The superior sensitivity for liquids is due to the possibility of adding a fluorescer to this type of sample.)

It is hard to overestimate the usefulness of low-level CL as a means of evaluating materials. The sensitivity makes it possible to look directly at degradation processes in various types of materials under very mild conditions approximating those of actual usage. Normally, the only way to evaluate materials in a reasonable period of time is to accelerate degradation by subjecting the material to very severe conditions. This evaluation procedure still takes more time than the CL method, and it requires the important assumption that degradation mechanisms are the same under normal and accelerated conditions. For example, scientists at Battelle have shown that CL measurements could predict the usefulness of a particular type of jet fuel. Previously, accelerated testing of this jet fuel did not give results correlating with actual performance. By varying the temperature of the CL measurement it was possible to demonstrate that the degradation mechanism did in fact change at high temperatures. With other types of materials, particularly organic polymers, CL measurements have been found to correlate well with the results of accelerated testing.

In addition to evaluating polymers and fuels, it has been shown that foods can be studied by CL. Foods, e.g., potato chips, give off weak CL which correlates closely with the tendency of the food to spoil. CL is much more sensitive than other methods for detecting rancidification.

These studies are only preliminary at this time. The power of the CL technique for evaluating materials is so great that it is sure to become widely used during the next few years.

6. Bioluminescence

Bioluminescence (BL) is chemiluminescence whose origin is either a living system or derived from one. Bioluminescing organisms are found in a wide variety of both genera and species, but wherever found add another dimension to nature's beauty. Bioluminescing organisms are found in all parts of the ecosphere; at great depths in the ocean as a bacterial passenger on the fish *Photoblepheron,* at quarter-mile depths in the Mediterranean where swims the fish* *Argyropelecus,* in the steaming jungles where the railroad worm *Phrixothrix tiemanrri* thrives while emitting its usual red light, in New Zealand in tubes hanging from the ceiling of the Waitomo Caves are the larvae of the fly *Arachnocampa luminosa,* in the jungles of Borneo where appears the fungi *Peromycena manipularis,* by day as a harmless mushroom but by night as a miniature lampshade, or, finally, the ubiquitous bioluminescing bacteria *Photobacterium phosphoreum.* But certainly the tiny crustacean *Cypridina hilgendorfii* found in the sandy shallows near the coast of Japan is outstanding for both its beauty and utility (the dried *Cypridina* when moistened and rubbed on the hands emits enough light so that a map can be read, a purpose to which it was put by the Japanese military during the Second World War).[129,130]

* Of the vertebrates the only fish that bioluminesces.

Table 7. Bioluminescing Organisms

Common name	Scientific name
Firefly	*Photinus*
	Photuris
	Luciola mingrelica
Ostracod crustacea	*Cypridina*
	Pyrocypris
Elaterid beetle	*Pyrophorus*
Marine fireworm	*Odontosyllis*
Decapod shrimp	*Systellaspis*
	Heterocarpus
Freshwater limpet	*Lafia*
Bacteria	*Achromobacter fischerii*
	Photobacterium fischerii
Clam	*Pholas dactylus*
Protozoa	*Gonyaulax polyedra*
(dinoflagellate)	
Fish	*Parapriacanthus*
	Apogon
	Aenomelopis
	Photoblepheron
Fungi	*Collybia velutipes*
	Armillaria mellea
Sea pansy	*Renilla reniformis*
Jellyfish	*Aequorea*
Sea pen	*Leioptilus*

Around 1885 Raphael Dubois, using the luminescing photogenic organ of the West Indian elaterid beetle *Pyrophorus*, discovered that when it was immersed in hot water until its light emission ceased, a heat-stable compound was extracted. It was also observed that when the cells of the photogenic organ were triturated in water at room temperature until the light emission ceased, a heat-labile compound was extracted. When the two oxygen-saturated extracts were mixed, light emission was immediately observed. The former extracted compound was named luciferin (LH_2) and is referred to as the substrate while the latter extracted compound was named luciferase (E) and is known to be an enzyme. It was at first thought that LH_2 was a protein; this is in general not true, although the *Aequorea* system is an exception in that the LH_2 is tightly bound to a protein matrix.[129,130] There are many different sources of many different luciferins and luciferases. A fairly representative collection of the sources is shown in Table 7.

Equal research attention has not been given all the entries in Table 7. The main reasons for selecting a given system are availability of the source of LH_2 and E, stability of the LH_2 and E, ease of handling, and cost. The

Table 8. Summary of Bioluminescence Reaction Types[a]

Reaction	Organism
1. Adenine nucleotide linked	
$LH_2 + ATP + O_2 \xrightarrow[Mg^{+2}]{E} h\nu$ (552–582 nm)	Firefly
$LH_2 + ado\ 3,5 - P_2 \xrightarrow[Ca^{+2}]{E} h\nu$ (485 nm) also (3,5 diphosphoadenosine) PAPS (3 phosphoadenosine-5-phosphosulfate)	Sea pansy (*Renilla reniformis*)
2. Pyridine nucleotide linked	
$NADH + H^+ + FMN \xrightarrow{E} FMNH_2 + NAD'E = $ oxidase and $FMNH_2 +$ $RCHO + O_2 \xrightarrow{E} h\nu$ (475–505 nm)	Bacterial
$NADH + H^+ + L \xrightarrow{E} LH_2 + NAD^> + LH_2 + O_2 \xrightarrow{E} h\nu$ (530 nm)	Fungi
3. Simple enzyme substrate system	
$LH_2 + O_2 \xrightarrow{E} h\nu$ (460 nm)	*Cypridina* *Gonyaulax* (protozoan) (470 nm) *Pholas* (clam) (480 nm) *Parapriacanthus* (fish) (460 nm) *Apogon* (fish) (460 nm) *Latia* (Limpet) (520 nm)
4. Activation of "precharged" systems	
precharged protein $\xrightarrow{Ca^{+2}} h\nu$ (460 nm)	*Aequorea*
precharged particle $\xrightarrow[O_2]{H^+} h\nu$ (470 nm)	*Gonyaulax*

[a] E, luciferase.

firefly and bacterial systems have been extensively studied over the past 75 years. Cypridina has been given almost as much attention, and the *Aequorea* and *Renilla* systems have in the more recent past been attracting considerable research attention.

The chemistry involved in BL reactions is widely studied and seems to be fairly well understood; however, certain mechanisms are still under discussion. Quantitative studies of these reactions show that extremely minute quantities (femptogram: 10^{-15} g) are involved as well as exceedingly small light fluxes (200 photons/s), as in the case of the endogenous emission of certain dino flagellates, and even lower fluxes from cells in a highly oxidized state. The latter BL is thought to be due to radical ion recombination. Some of what is known regarding BL reaction types is summarized in Table 8. Only those reaction types which have been or are being established as analytically significant will be dealt with in this chapter.[131-141]

7. Firefly Reaction

7.1. Mechanism

Of all the systems studied, certainly the firefly system is the most popular. The following schematic summarizes the firefly mechanism as presently proposed.[132,134,142] (Though not shown, this reaction is best carried out in a glycine buffer at pH 7.8 and at 25°C.)[134] Abbreviations used: ATP, adenosine triphosphate; PP, pyrophosphate; L=O, oxyluciferin; and L, dehydroluciferin:

$$LH_2 + E + ATP \xrightleftharpoons{Mg^{+2}} E{\cdot}LH_2{\cdot}AMP + PPMg^{+2} \qquad (38)$$

$$E{\cdot}LH_2{\cdot}AMP + O_2 \rightarrow E + L{=}O + CO_2 + AMP + h\nu \qquad (39)$$

$$\phi_{BL} \simeq 1 \qquad h\nu(max) = 562 \text{ nm}$$
Other equilibria

$$E{\cdot}L{\cdot}AMP + PP \xrightleftharpoons{Mg^{+2}} E + L + ATP \qquad (40)$$

$$E{\cdot}LH_2{\cdot}AMP \rightleftharpoons E + LH_2{\cdot}AMP \qquad (41)$$

The equilibrium constant for the dissociation of the Mg–ATP^{+2} complex has been shown to be about 2.0×10^{-5} M at 20°C and was determined by measuring the intensity of light emitted by the firefly BL reaction.[143]

The following structures have been proposed for the firefly system[135]:

(a) Luciferin (LH$_2$)

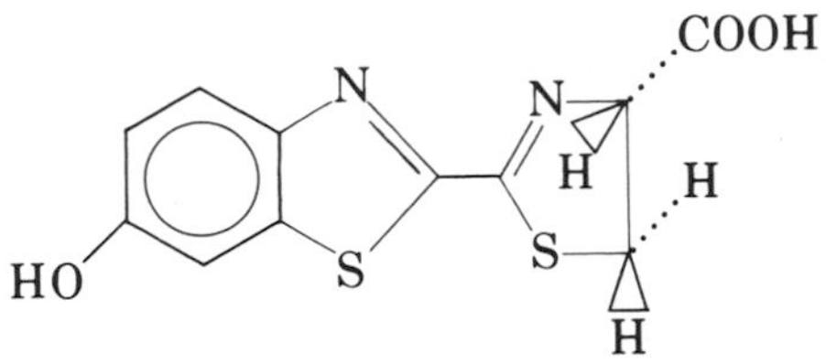

(b) Oxyluciferin (thought to be the product-emitter)

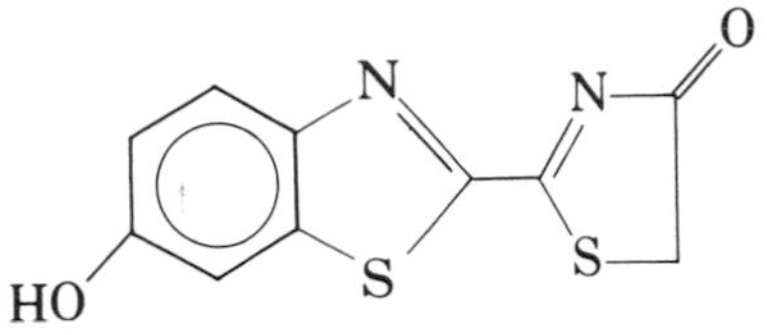

(c) Dehydroluciferin (L)

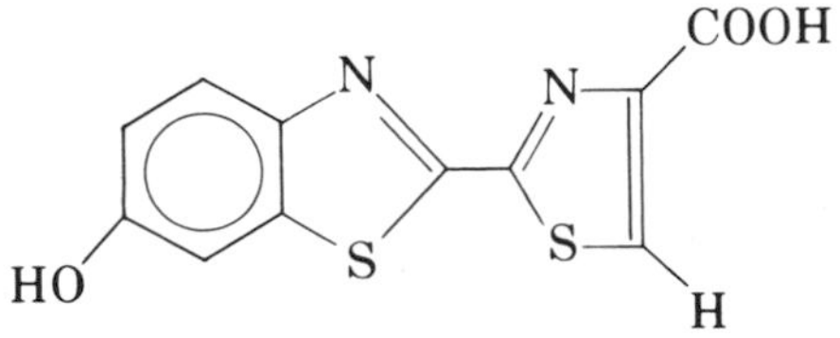

7.2. Analytical Characteristics

ATP, molecular oxygen, and Mg^{+2} are required for the reaction. This reaction can be used to assay any component of the reaction since the light output can be made to be a function of any desired component by making the rest of the reaction components in excess. However, most frequently the reaction is used for the analysis of ATP. By the addition of the appropriate coupling enzymes or enzyme systems, the following substances can be detected: ADP, AMP, creatine phosphate, and glucose, as well as several others.

Sensitivity and specificity are characteristics usually considered in the description of an analytical method. Sensitivity and lower levels of detection are not usually the same when various works are compared because of the variety of sample types to which BL analysis is applied. We therefore leave these data in the experimental setting when they are reported here. Linearity of response, however, is often the same from one report to the next; therefore, we generalize by suggesting that linear response can be expected over a minimum of six decades of concentration of the analyte, starting at the lower level of detection.[145–147]

Specificity studies of the firefly BL reaction have been made from two points of view: specific-ion inhibition and ATP analysis in the presences of other adenine nucleotides.

In regard to the specificity of firefly luciferase, inhibition of its activity has been reported for certain anions. In solutions of ATP and Mg^{+2} in concentrations 10 times that normally used, specific-ion inhibition was observed. The order of decreasing effectiveness in inhibition by the anion was: SCN^-, I^-, NO_3^-, Br^-, Cl^-. It was suggested that a small conformational change occurred near the active site in the luciferase molecule as a result of the binding of the anion.[148]

It has been stated categorically that the firefly luciferin and luciferase lacked specificity; however, this comment must be defined as to the nature of specific interferents. It has been shown, for example, that cytidine-5′-triphosphate (CTP) and inosine-5′-triphosphate (ITP) can stimulate light emission to the same extent as ATP. Some researchers have reported that the presence of adenosine diphosphate (ADP), uridine triphosphate (UTP), and guanine triphosphate (GTP) do not affect the light output of the BL system with respect to the ATP concentration; however, others have reported that the presence of these compounds does indeed reduce the light output of this BL system.[149–151]

One important factor affecting ATP analysis is the method of luciferase preparation. The simplest approach is to grind vacuum-dried firefly lanterns in an arsenate or phosphate buffer, filter, and add some magnesium sulfate. However, the resulting preparation is unstable and can be used for a few days at most.[144,145,152] Another rapid method for luciferase preparation involves grinding followed by extraction with cold water. The mixture is centrifuged and the supernatant is saved. This preparation also has poor stability.[144,145,152]

A good method for getting pure, stable enzyme is to do repeated fractional precipitations with ammonium sulfate after grinding and extracting the firefly lanterns.[144,145,152] The resulting preparation may be stored for years at $-20°C$ without loss of activity. Gel filtration has also been successfully used to prepare pure luciferase.[153,154]

It has been reported that bovine serum albumin, sucrose or ascorbic acid with mercaptoethanol stabilize enzyme activity.[155,156]

ATP is determined by adding it to an excess of reactants and measuring the BL intensity vs. time after addition. Calibration curves are generated from measurements of either maximum intensity or integrated intensity as a function of ATP concentration. Because ATP is present in all living cells, whether photosynthetic or heterotrophic (and for a short time in the debris of dead cells), the firefly ATP reaction can be applied as a biodetector.[157-159]

Minimum levels of detection vary with the care taken in the extraction of the ATP from the sample as well as the extraction of the luciferin and the luciferase from the firefly lanterns. A minimum level of detection for ATP has been reported at 1×10^{-13} g with linearity from 1×10^{-12} to 1×10^{-5} g.[149,160] Because the quantum efficiency of the firefly reaction is approximately unity, and in consideration of modern, fast, high-efficiency photomultiplier tubes (PMT) coupled with low-noise electronics, extraordinarily high sensitivities are possible for the components of this reaction. From 0.1 to 1 pmol is claimed for a minimum detection limit for routine analysis of ATP with crude extracts. Where great care is taken to avoid contamination and interferents such as AMP and endogenous ATP which limit the minimum detectable level, $2-\times-10^{-5}$-pg minimum detection limit has been reported, or approximately that amount of ATP in a single bacterium. Optimum detection levels for ATP are only achieved where great care is exercised in removing or masking interferents.

Kinetic studies involving ATP, glucose, and hexokinase have been made. The kinetic curves associated with this chemical system were independent of the amount of enzyme present. The shape of the curves were determined by the proportions of ADP and inorganic phosphate present in the samples of ATP. At certain levels of ADP and phosphate the intensity of the BL exceeded that which was expected for the level of ATP. It was suggested that inorganic phosphate inhibits the hydrolysis of ATP and also increases the level of ATP in the sample.[161]

Before an ATP assay can be performed the ATP must be extracted from the sample of interest. Solid samples, e.g., seeds and freeze-dried materials, must be ground before extraction. For bacteria, it is usual to work with cell suspensions. Commonly used extracting media include cold trichloroacetic acid,[162] 30% DMSO,[163] boiling ethanol or butanol,[164-169] and hydrochloric or perchloric acid.[166] In all cases, it is essential to verify that the extraction procedure does not destroy any ATP. If complete ATP recovery cannot be demonstrated, quantitative analysis is still possible using the method of standard additions. The samples with and without added standard should be treated identically so that the percent recovery remains constant.

7.3. Applications

7.3.1. Analysis for ATP

Estimates of both marine and terrestrial biomass constitute the most important class of measurements to which the firefly assay of ATP is applied. Biomass measurements contribute to the quantitative delineation of complex life cycles and the interrelationships of heterogeneous microorganism populations. In the management of both industrial and human waste, biomass estimates are of great utility. Biological oxygen demand (BOD), which may be measured directly by means of the ATP analysis or inferred from biomass estimates, may be used to determine the extent of pollution in water systems as well as the viability of the system to support a given level and diversity of life.

Firefly ATP assay is applied to biomass measurement because ATP is found in all living cells and is not associated with nonliving matter, the ratio of the concentration of ATP per organism to the concentration of cellular carbon per organism is nearly a constant, and ATP can be quantitatively measured in extremely low concentrations by the dependence of the firefly BL reaction on ATP.

While estimates of biomass may be expressed in terms of fresh weight, dry weight, volume, mass, etc., the most significant physiological parameter is cellular organic carbon (COC). Expressions of biomass involving COC minimize errors arising from samples which contain large vacuolated cells or cells with thick inorganic frustules such as diatoms. In studies of fresh water and marine bacteria, algae, protozoans, microzooplankton, and macrozooplankton, it has been shown that the ratio of the concentration of ATP per organism to COC per organism is a constant as shown in Figure 17. This ratio is also independent on the growth phase of the microorganism. From the data shown in this figure it can be seen that, given the ATP concentration of a sample, the COC can be computed from the ATP concentration. By means of total carbon analysis and with a knowledge of the ATP concentration, live/dead cell ratios can be established.[170,171]

If the microorganisms in the sample are in low concentration, they may be concentrated by filtration or some other suitable means. They are then lysed by a standard lysing technique such as treatment with perchloric acid, butanol, trichloroacetic acid, arsenate buffer, and DMSO. With samples in which the microorganisms are abundant, the sample may be introduced directly into boiling Tris buffer, pH 7.7, to bring about the release of the cellular ATP. In all cases, however, care must be taken to deactivate the ATPase which is released along with the ATP when the cell wall is ruptured so that concentrations of ATP measured are not erroneously low. After both lysing and separation of the ATP from the cell debris, an aliquot of the sample is combined with firefly luciferin and luciferase for measurement.

Many microbiological studies require measurement of the number of cells in a given sample. Commonly this is done by the laborious technique

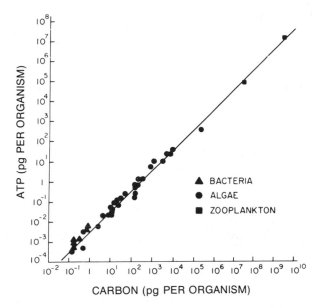

Figure 17. ATP concentration (pg/organism) vs. concentration of organic carbon (pg/organism) for a variety of different organisms.

of plate counting, i.e., applying a sample to a plate, adding nutrients, and counting the number of colonies grown.[172] ATP analysis provides a much more rapid method for estimating the number of cells. However, there are still questions as to the amount of ATP per cell as well as to the effect of growth stage on ATP levels.[160,173–175]

Table 9 lists some applications of the ATP assay to a wide variety of problems. The length of this list, which is by no means complete, is evidence of the great utility of the ATP assay and helps to explain why commercial instruments have been developed specifically for this analysis.

7.3.2. Analysis of Species Other Than ATP Based on Coupling Reactions

By the nature of the ATP assay by luciferin and luciferase it was proposed at an early stage in the study of the assay that by using the appropriate coupling enzymes or enzyme systems that, besides ATP, other compounds could also be assayed by this means; they include ADP, AMP, creatine phosphate, and glucose- and ATP-synthesizing systems such as phosphorulating mitochondria or phosphorulating chloroplasts. The enzyme activity of hexokinase, myokinase, various ATPases, or apyrases can also be estimated by this assay.[168,169,176–180]

Table 9. Some Systems Studied Using ATP Assay

System	Reference
Biomass distribution and turnover rates in the ocean	254, 255, 264, 274, 275
Bacterial content in lake sediments and terrestrial soils	258, 259, 260
Sludge activity in biological waste treatment	146
Process control and optimization of fermentation processes	261
Bacterial contamination of food and water	160, 173, 253
Bacterial infections of the urinary tract	272, 250, 249
Organic pollution in water	262, 273
Contamination of lubricating and cutting oils	263
Virus-containing cells	265
Sugar inhibition of root growth	266
Salamander limb regeneration	267
ATP synthesis by mitochondria	177
Response of tumor cells to treatment	268
Cancer detection	269, 271
Drug potency	270
ATP at nerve endings	251
Growth of leaves of green and etiolated plants	252

ADP. The presence of ADP in a given sample can in fact be quantitatively shown in the following way. After half of the extracted sample is first assayed for ATP and the value recorded, and the other half of the sample is treated with adenylate kinase, any ADP in the sample will be reversibly converted to ATP. If an increased level of light emission is observed in the second half of the sample extract, then twice the difference in that emission level and that of the first half of the sample extract can be taken as a measure of the ADP in the sample.

AMP. A microassay for AMP has reported a minimum level of detection of 5×10^{-14} mol of cyclic AMP (cAMP) which utilizes the conversion of the cAMP to ATP by the enzyme system previously described.[180] In this report it is proposed that as little as 1 mg of brain tissue could be assayed for cAMP by this method. The concentrations of cAMP found in various tissues have been reported.[181,182]

The study of hormone action depends to some extent on the analysis of cAMP, adenosine 3′,5′-monophosphate. Cyclic AMP has been converted enzymatically to ATP through the use of 3′,5′-cyclic nucleotide phosphodiesterase, myokinase, and pyruvate kinase.[183,184] For the analysis of cAMP in urine, the urine sample was first buffered and then filtered to remove insoluble material as well as any cells present. The minimum level of detection of cAMP was given as 7.2×10^{-9} M. This analysis was compared with the myosin–ATPase cycling assay of cAMP and was found to correlate well.[155,185,186]

Creatine Phosphate (CP). Creatine phosphate's role in muscle function has been well established as a source of quick energy, and a rapid, simple method for its analysis involves the use of the firefly BL assay of ATP. Creatine phosphate in an excess of creatine phosphokinase and AMP produces an amount of ATP which is proportional to the amount of CP initially present. Thus by measuring ATP formed in this way, the concentration of an unknown amount of CP can be estimated.[245,187] For the analysis of CP in samples containing appreciable amounts of ATP and ADP, the ATP and ADP can be converted to AMP by treating the sample with an ATPase and myokinase. The sample is then boiled to inactivate the ATPase.[145]

Creatine Phosphokinase (CPK). A recently proposed screening test for muscular distrophy depends on the assay of CPK in blood samples. It was previously mentioned that the analysis of CP was carried out by means of converting AMP to ATP with CP and CPK where CP was in limiting concentration with respect to the other reactants. The amount of ATP produced by this reaction was proportional to the unknown amount of CP in the sample. However, in the case of the analysis of CPK, if its concentration is limiting with respect to the other reactants then the amount of ATP produced by the reaction is proportional to the amount of CPK present in the sample.[188] Some researchers have shown that the amount of CPK found in the blood of victims of heart attack is an indication of the severity of the heart attack as well as the success of the prescribed therapeutic regime.

Delayed Bioluminescence. Delayed-bioluminescence analysis of purine and pyrimidine ribose and deoxyribose nucleotide triphosphates including guanosine triphosphate has been carried out using the firefly reaction.[150,189] Adenosine tetraphosphate has also been studied by its delayed bioluminescence with firefly luciferin and luciferase.[190]

Pyrophosphate. Recently a report of a coupled-enzyme method for the analysis of pyrophosphate using ATP-sulfurylase appeared and is illustrated by the following reaction:

$$\text{ATP} + \text{sulfate} \xrightleftharpoons{\text{ATP-sulfurylase}} \text{APS} + \text{pyrophosphate} \qquad (42)$$

APS is adenylylsulfate-3'-phosphate. This reaction lies hard to the left under normal conditions. Because the reaction relies on an enzyme reaction, it is affected by various inhibitory compounds, therefore the use of internal standards is recommended. Obviously ATP-sulfurylase activity or quantity can be assayed in a similar manner. The experimental scheme proposed started with the combination of 1 mliter of 5×10^{-2} M Tris-HCl buffer, 1 mliter of 1.5×10^{-2} M sodium arsenate, and 1 mliter of 5×10^{-3} M $MgCl_2$, which was allowed to rest for several minutes. Appropriate amounts of ATP-sulfurylase and firefly extract were introduced along with 2×10^{-9} mol of APS, and the background light level was determined. The sudden addition of from 1 to 100×10^{-12} mol of pyrophosphate caused an immediate increase in the light output and was monitored for 2 min. The maximum levels of emission were selected for purposes of quantitation and compared with a calibration curve generated in a similar manner with known amounts of the pyrophosphate. Internal standards were used when necessary.[192,193]

Arginine Phosphate. An enzymatic method for the determination of arginine phosphate has been described, in which arginine phosphate is treated with arginine kinase in the presence of ADP. The resulting ATP concentration is proportional to the concentration of arginine phosphate. The method was compared with an optical assay and the limits of detection were found to be 5×10^{-11} and 1×10^{-8} mol, respectively.[161,191]

Others. Other examples include the analysis of acetate by means of following the disappearance of ATP when the enzyme acetate kinase is employed, or the analysis of glucose by following the disappearance of ATP when treated with hexokinase. Phosphoenolpyruvate and ADP in the presence of the enzyme pyruvate kinase produces ATP whose concentration is proportional to that of either ADP or phosphoenolpyruvate, whichever is in limiting concentration. Also 3-phosphoglycerate in the presence of 3-phosphoglycerase consumes an amount of ATP that is proportional to the concentration of either the substrate or the en-

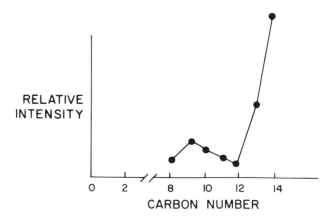

Figure 18. Relative intensity vs. carbon number of the alkylaldehyde.

zyme.[166] These are a few of the possible analyses which can be based on the amount of ATP present as the result of reactions which either consume or product it.

8. Bacterial Bioluminescence

8.1. Mechanism

The bacterial system follows the firefly system in popularity for both research and application. The following reactions schematically represent the bacterial system[132,133,194–199]:

$$FMNH_2 + O_2(g) \rightarrow FMN + H_2O_2 \tag{43}$$

$$2FMNH_2 + E + 2O_2 \rightarrow \underset{(I)}{E \cdot (FMNH_2 \cdot O_2)} \tag{44}$$

$$I + RCHO \rightleftharpoons \underset{(II)}{(I) \cdot RCHO} \tag{45}$$

$$II \rightarrow \underset{(III)}{E \cdot (FMN)_2{}^*} + RCOOH + H_2O \tag{46}$$

$$III \rightarrow E + 2FMN + h\nu \ (492 \ nm) \tag{47}$$

$$\phi_B = 1 \ \text{quanta of light/RCHO}$$

The structures of FMN and FMNH$_2$ are as follows:

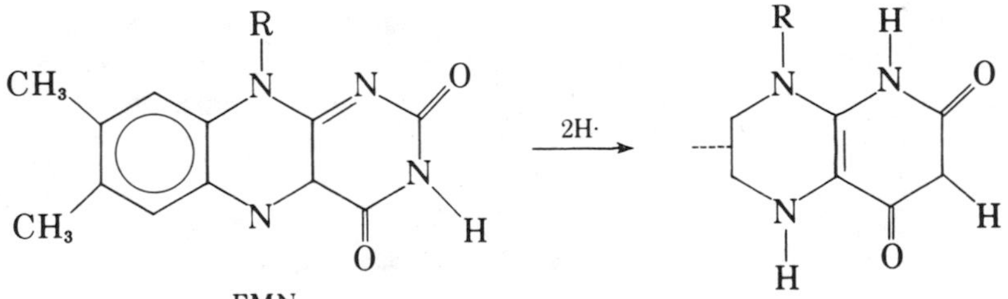

$$R = -(CH_2)_4CH_2-O-\overset{\displaystyle O}{\underset{\displaystyle OH}{\overset{\|}{P}}}-OH \tag{48}$$

The stoichiometry of the reaction is, however,

$$2FMNH_2 + 2O_2 + RCHO + E \rightarrow 2FMN \tag{49}$$

$$+ H_2O + H_2O_2 + RCOOH$$

This reaction is not catalyzed by FADH$_2$ (the reduced form of flavin adenine dinucleotide); however, $FMNH_2$ is required for this reaction to proceed.[196]

The dependence of the bacterial luciferase on the presence of alkylaldehyde has been well established. Dodecanal complexed with bisulfite is widely used; however, many long-chain alkylaldehydes may be used and in fact it has been shown that as the chain length increases so does the relative intensity of the reaction. Both the BL intensity maximum and ϕ_B depend on the chain length of the aldehyde used. Figure 18 summarizes the dependence of the light intensity maximum on chain length.[140] The absence of aldehyde does not lend to favorable energetics for BL. Emissions of 490 nm ($\equiv 60$ kcal/einstein) require at least 60 kcal/einstein excitation energy. In the transition from II to III 95 kcal are involved; without the aldehyde, though, only 27 kcal are available. Thus, with the aldehyde present adequate energy exists to account for BL observed.[194,200-202]

It has been reported that an aldehyde analog can be substituted for the aliphatic aldehyde in the bacterial BL system. It was found that decylnitrite was active as a substitute for the aldehyde, although two other analogs, decylformate and decylformamide, were shown to be inactive. In the case of the active analog, the light intensity and the quantum efficiency were lower than that found when the aldehyde was used. However, it was observed that the rate of decay of the emission was much slower in the case of the analog than with the aldehyde.[200]

Bacterial luciferase can be obtained from several different bacteria.[194,199] Different sources of luciferase tend to behave differently with respect to BL efficiency, kinetics, and specificity. Most commonly the luciferase is obtained from either *Photobacterium fischerii* or *Achromobacter fischerii*. A relatively crude preparation is available commercially.

Table 10. FMN Concentration in Certain Biological Materials

Material	μg per cell or mliter
Escherichia coli	7×10^{-11}
Serratia marcescens	6×10^{-12}
Bacillus globigii	3×10^{-11}
Bacillus globigii spores	
Urine	1×10^{-1}
Whole blood (5.5×10^6 rbc/mm^3)	5
Serum	1.1

Table 11. Sensitivity of Methods for FMN Assay

Method	mg/mliter
Paper chromatography	10^{-2}
Cytochrome c reductase	10^{-2}
Lactic oxidase	1
Fluorometry	10^{-4}
Bacterial bioluminescence	10^{-5}

8.2. Analysis for FMN

To perform an analysis, photobacterium extract is commonly used at a concentration of 1 mg/cm^3 in 0.05 M Tris buffer (pH 7.4). Dodecylaldehyde complexed with bisulfite is used, and the FMN is reduced to FMNH$_2$ by using NaBH$_4$, 10 mg/10 mliter FMN, with PdCl$_2$ as a catalyst. The relationship between light output and FMNH$_2$ concentration is linear from 10^{-1} to 10^{-5} μg/0.1 mliter sample (10^{-15} g/I). Endogenous light may be observed in the absence of reducing agent, probably due to the presences of NADH (diphosphopyridine nucleotide, reduced form) and NADH dehydrogenase. For natural samples such as those seen in Table 10, one of the best extraction media is boiling 6% butanol in 0.01 M Tris containing 10^{-3} EDTA. The extraction of the FMN is complete in less than 2 min.

Linearity of response for the FMN luciferase assay has been reported over seven orders of magnitude regardless of the means selected for the reduction of FMN. These methods include the enzymatic reduction with NADH and NADH dehydrogenase, PdCl$_2$-catalyzed NaBH$_4$, or photoreduction of FMN in the presence of 2×10^{-2} M EDTA, pH 7, with care taken to eliminate molecular oxygen in the last two cases due to the autooxidation of FMNH$_2$ by oxygen.[194,203]

The lower level of detection for one analysis was reported as 1×10^{-11} g of FMN; however, others have reported lower levels where great care is taken in the purification of the enzyme and the particular bacterial source is selected for its enzyme's high specific activity.[204]

The lower limit of detection for FMN (reduced form) by several different techniques are shown in Table 11. Although fluorometry can be seen to be very good, it suffers from a lack of specificity, since such compounds as chlorophyll, pterines, alloxan and benzo(a)-pyrene absorb and emit at wavelengths similar to those of FMN.

8.3. Analysis for Other Compounds Based on Coupling to FMN

8.3.1. Systems Involving NADH and NADPH

As previously mentioned, coupled-enzyme reactions present wide analytical possibilities with regard to the firefly ATP system; certainly the same can be said for the bacterial BL systems, owing to their dependence on the reduced form of flavin mononucleotide ($FMNH_2$), particularly when it is realized that the oxidized form of $FMNH_2$ can be converted to the reduced form by NADH and H^+ or NADPH in the presence of NADH dehydrogenase. As a consequence many compounds which are either convertible in dehydrogenase reactions or possibly lead to the dehydrogenase step can be the subject of this type of analysis. The role played by these coenzymes is virtually ubiquitous in biochemical systems. Examples of coupled-enzyme systems will serve to illustrate the point.[195,203,205–208]

In the study of the fate of the long-chain alkylaldehyde required for the bacterial luciferase $FMNH_2$ system, a coupled enzyme system was employed where ^{14}C-labeled decanol was enzymatically converted to decanal with alcohol dehydrogenase and NAD^+. The resulting NADH with NADH dehydrogenase was used to reduce the FMN, which then reacted with the decanal and oxygen in the presence of the bacterial luciferase to give light and the corresponding carboxylic acid. The reactions are shown below:

$$NAD^+ + \text{decanol} \xrightarrow{\text{alcohol dehydrogenase}} NADH + H^+ + \text{decanal} \qquad (50)$$

$$FMN + NADH + H^+ \xrightarrow{\text{NADH dehydrogenase}} NAD^+ + FMNH_2 \qquad (51)$$

$$FMNH_2 + \text{decanal} + O_2 \xrightarrow{\text{luciferase}} \text{decanoic acid} + h\nu \qquad (52)$$

An interesting sidelight of this study was that the level of light emission for the coupled system was relatively high for as long as half an hour; however, when pure $FMNH_2$ was used to catalyze the reaction the light level, although higher in the beginning with respect to the coupled system, diminished within seconds to the background level, as shown in Figure 19. The reason for this interesting behavior has not been explained experimentally, but it might be speculated that the rate-limiting reaction is not the $FMNH_2$–luciferase–aldehyde reaction. It might be expected that this sort of behavior will be seen in other coupled systems.[201]

Ethanol can also be measured using bacterial BL in a sodium pyrophosphate buffer along with the enzyme alcohol dehydrogenase. The amount of NADH formed is measured as before.

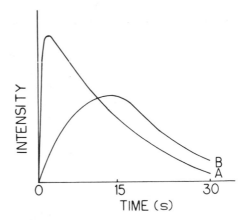

Figure 19. Log of relative CL intensity vs. time (min). The exponential curve is due to the use of highly purified FMNH$_2$ while the slower rising curve is due to the production of FMNH$_2$ by a coupled reaction.

Malate in Tris buffer in combination with 3-acetylpyridine-NAD (AP-NAD) and hydrazine sulfate, when treated with the enzyme malate dehydrogenase, produces an amount of AP-NADH which is proportional to the beginning concentration of the malate. As was previously described, AP-NADH can also be used to reduce FMN and thus is detected quantitatively by the bacterial luciferase. By this method 1 to 1.5 pmol could be estimated.

In another coupled enzyme reaction micro amounts of ammonia have been determined by using glutamic dehydrogenase and α-ketoglutamic acid in the presence of NADH. In this analysis the disappearance of NADH was the analytical objective, for its disappearance was reflected in a diminishing of the light intensity due to the decrease in the rate at which FMNH$_2$ is formed. In a typical analysis, phosphate mercaptophenol buffer is combined with tetradecanal, FMN bacterial luciferase, NADH, and glutamic dehydrogenase. A stable light intensity is reached in less than one minute at which time a sample of unknown ammonia concentration and ketoglutaric acid is suddenly added to the reaction mixture. The change in the intensity of light is measured and is proportional to the concentration of ammonia present in the sample. A standard commercial liquid scintillation counter was used for these measurements. The linear range for this measurement was reported as 0.01–0.20 μmol; however, it was stated that the precision was better than any other micro method for the analyte.[209]

Clearly NAD$^+$ could also be quantitatively measured and in fact has been in the concentration range of from 0.2 to 2.4 pmol. In all of the coupled-enzyme reaction assays mentioned, linearity of response with respect to the analyte has been the rule over the concentration ranges cited.

In samples which contain NAD$^+$, NADH, and NADPH where the concentration of NAD$^+$ is of interest, the NADH and the NADPH must be eliminated. This is accomplished by treatment of the sample with HCl.[206,207]

Aside from being able to assay native NADH samples by means of the bacterial luciferase, NADPH can also be assayed in a manner similar to that of NADH. However, the overall decay was slower and the peak intensity was about $\frac{1}{3}$ the peak intensity observed for NADH. The interest in using NADPH stems from the existence of specific NADP-requiring dehydrogenase enzymes. The NADH analog AP-NADH can also be used to reduce the FMN required by the bacterial luciferase. It has been observed that the maximum intensity is greater and the rate of decay slower than that seen for either NADH or NADPH. The reason for this behavior is as yet unclear. Glucose in the concentration range of 0.5–6.0 pmol has been assayed by the coupled-enzyme method. Glucose in Tris buffer along with NADP, ATP, and mercaptoethanol is combined with the enzymes hexokinase and glucose-6-phosphate dehydrogenase and the formation of NADPH is followed, measured by the $FMNH_2$ it generates in combination with FMN. The reaction time was 30 min at 38°C.[210]

$$\text{Glucose} + \text{ATP} \xrightarrow{\text{hexokinase}} \text{glucose-6-phosphate} + \text{ATP} \quad (53)$$

Glucose-6-phosphate + NADP

$$\xrightarrow{\text{G-6-P dehydrogenase}} \text{6-phosphogluconate} + \text{NADPH} \quad (54)$$

In samples containing both NADH and NADPH, if the concentrations are within the proper range, it is possible to measure each separately by means of selective enzymes and bacterial luciferase. From a lyophilized tissue sample the unwanted oxidized forms of pyridine nucleotides may be eliminated with 0.1 N sodium hydroxide, 0.5 mM cysteine hydrochloride, and heat. After this operation, the NADPH is oxidized with glutathione and glutathione reductase after a 30-min incubation period. The NADH can then be measured independently of the NADPH. The NADH can be oxidized with 3-hydroxybutyrate dehydrogenase and acetoacetate. The NADPH is unaffected by these compounds and may be measured independently of the NADH. This type of sample can be dealt with by selective oxidation, provided the concentration of either enzyme does not exceed 5×10^{-5} M.[207]

8.3.2. FAD

Since it is possible to convert flavin adenine dinucleotide (FAD) to FMN, it is therefore feasible to quantitatively assay FAD even if the sample nuclei contain FMN. FMN and FAD can be extracted from cellular material by treatment with cold perchloric acid and lumiflavin (the presence of the latter protects the FMN and FAD from photooxidation). After reduction of FMN with $NaBH_4$ in the presence of the catalyst $PdCl_2$, the concentration of FMN is determined by measuring the maximum intensity of the light emitted from an aliquot of the sample solution added to a mixture of bacterial luciferase and dodecylaldehyde. The total phosphoflavin (FMN plus FAD) content of the extract is measured after boiling the perchloric acid extract for 30 min, which converts the FAD to FMN, and assaying as before. Thus the concentration of FAD is computed as the difference in the former and latter determinations.[208]

8.3.3. Flavoproteins

It was pointed out that when three FAD flavoproteins (diaphorase, glucose oxidase, and L-amino acid oxidase) were treated with cold (2°C) perchloric acid, less than 1% of the FAD was converted to FMN; however, when boiled in perchloric acid for 30 min, 100% conversion was observed. There was no measurable destruction of FMN under the conditions cited. Spontaneous hydrolysis of FAD to FMN is thought to account for the slight BL catalysis observed for FAD alone.[208]

Other flavoproteins have been studied for their activity in the presence of NADH reaction and no appreciable activity was observed that could not be explained by the free FMN present in the materials. The flavoproteins studied in this way were flavodioxin, dihydro-orotic dehydrogenase, and glycollate oxidase. However, one of the fractions taken during the chromatographic purification which was determined to be a flavoprotein and known as the yellow fraction did show appreciable activity in the NADH reaction. In this study assays for both FMN and NADH were performed under these optimum conditions. The lower levels of detection for FMN and NADH were 1×10^{-14} mol and 1×10^{-16} mol respectively and it was suggested that the lower level of detection for NADH can be lowered another two level orders of magnitude. Some of the practical implications of this study include care in the preparation of the enzyme and care in selection of the bacterial source of the luciferase when the peak intensity is used analytically.[194]

8.3.4. In Situ Application

Applications of in situ bacterial BL includes the effects of anesthetics on luminous bacteria. Short-lived concentration-dependent changes in endogenous light brightness has been observed for the marine bacteria *Photobacterium phosphoreum* and *Photobacterium fischerii*, on exposure to low concentrations of a series of anesthetics such as halothane (trifluorochlorobromomethane) chloroform, ether, freon 22, trichloroethylene, etc.

It has been proposed from the above study that luminous bacteria be used for experiments on the mechanism of narcosis as well as to estimate anesthetic potency.[211-214]

The effect of the vapors of various solvents on the *in situ* bacterial BL was studied by both early and later researchers.[129,215] More recently researchers at Beckman Institute, Inc., have utilized luminescing marine bacteria to make a device designed for the purpose of detecting the vapors of ethyl alcohol found or not found on the breath of suspected drunk drivers. This remarkable device has both an audio and meter readout, is hand held, and has sufficiently low detection level so that alcohol on one's breath can easily be detected. The fraction of the bacteria's lifetime during which they may be utilized for this purpose is about two weeks and their replacement cost is nominal. Their application to air pollution problems has not been overlooked.[216-218] Other applications of *in situ* bacterial BL depends on the enhancement of light output rather than its decrease.[219,220]

9. Other Bioluminescence Reactions

9.1. *Aequorea*

Another interesting BL system is that of *Aequorea*, a hydromedusid. Of particular interest is that on the surface the requirements for luminescence do not seem to include the usual luciferin and luciferase. Study of the system has shown that luminescence involves only a protein, named aequorin, and Ca^{+2} not O_2 (gas). (Virtually every BL system requires O_2.)[221,222]

Spectroscopic studies have shown that the product of the aequorin–Ca^{+2} reaction is stable and has a spectrum which resembles the reduced form of the nicotin-amide dinucleotide NADH. The photoprotein, aequorin, is then a matrix for the luciferin, luciferase, and O_2, thus forming a complex which is the basis for Ca^{+2}-triggered, EDTA-inhibited, BLing system.[223] Note that the complex may contain NAP. Such a system is referred to as a "precharged" system.[224-226]

$$aequorin + 3Ca \rightarrow (\text{blue-fluorescing protein}) + h\nu \ (469 \ nm) \qquad (55)$$

blue-fluorescing protein

Early research proposed this reaction to be specific for Ca^{+2}; however, later work has shown that over a dozen cations (such as Cu^{+2}, Co^{+2}, Pb^{+2}, and Yb^{+3}, to list a few) were capable of exciting light emission from aequorin. These cations are not normally present in significant amounts in biological fluids, therefore the aequorin reaction may still serve as a means of Ca^{+2} trace analysis.[227-230] In spite of this apparent lack of specificity, numerous suggestions for analysis of Ca^{+2} by these means have been made.[221,231] The use of EDTA as a means of eliminating interferences has also been suggested.[223,229]

9.2. *Cypridina*

Cypridina hilgendorfii, the ostracod, is the source of a BLing luciferin–luciferase system of greater simplicity relative to those BL reactions discussed thus far. Referred to as an enzyme–substrate system, the reaction goes as follows[132,232-234]:

$$L + 2H^+ + O_2 \text{ (g)} \rightarrow LH_2 \cdot O_2 \tag{56}$$

$$LH_2 \cdot O_2 \xrightarrow{\text{E, Na}^+ \text{ or K}^+} \text{products} + h\nu \text{ (460 nm)} \quad \phi = 0.15 \tag{57}$$

E, the luciferase, has been found to be a rather simple protein with a molecular weight of approximately 50,000 and an isoelectric point at 4.35. Alkali metal cations are required for activity. Reports of cross-reactions between luciferins and luciferases of *Cypridina hilgendorfii* and the two fishes *Parapricanthus ransonneti* and *Apoyou ellioti* raised interesting evolutionary questions and thus gave rise to intensive research to determine the morphological origin of the LH_2s and Es.

Cypridina luciferin

Dried specimens of cypridina will BL if moistened in an O_2-containing environment, and were thus utilized by the Japanese during World War II as portable, wireless light sources for such purposes as map reading.

Table 12. Nucleotide Activators

Compound	Relative activity (%)
3',5'-diphosphoadenosine	100
2',5'-diphosphoadenosine	1
Coenzyme A	7
3'-phosphoadenosine-5'- phosphosulfate (PAPS)	15 or 98[a]
ATP	0

[a] 15% with no acid hydrolysis, 98% with acid hydrolysis.

9.3. *Renilla*

Renilla reniformis, commonly referred to as the sea pansy, produces a remarkable BL taking the form of blue-green concentric waves emanating across its surface. The adenine-nucleotide-linked reactions that lead to BL of the renilla system with an emission maximum at 485 nm are[132,139,141,235]

$$LH_2 + \text{nucleotide activator} + Ca^{+2} + E \rightarrow \text{activated } LH_2 \quad (58)$$

E, probably a sulfokinase

$$\text{activated } LH_2 + O_2 \rightarrow \text{product} + h\nu \quad (59)$$

E', luciferase

9.4. Fungal

The fungal BLing systems are pyridine-nucleotide linked and may follow the path shown in producing light:

$$NADH + H^+ + X \xrightarrow{\text{oxidase}} XH_2 + NAD^+ \quad (60)$$

$$XH_2 + O_2 \xrightarrow[\text{luciferase}]{\text{particulate}} \text{products} + h\nu \ (528 \text{ nm}) \quad (61)$$

The linkage of this reaction to the triphospho form of the pyridine nucleotide make the fungal system potentially important.[132,236]

10. Instrumentation

10.1. Introduction

The apparatus used for BL assays includes adaptations of well-known and readily available laboratory instrumentation as well as instruments designed specifically for the purpose. This last category includes both commercial instruments and those designed by the individual investigator. The commercial instruments are the more recent and are usually designed around the ATP assay, which employs firefly luciferin and luciferase. These same instruments could also be used with other BL systems, as well as many of the CL systems discussed earlier, however, the techniques have not yet been worked out in detail.

Before discussing the merits of specific instruments, an understanding of typical CL and BL response functions is necessary. Most CL and BL analyses are initiated by mixing the analyte with the necessary reagents. The resulting intensity–time curve is recorded and analyzed to determine concentration. The experimental intensity–time curve will reflect not only the kinetics of the reaction but also the rate at which mixing occurs. This is illustrated by Figure 20, which shows mixing and reaction functions individually and their combination into an overall response function, for two cases; one where k_0 equals k_1 and the other where k_0 exceeds k_1 by a factor of 20. This figure illustrates several important points. First the mixing must be carefully controlled to get reproducible intensity–time

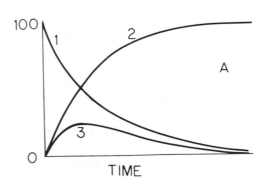

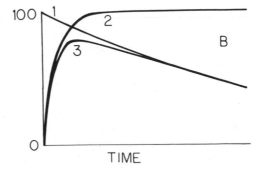

Figure 20. Effect of mixing rate on CL and BL intensity–time curves: curve 1, Hypothetical intensity–time curve with instantaneous mixing; curve 2, Percent mixed as a function of time; curve 3, Observed intensity–time curve. (A) The rates of mixing and luminescence are equal; (B) mixing is 20 times faster than luminescence.

curves. The time and magnitude of maximum BL intensity is particularly dependent on mixing, which accounts for the observation that better precision is usually obtained by integrating the intensity–time curve rather than using the peak maximum as a measure of concentration. A second important conclusion is that mixing is most important when k_0 and k_1 are similar in value and becomes progressively less important as k_0 becomes large relative to k_1. In other words, best precision can be obtained either by working with reactions which are relatively slow or by using instrumentation which provides for rapid mixing.

In flow systems, since a steady-state signal is being observed, the effect of mixing is different. Fluctuations in mixing lead to fluctuations in steady-state signal. This effect is most serious when k_0 and k_1 are similar in value.

10.2. Adaptation of Laboratory Instruments Designed for Other Purposes for Measurement of CL and BL

10.2.1. Liquid Scintillation Counter

The liquid scintillation counter (LSC) was the first instrument to be used for monitoring BL and is still the most sensitive commercial instrument for measuring low-level photon fluxes.

The problem solved by the LSC in its typical application (the quantitative determination of beta-emitting radioisotopes), is that of detecting ultra-low-level photon fluxes. The photon fluxes are the result of the beta energy being transferred to the solvent by collisional processes and its subsequent transfer to fluorescent compounds dissolved in the solvent along with the radioisotope-labeled sample.

Figure 21, the block diagram of a basic LSC, shows the top view of the sample vial located between and near two photomultipliers (PMTs). During a given beta event many fluorescent molecules are excited so that both photomultipliers normally will detect photons. The electronics are set up so that a count is recorded only when both PMTs see a signal within a short time interval (20 ns). This coincidence requirement lowers detection limits by discriminating against noise pulses from the PMTs. However, for CL and BL, the coincidence requirement is undesirable since it causes a decrease in sensitivity.

In the case of most LSCs it is possible to disable the coincident requirement but this results in the use of only one PMT to detect the light emitted from the BL reaction and in the presence of its singles count rate. Detection of the emission with one PMT reduces by at least one-half the optical collection efficiency unless the side of the sample vial away from the PMT in use is provided with a reflector. The use of coatings itself

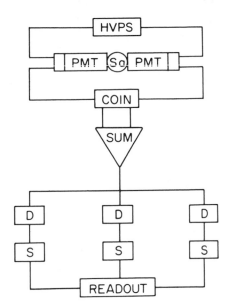

Figure 21. Block diagram of a typical three-channel LS counter. High-voltage power supply (HVPS), photomultiplier tube (PMT), sample (Sa), coincidence amplifier (COIN), summing amplifier (SUM), discriminator (D), and scaler (S).

suggests, however, that care must be taken in their application. Flash-evaporated aluminum or silver make good reflectors, although almost equally good reflection at considerably less cost can be obtained from the back surface of the sample vial by applying $MgSO_4$ to it. This author has achieved considerable success in improving optical efficiency by applying the white correction fluid commonly used by typists. When using an LSC for these purposes, the data are improved by selecting the PMT for use which exhibits the lowest singles count rate, i.e., the lowest background.

When used for its usual purpose, the LSC sample is prepared in glass vials an inch in diameter and 22 mliter in volume. Either in a series or singly the samples are placed in a sample-vial conveyor belt and automatically fed into the instrument. By disabling the sample conveyor mechanism and the shutters in the counting well (they protect the PMTs from being light shocked) it is possible to introduce a flowcell probe as is shown in Figure 22. The typical use to which this device is put is that of monitoring radioisotopes in a flowing stream, either by causing the sample to pass over scintillating crystals or by externally mixing a fraction of the sample stream with scintillator and pumping the mixture into and out of the cell at predetermined times automatically. In the case of the BL assay the reagents can be first mixed in the counting vial and introduced into the instrument or by using a probe as shown in Figure 22 the analyte can be introduced into the reagents already in place in the vial from a syringe attached to one of the feed-through tubes. In the former case care must

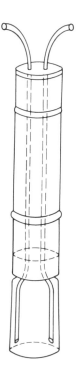

Figure 22. An illustration of a sampling probe that may be used with a standard LS counter for measuring CL intensity. This probe is inserted in the counting well and light tightness is assured by a double set of O rings. Solvents and reactants are conducted in and out of the probe by the tubing shown. The reaction vessel is a standard LS vial.

be taken to initiate the count for each sample of the set after a precise time delay after the introduction of the analyte into the reagents. During this time delay good mixing can be achieved, and the sample vial can be manually manipulated as necessary in order to get it placed before the PMTs; the count of the sample may proceed for as long after the time delay as is desired or thought necessary.[152,159,192,193,241–243] In the latter case, with the probe in place, the reagents are introduced into the cell by means of a syringe or a pump and the analyte is then quickly and forcefully injected to promote mixing. The count may be initiated at any time and proceed for as long as is necessary. If the count is started just prior to the injection of the analyte the count should be continued long enough so that the count rate is less than 1% of the maximum observed count rate so that nonreproducible mixing does not degrade the precision to a great extent.[180,237,244]

Even though many LSCs have, as an integral part of the instrument, a rate meter and therefore an analog output, the LSC is not easily adapted to a continuous or flowing type of BL analysis for mechanical reasons, primarily because of the location and configuration of the counting well and the restrictions that it places on a flowcell probe, as well as the problem of maintaining light tightness around the probe. The dead volumes in the feed-through tubes may be large as compared with the sample volume and complete removal of the spent reagents and product by aspiration presents a special problem since the removal does not depend on gravity but rather opposes it. Since the flowcell is not easily viewed, visual confirmation of the cell's emptiness is not possible. LSCs have been widely used in a variety of ways for the analysis of ATP.[238-240,246] The use of the LSC in conjunction with an autoanalyzer for the analysis of ATP from hemolyzates of erythrocytes has been described. The ATP-containing sample is air-segmented in the usual way and mixed with the reagents just prior to entering a helix of glass tubing which was located before the PMT of the LSC. Flow rates and dilution ratios were carefully controlled so that reproducibility was ±3%. Recovery studies showed a range of from 98.2 to 100% recovered ATP within the concentration range of 1×10^{-9} to 1×10^{-4} g/mliter ATP.[180,181,241,244]

A rapid, 200-samples-per-hour, automated system for ATP analysis has been described which utilizes an auto pipetter for accurate and reproducible delivery of the luciferin–luciferase mixture into a glass scintillation vial containing the ATP-bearing sample. The vial is located in the LSCs vial conveyor and after the pipetter operates, positions the vial before the LSC's PMT in the counting well. The time delay between the pipetter's operation and the beginning of the count is thus reproduced from sample to sample with good precision, ±0.1 s. In this study the coincidence requirement was disabled and internal standards were used, randomly located within the sample set. The integral discriminator was set so that 50,000 counts/0.1 min was equivalent to 1×10^{-10} mol ATP. By this means linear response of the LSC of from 10 to 300×10^{-12} mol ATP was observed without overflowing the counter's scalers.[182,241,245]

In general the cost of a LSC precludes its use for the purpose of BL assay alone, for the larger instrument's cost exceeds ten thousand dollars. However, several of the scientific instrument companies are marketing lower priced instruments which differ from the larger ones only in that they lack the automated multiple sample capability. This latter condition in no way affects their application to the BL assay, in general, since the multiple sample feature is not employed. The LSC should also be considered to be less than ideal for flow, or continuous BL measurements due to the limitation imposed by the sample counting well. However as a multipurpose instrument, i.e., BL assay and radioassay, the LSC is the best choice of instrument. The many biochemical laboratories which already possess a LSC need not purchase additional instrumentation for the purpose of performing BL assays; moreover, as will be seen in the next section a simple photometer or fluorometer can also be used for BL assays.

10.2.2. Photometers, Fluorometers, and UV–Vis Spectrophotometers

The fluorometer used as a photometer, the filter photometer and the UV–vis spectrophotometers came into use for the purpose of BL assay about the same time as the LSC.[247] If for no other reasons their simplicity, availability, and cost (as well as cost of operation) recommend them to the experimentalist for use in monitoring BL reactions. Each instrument is comparable in complexity and sensitivity although, as was the case with the LSC, since neither the fluorometer nor the filter photometer was designed specifically for BL monitoring they impose certain restrictions on their use for that purpose. In general the models of these instruments which utilize the double-beam principle are not well suited to monitoring BL emission due to the use of the null principle of measurement. Also those which detect radiation by means of a photocell as opposed to a PMT will have poor sensitivity (response) for those BL reactions whose level of emission is low.

Since many photometers are analog devices, their output can be monitored by means of a potentiometric recorder, which is particularly desirable in the case of BL measurements. Recorders which are equipped with an integrator add additional ease to the analysis of data. When this instrument is used for monitoring a BL reaction, the light source is disabled in some way and the detector current is monitored as a function of time. The record of the output resembles $C(t)$ shown in Figure 20 and can be analyzed as previously described. If the reactants are not mixed with the sample cell in place, a delay in the time that the measurement is commenced will necessarily result, and it must be carefully reproduced from sample to sample. This will not be the case when the sample and the reactants are mixed in the sample cell with it in place before the PMT. The optical efficiency of this arrangement can be optimized by the addition of a reflective surface on the side of the sample cell away from the PMT, but the ultimate optical efficiency will depend upon the distance separating the detector and the sample and will decrease as the square of the distance as it increases. It is therefore desirable to place the sample as close to the PMT as possible, for maximum optical efficiency.

All three of the instruments discussed (i.e., filter photometer, filter fluorometer, and the single-beam spectrophotometer) are considerably better suited for flow measurements than the LSC first discussed. With all three, however, neither optical efficiency nor the efficiency for removal of the spent reagents and product are high maximum. Systems designed specifically for the purpose of a flowing measurement maximize these parameters.[245,248]

One flowing type of system that has been used for monitoring BLing reactions is the Technicon Autoanalyzer. As was the case with the filter photometer, filter fluorometer, and the single-beam spectrophotometer, the autoanalyzer's light source must be disabled and the backside of the absorption cell should be provided with a reflective surface. The autoanalyzer does have an efficient means of disposing of the spent reagent and product thus minimizing cross contamination between samples; however, it suffers from the limitation of being designed for absorption measurements.[244]

10.3. Commercial Instruments

Due primarily to the increased use of and diversity in the applications of the firefly ATP analysis, a number of commercial instruments designed specifically for this analysis have become available during the last few years. In general these instruments incorporate a PMT and a well-placed reaction cell along with associated electronics. A recorder output is provided as well as a meter or digital readout. *In situ* mixing is the rule; however, simple modifications would make these instruments applicable to the flow type of measurement. In the case of one of the instruments the linear calibration factor is electronically stored in the instrument and the emission level of the BL reaction is read out directly in units of ATP concentration. The cost of these single-purpose instruments is between $1,000 and $5,000, not including the recorder. Most of the instrument manufacturers include as accessories reagent kits, which contain sufficient luciferin, luciferase, and buffer for several hundred analyses, bacteriological kits for extracting the ATP from a titer of some bacteria of interest, and other items of the hardware type (e.g., digital readout tubes, etc.).

The Lab-Line ATP photometer and the JRB ATP photometer appear to be the same instrument and therefore their characteristics will be presented here as though they were. It is a small, portable, self-contained instrument using modern solid-state electronics. A means is provided of selecting the wavelength range from 560 to 580 nm from the total emission of the sample for measurement. The BL in this range may be integrated for 60 s after a 15 s delay, and a 6-s integration mode with no delay is also provided so that a measure of the peak maximum may be preserved. The 15 seconds provide for sample placement. The readout is either a light-emitting diode (LED) digital display, or a potentiometric record. The sample and the reagents are mixed in a 22-mliter disposable glass scintillation vial and provide for sample sizes of up to 10 mliter, a feature unique to this particular instrument. The cost of operation is suggested to include only the cost of the reagents and is estimated to be from 3 to 4 cents per analysis. The instrument's cost is approximately $5,000. The fixed wavelength feature, if considered to be literally fixed, seems to be the chief shortcoming of this instrument when considered in terms of other possible BLing systems, for nearly all of the other BLing systems (i.e., the bacterial, fungal, cypridina, and aequoria) exhibit emission at wavelengths considerably shorter than the 560–580 nm range. The literature associated with this instrument also proposes its use with the Luminol CL system, which seems unlikely since Luminol's spectrum is centered at about 420 nm. Although it is not clear from the preliminary information on this instrument, it may be a simple matter to alter the bandpass of the instrument to complement the particular BL system chosen for use. One of the chief advantages of this instrument when compared with the others is that it can accommodate a rather large sample, 10 mliters, as well as ones as small as 10 μliters. This instrument is also equipped with a foot-pedal switch so that the 15-s delay of the measurement can be initiated as soon as the sample is in place or prior to placement. In general, *in situ* mixing is not done when this instrument is used for BL assay.

The Du Pont Biometer is basically a sensitive photometer equipped with a digital readout and recorder readout capabilities. Signal collection takes place over a 3-s interval without any delay. The sample size is limited to 10 μliters and the reaction volume 100 μliters. Mixing takes place in front of the PMT. A wavelength range of from 350 to 600 nm is detected by the PMT; thus, the Biometer's use with other BL systems is possible. The Biometer is equipped with an electronic means of treating the signal with a linear calibration coefficient so that concentration of ATP can be read out directly. The cell containing the luciferin–luciferase reaction mixture is introduced into the instrument and rotated 180° so that the final position of the cell is in front of the PMT and directly beneath a septum through which the ATP-containing sample is introduced by means of a syringe. The force of the injection is relied upon to cause mixing and in view of the small sample size (10 μliters), and small reaction volume (100 μliters) mixing is efficiently carried out in this way. The output of the PMT is stored as an analog signal (a charge on a capacitor) and when the count button is displaced the charge on the sampling capacitor is passed through an oscillator–gate–ramp circuit to provide a number of counts which are proportional to the charge on the capacitor. These counts are then displaced in the digital readout. A wide range of light fluxes are handled by an automatic ranging feature on the instrument; a total of 5 decades of linear response is thus provided for. When the optional potentiometric readout is used the current output of the PMT is sampled directly as a function of time. The cost of analysis is estimated to be from 3 to 4 cents per analysis and the cost of the instrument about $5,000. It has been pointed out that the chief shortcoming of the Biometer, particularly in the case of certain biomass measurements, is that the maximum sample size is 10 μliters.[254,255] It has also been reported that the Biometer gave erroneously high counts when applied to the microbiological study of food extracts, and particularly when molds, yeasts, and cocci bacteria were present. It was claimed the heat-killed bacteria still showed the presence of ATP. It was therefore concluded that it was not possible to differentiate between living and dead cells. It was stated that it was not possible to use ATPase to reduce the background since it would also attack that released from the bacteria during analysis. Unless the source of the high background was electronic, the difficulties encountered from these analyses with the Biometer are to be expected with any instrument of this type. It should be added that the researchers in the study mentioned did not identify the source of the apparent high background.[256,257]

The Biometer is backed up by considerable applications assistance as well as accessories. These include kits containing purified luciferin, luciferase, and the required buffers, a bacteriological kit for the extraction of the ATP, and filter assemblies.

The Aminco Chem-glow photometer, of the instruments thus far described, is the lowest in cost, less than $1,000. The recommended sample volume is 10 μliters and the reaction volume 100 μliters. The reaction cell is positioned in front of a mirror, in a ring that rotates so that the cell is in front of the PMT and beneath the sample injection port. In this manner ambient light is never incident upon the PMT. The Chem-glow photometers are available for use with either 6-×-50-mm reaction cells or 12-×-35-mm reaction cells. The Chem-glow photometer relies on the force of the injection of the sample or the reagents to cause mixing and due to the size of the sample and reaction volumes this type of mixing is probably reasonably efficient. The Chem-glow photometer utilizes an RCA 931 A PMT which exhibits S-4 response or covers the 300–650 nm range with a maximum quantum efficiency of 10+% at about 400 nm. The instrument scales the current from the PMT with a galvanometer to units of relative intensity. The wavelength range of the PMT used is adequate for virtually all of the BLing systems, including the metal-catalyzed luminol–H_2O_2 CL system. The instrument features seven ranges of sensitivity through three orders of magnitude; however, their selection is manual. The relative intensity of the BL emission is displayed on a meter, but recorder output is also an option. A unique feature of the Chem-glow photometer is that through the use of a coil type of flowcell it can be used as a detector for BLing samples as they emerge from a Technicon Autoanalyzer. In this way continuous monitoring of a flowing sample can be accomplished. A shortcoming of this instrument seems to be the size of the sample. As was previously mentioned, for applications such as certain biomass measurements this size is very much too small.

From the point of view of one who wishes to do either accurate and/or routine analysis using the Chem-glow photometer, a meter readout without a recorder readout as well can give rise to difficulties in estimating the meter's maximum deflection. The maximum deflection point would be the natural point for the operator to take his reading since that is the only point during measurement and the meter's movement at which the needle stops. If the pause is a short one, considerable error can arise in the estimation of the point of maximum deflection. These problems would be relieved by the use of a potentiometric recorder. Since the instrument is not switch activated to start a measurement but rather is continuously activated, the experimentalist has the advantage of seeing the reaction from the beginning to the end, particularly if a recorder is being used.

In summary, the commercially available instrumentation seems to adequately fill the needs of most investigators of BL reactions where their studies are directed toward applications of assay. For the needs of those investigators who are concerned with the nature of BL reactions, these instruments must serve in conjunction with others in their studies. None of the commercial ATP photometers employ true photon counting as does the LSC.

10.4. Rapid-Scan Spectrometers

If CL or BL intensity changes significantly while a spectrum is being scanned, the resulting spectrum will be biased in favor of those wavelengths that were measured while intensity was greatest. One solution to this problem is to use a rapid-scan spectrometer to measure the spectrum. Two different types of rapid-scan spectrometers have become commercially available in the last few years.

10.4.1. Silicon Vidicon Rapid-Scan Spectrometer

In the rapid-scan spectrometers marketed by Tektronix and SSR Instruments Co., source radiation is dispersed by a grating and focused

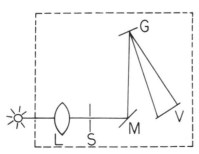

Figure 23. Block diagram of optics for Vidicon rapid-scan spectrometer. L, lens; S, slit; M, mirror; G, grating; and V, silicon vidicon detector surface.

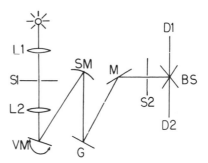

Figure 24. Block diagram of optics for Harrick rapid scan spectrometer. L_1, L_2, lenses; S_1, S_2, slits; VM, vibrating mirror; SM, spherical mirror; M, plane mirror; G, grating; BS, beam splitter; D_1, D_2, detectors.

onto a silicon vidicon as shown in Fig. 23. The silicon vidicon consists of an array of photodiodes. When light strikes a photodiode, it stores a positive charge which is then neutralized by scanning the back of the silicon target with a focused beam of electrons. The resulting current is proportional to the intensity of radiation striking that particular photodiode. Altogether there are 500 separate resolution elements, each corresponding to a different wavelength. The spectrum is displayed on a cathode ray tube as counts vs. channel number. The total wavelength range displayed depends on the dispersion of the grating. For a given grating position, the channel numbers must be calibrated in terms of wavelength using a known emission spectrum. Silicon vidicon targets give useful response from 350 to 1000 nm with maximum sensitivity in the visible.

In normal operation the electron beam completes a scan of the silicon vidicon target every 32 ms. However, the electron beam can be turned off while a spectrum accumulates on the silicon target, and then after the emission is over the spectrum can be scanned. Using this mode, spectra from any reaction can be measured no matter how fast, provided enough intensity reaches the silicon target.

10.4.2. Vibrating Mirror Rapid-Scan Spectrophotometer

The other type of rapid-scan spectrophotometer, available commercially from Harrick Scientific Corp., more nearly resembles a conventional scanning spectrophotometer. A diagram is shown in Figure 24. The light from the source goes through a slit and is focused on a plane mirror. The light reflected off the plane mirror strikes a spherical mirror which images the luminescence on a grating. The position of the plane mirror is varied electromagnetically. As it varies, light from the source strikes a different part of the spherical mirror. Because the mirror is spherical, the light strikes the grating at the same place at all times. However, as the plane mirror moves, the angle of incidence or the grating changes and, as a result, a different part of the spectrum passes through the exit slit. The scan rate is varied by varying the frequency at which the plane mirror moves. Maximum scan rate is 100 spectra/s covering a range of 100 nm. By appropriate choice of grating and detector, any wavelength from 200 nm to the IR can be detected. When scanning rapidly, spectra are accumulated and displayed on a cathode ray tube.

The spectrophotometer comes equipped with a source and double-beam optics for absorbance. To measure CL and BL spectra it would be necessary to disable the double-beam feature and to perform the luminescing reaction in the location normally occupied by the source.

10.4.3. Comparison

The authors are not in a position to make a critical comparison of the two types of rapid-scan spectrophotometers. The Harrick instrument is useful over a wider wavelength range, but because of the moving mirror it is a more delicate instrument than the silicon vidicon spectrophotometers. Both types of instruments require fairly intense luminescence to get a good spectrum. Both instruments offer a capability not previously available commercially.

11. References

1. M. A. A. Clyne, B. A. Thrush, and R. P. Wayne, *Trans. Faraday Soc.*, 60, 359–370 (1964).
2. C. P. Fenimore and G. W. Jones, *Combustion and Flame*, 8, 133–137 (1964).
3. K. D. Gundermann, *Chemilumineszenz Organischer Verbindungen*, pp. 6–90, Springer-Verlag, New York (1969).
4. A. U. Khan and M. Kasha, *J. Am. Chem. Soc.*, 92, 3293–3300 (1970).
5. F. E. Lytle and D. M. Hercules, *J. Am. Chem. Soc.*, 91, 253–257 (1969).
6. E. A. Chandross, J. W. Longworth, and R. E. Visco, *J. Am. Chem. Soc.*, 87, 3259–3260 (1965).
7. M. M. Rauhut, B. G. Roberts, and A. M. Semsel, *J. Am. Chem. Soc.*, 88, 3604–3617 (1966).
8. W. D. McElroy, H. H. Seliger, and E. H. White, *Photochem. Photobiol.*, 10, 153–170 (1969).
9. W. R. Seitz and M. P. Neary, *Anal. Chem.*, 44, 188A–202A (1974).
10. U. Isacsson and G. Wettermark, *Anal. Chim. Acta*, 68, 339–362 (1974).
11. D. T. Bostick and D. M. Hercules, *Anal. Chem.*, 47, 447 (1975).
12. J. P. Auses, S. L. Cook, and J. T. Maloy, *Anal. Chem.*, 47, 244 (1975).
13. B. L. Strehler in *Methods of Biochemical Analysis*, Vol. 16 (D. Glick, ed.), Interscience, New York, (1968), pp. 154–5.
14. J. A. Hodgeson, W. A. McClenny, and P. L. Hanst, *Science*, 182, 248–258 (1973).
15. A. Fontijn, A. Sabadell, and R. J. Ronco, *Science and Technology*, 26, 231–237 (1970).
16. R. K. Stevens and J. A. Hodgeson, *Anal. Chem.*, 45, 443A–449A (1973).

17. J. A. Hodgeson, submitted to *Toxicol. Environ. Chem. Rev.*
18. A. Fontijn, D. Golomb, and J. A. Hodgeson in *Chemiluminescence and Bioluminescence,* (M. J. Cormier, D. M. Hercules, and J. Lee, eds.), pp. 393–426, Plenum Press, New York (1963).
19. H. J. Bernanose and M. G. Rene in *Ozone Chemistry and Technology,* (American Chemical Society), pp. 7–12, Washington, D.C. (1959).
20. V. H. Regener, *J. Geophys. Res., 65,* 3975–3977 (1960).
21. V. H. Regener, *J. Geophys. Res., 69,* 3795–3800 (1964).
22. J. A. Hodgeson, K. J. Krost, A. E. O'Keffe, and R. K. Stevens, *Anal. Chem., 42,* 1795–1802 (1970).
23. E. H. Steinberger, J. Sivan, J. Neumann, and N. W. Rosenberg, *J. Geophys. Res., 72,* 4519–4524 (1967).
24. R. Guicherit, *Z. Anal. Chem., 256,* 177–182 (1971).
25. D. Bersis and E. Vassiliou, *Analyst, 91,* 499–505 (1966).
26. G. W. Nederbragt, A. Van der Horst, and J. Van Duijn, *Nature, 206,* 87 (1965).
27. G. J. Warren and G. Babcock, *Rev. Sci. Inst., 41,* 280–281 (1970).
28. R. Atkinson, B. J. Finlayson, and J. N. Pitts, Jr., *J. Am. Chem. Soc., 95,* 7592–7599 (1973).
29. D. H. Stedman and H. Niki, *J. Phys. Chem., 77,* 2604–2609.
30. D. H. Stedman, E. E. Daby, F. Stuhl, and H. Niki, *J. Air Poll. Control Assoc., 22,* 260–263 (1972).
31. A. Fontijn, A. J. Sabadell, and R. J. Ronco, *Anal. Chem., 42,* 575–579 (1970).
32. P. N. Clough and B. A. Thrush, *Trans. Faraday Soc., 63,* 915–925 (1967).
33. John E. Sigsby, Francis M. Black, Thomas A. Bellar, and Donald L. Klosterman, *Environ. Sci. Technol., 7,* 51–54 (1973).
34. L. P. Breitenbach and M. Shelef, *J. Air Poll. Control Assoc., 23,* 128–131 (1973).
35. D. H. Stedman and G. Kok, Status Report on NASA grant No. NGR 23-005-559 (1974).
36. A. Fontijn, C. B. Meyer, and H. I. Schiff, *J. Chem. Phys., 40,* 64–70 (1964).
37. F. M. Black and J. E. Sigsby, *Environ. Sci. Technol., 8,* 149–152 (1974).
38. W. A. McClenny, J. A. Hodgeson, and J. P. Bell, *Anal. Chem., 45,* 1514–1518 (1973).
39. M. F. R. Mulcahy and D. J. Williams, *Chem. Phys. Lett., 7,* 455–458 (1970).
40. A. Fontijn, *J. Chem. Phys., 44,* 1702–1707 (1966).
41. B. Krieger, M. Malki, and R. Kummler, *Environ. Sci. Technol., 6,* 742–744 (1972).
42. W. A. Kummer, J. N. Pitts, Jr., and R. P. Steer, *Environ. Sci. Technol., 5,* 1045 (1971).
43. J. N. Pitts, Jr., H. Fuhr, J. S. Gaffney, and J. W. Peters, *Environ. Sci. Technol., 7,* 550–552 (1973).
44. D. H. Fine, F. Rufeh, and B. Gunther, *Anal. Lett., 6,* 731 (1973).
45. D. H. Fine, F. Rufeh, and D. Lieb, *Nature, 247,* 309 (1974).
46. W. L. Crider, *Anal. Chem., 37,* 1770–1773 (1965).
47. A. Syty and J. A. Dean, *Appl. Opt., 7,* 1331–1336 (1968).
48. T. M. Sugden and A. Demerdache, *Nature, 195,* 596 (1962).
49. R. M. Dagnall, K. C. Thompson, and T. S. West, *Analyst, 92,* 506–512 (1967).
50. S. S. Brody and J. E. Chaney, *J. Gas. Chromatogr., 2,* 42 (1966).
51. M. C. Bowman and M. Beroza, *Anal. Chem., 40,* 1449 (1968).
52. H. W. Grice, M. L. Yates, and D. J. David, *J. Chrom. Sci., 8,* 90–94 (1970).
53. R. K. Stevens, J. D. Mulik, A. E. O'Keefe, and K. J. Krost, *Anal. Chem., 43,* 827–831 (1971).
54. R. E. Pescar and C. H. Hartmann, *J. Chromatogr. Sci., 11,* 492–501 (1973).

55. R. K. Stevens, A. E. O'Keefe, and G. C. Ortman, *Environ. Sci. Technol.*, *3*, 652 (1969).
56. K. M. Aldous, R. M. Dagnall, and T. S. West, *Analyst*, *95*, 417–424 (1970).
57. R. F. Maddalone, G. L. McClure, and P. W. West, *Anal. Chem.*, *47*, 316 (1975).
58. G. L. Everett, T. S. West, and R. W. Williams, *Anal. Chim. Acta*, *68*, 387 (1974).
59. R. Belcher, S. L. Bogdanski, Thi. Kouimtzis, and A. Townshend, *Anal. Chim. Acta*, *68*, 297 (1974).
60. R. Belcher, S. L. Bogdanski, S. A. Ghonaim, and A. Townshend, *Anal. Lett.*, *7*, 133 (1974).
61. R. Belcher, S. L. Bogdanski, and A. Townshend, *Anal. Chim. Acta*, *67*, 1 (1973).
62. C. P. Fenimore and G. W. Jones, *Combust. Flame*, *8*, 133–137 (1964).
63. R. M. Dagnall, K. C. Thompson, and T. S. West, *Analyst*, *93*, 72–78 (1968).
64. M. J. Prager and W. R. Seitz, *Anal. Chem.*, *47*, 148 (1975).
65. A. Syty, *Anal. Lett.*, *4*, 531–536 (1971).
66. A. Syty, *At. Absorpt. Newsl.*, *12*, 1–2 (1973).
67. W. N. Elliott, C. Heathcote, and R. A. Mostyn, *Talanta*, *19*, 359–363 (1972).
68. D. R. Campbell and W. R. Seitz, *Anal. Lett.*, *9*, 543 (1976).
69. K. J. Krost, J. A. Hodgeson, and R. K. Stevens, *Anal. Chem.*, *45*, 1800–1804 (1973).
70. F. M. Zado and R. S. Juvet, Jr., *Anal. Chem.*, *38*, 569–573 (1966).
71. R. S. Braman, *Anal. Chem.*, *38*, 734–742 (1966).
72. C. V. Overfeld and J. D. Winefordner, *J. Chromatogr. Sci.*, *8*, 233 (1970).
73. P. T. Gilbert, *Anal. Chem.*, *38*, 1920 (1966).
74. B. Versino and G. Rossi, *Chromatographia*, *4*, 331 (1971).
75. R. F. Moseman and W. A. Aue, *J. Chromatogr.*, *63*, 229 (1971).
76. B. Gutsche and R. Herrmann, *Z. Anal. Chem.*, 253–257 (1971).
77. B. Gutsche and R. Herrmann, *Z. Anal. Chem.*, *249*, 168–171 (1970).
78. B. Gutsche and R. Herrmann, *Z. Anal. Chem.*, *245*, 274 (1969).
79. M. C. Bowman and M. Beroza, *J. Chromatogr. Sci.*, *9*, 44 (1971).
80. B. Gutsche, R. Herrmann, and K. Ruediger, *Z. Anal. Chem.*, *241*, 54 (1968).
81. B. Gutsche and R. Herrmann, *Z. Anal. Chem.*, *242*, 13 (1968).
82. B. Gutsche and R. Herrmann, *Analyst*, *95*, 805 (1970).
83. R. M. Dagnall, K. C. Thompson, and T. S. West, *Analyst*, *94*, 643 (1969).
84. M. M. Rauhut, A. M. Semsel, and B. G. Roberts, *J. Org. Chem.*, *31*, 2431 (1966).
85. W. R. Seitz and D. M. Hercules, *J. Am. Chem. Soc.*, *96*, 4094 (1974).
86. W. R. Seitz, *J. Phys. Chem.*, *79*, 101 (1975).
87. P. B. Shevlin and H. A. Neufeld, *J. Org. Chem.*, *35*, 2178 (1970).
88. H. Ojima, *Nippon Kagaku Zasshi*, *82*, 973 (1961).
89. A. K. Babko and L. I. Dubovenko, *Ukr. Khim. Zh.*, *29*, 1083 (1963).
90. T. G. Burdo and W. R. Seitz, *Anal. Chem.*, *47*, 1639 (1975).
91. A. K. Babko and N. M. Lukovskaya, *Zavod. Lab.*, *29*, 404 (1963).
92. A. K. Babko and N. M. Lukovskaya, *Zh. Anal. Khim.*, *17*, 47 (1962).
93. I. E. Kalinichenko and O. M. Grishchenko, *Ukr. Khim. Zh.*, *36*, 610 (1970).
94. I. E. Kalinichenko, *Ukr. Khim. Zh.*, *35*, 755 (1969).
95. L. I. Dubovenko and T. A. Bogoslovskaya, *Ukr. Khim. Zh.*, *37*, 1057 (1971).
96. A. K. Babko and N. M. Lukovskaya, *Zh. Anal. Khim.*, *20*, 1153 (1965).
97. A. K. Babko, L. I. Dubovenko, and L. S. Mikhailova, *Metody Anal. Khim. Reakt. Prep.*, *13*, 139 (1969).
98. L. I. Dubovenko and Chan Ti Huu, *Ukr. Khim. Zh.*, *35*, 637 (1969).
99. L. I. Dubovenko and Chan Ti Huu, *Ukr. Khim. Zh.*, *35*, 957 (1969).
100. A. A. Ponomarenko and L. M. Amelina, *Zh. Obshch. Khim.*, *35*, 750 (1965).
101. A. A. Ponomarenko and L. M. Amelina, *Zh. Obshch. Khim.*, *35*, 2252 (1965).

102. A. A. Ponomarenko and B. I. Popov, *Zh. Anal. Khim., 19,* 1397 (1964).
103. A. A. Ponomarenko and L. M. Amelina, *Zh. Anal. Khim., 18,* 1244 (1963).
104. A. A. Ponomarenko, B. I. Popov, L. M. Amelina, L. V. Grishchenko and R. E. Shindei, *Zh. Obshch. Khim., 34,* 4118 (1964).
105. A. K. Babko and N. M. Lukovskaya, *Zh. Anal. Khim., 17,* 50 (1962).
106. W. A. Armstrong and W. G. Humphreys, *Can. J. Chem., 43,* 2576 (1965).
107. N. M. Lukovskaya, A. V. Terletskaya, and N. I. Isaenko, *Zavod. Lab., 37,* 897 (1971).
108. J. Kubal, *Chem. Listy, 62,* 1478 (1968).
109. K. Weber, D. Prpic-Majic, and B. Svetlicic, *Arch. Exp. Veterinaermed., 23,* 935 (1969).
110. W. R. Seitz, W. W. Suydam, and D. M. Hercules, *Anal. Chem., 44,* 957 (1972).
111. W. R. Seitz and D. M. Hercules, *Anal. Chem., 44,* 2143 (1972).
112. W. R. Seitz and D. M. Hercules in *Chemiluminescence and Bioluminescence* (M. J. Cormier, D. M. Hercules, and J. Lee, eds.), Plenum, New York, 1973, pp. 427–449.
113. R. Li and D. M. Hercules, *Anal. Chem., 46,* 916 (1974).
114. M. P. Neary, R. Seitz, and D. M. Hercules, *Anal. Lett., 7,* 583 (1974).
115. A. Hartkopf and R. Delumyea, *Anal. Lett., 7,* 79 (1974).
116. T. Sheehan and D. M. Hercules, *Anal. Chem. 49,* 446 (1977).
117. D. T. Bostick and D. M. Hercules, *Anal. Lett., 7,* 347 (1974).
118. D. C. Williams, G. F. Huff, and W. R. Seitz, *Clin. Chem., 22,* 372 (1976); D. C. Williams and W. R. Seitz, *Anal. Chem., 48,* 1478 (1976).
119. I. E. Kalinichenko and O. M. Grishchenko, *Ukr. Khim. Zh., 35,* 755 (1969).
120. J. Bognar and L. Sipos, *Mikvochim. Ichnoanal. Acta, 3,* 442 (1963).
121. A. K. Babko, A. V. Terletskaya, and L. I. Dubovenko, *J. Anal. Chem. USSR, 23,* 809 (1968).
122. L. I. Dubovenko and A. P. Tovmasyan, *Ukr. Khim. Zh., 37,* 1057 (1971).
123. L. I. Dubovenko and L. D. Guz, *Ukr. Khim. Zh., 36,* 1264 (1970).
124. L. I. Dubovenko and E. Ya Khotinets, *Zh. Anal. Khim., 26,* 784 (1971).
125. J. T. Brown, University of Georgia, unpublished data.
126. W. Nonidez, University of Georgia, unpublished data.
127. R. L. Bowman and N. Alexander, *Science, 154,* 1454 (1966).
128. G. D. Mendenhall and R. A. Nathan, Battelle Memorial Institute, unpublished work.
129. E. N. Harvey, *Bioluminescence,* Academic Press, New York (1952).
130. W. D. McElroy, B. Glass, eds., Light and Life, Johns Hopkins Press, Baltimore, 1961.
131. W. D. McElroy and H. H. Seliger, The chemistry of light emission, *Advances in Enzymology, 25,* 119 (1963).
132. M. J. Cormier, J. E. Wampler, and K. Hori, Bioluminescence: Chemical aspects in *Progress in the Chemistry of Organic Natural Products* (W. Herz, H. Grisebach, and G. W. Kirby, eds.), Vol. 30, Springer-Verlag, New York, (1973).
133. J. Lee, D. M. Hercules, and M. J. Cormier in Mechanisms of chemiluminescence and bioluminescence, in *Chemiluminescence and Bioluminescence* (J. Lee, D. M. Hercules, and M. J. Cormier, eds.), Plenum Press, New York (1973).
134. H. H. Seliger, R. A. Morton, *Photophysiology, 4,* 253 (1968).
135. W. D. McElroy and M. DeLuca, Chemical and enzymatic mechanisms of firefly luminescence, in *Chemiluminescence and Bioluminescence* (J. Lee, D. M. Hercules, and M. J. Cormier, eds.), Plenum Press, New York (1973).
136. F. McCapra, M. Roth, D. Hysert, and K. A. Zaklika, Model compounds in the study of bioluminescence, in *Chemiluminescence and Bioluminescence* (J. Lee, D. M. Hercules, and M. J. Cormier, eds.), Plenum Press, New York (1973).

137. T. Goto, I. Kubota, N. Suzuki, and Y. Kishi, Aspects of the mechanism of bioluminescence, in *Chemiluminescence and Bioluminescence* (J. Lee, D. M. Hercules, and M. J. Cormier, eds.), Plenum Press, New York (1973).
138. M. DeLuca and M. Dempsey, Mechanism of bioluminescence and chemiluminescence elucidated by use of oxygen-18, in *Chemiluminescence and Bioluminescence* (J. Lee, D. M. Hercules, and M. J. Cormier, eds.), Plenum Press, New York (1973).
139. K. Hori and M. J. Cormier, Structure and synthesis of a luciferin active in the bioluminescent systems of renilla and certain other bioluminescent coelenterates, in *Chemiluminescence and Bioluminescence* (J. Lee, D. M. Hercules, and M. J. Cormier, eds.), Plenum Press, New York (1973).
140. J. Lee and C. L. Murphy, Effects of aldehyde chain length and type of luciferase on the quantum yields of bacterial bioluminescence, in *Chemiluminescence and Bioluminescence* (J. Lee, D. M. Hercules, and M. J. Cormier, eds.), Plenum Press, New York (1973).
141. J. M. Anderson and M. J. Cormier, Lumisomes: A bioluminescent particle isolated from *Renilla reniformis*, in *Chemiluminescence and Bioluminescence* (J. Lee, D. M. Hercules, and M. J. Cormier, eds.), Plenum Press, New York (1973).
142. E. H. White, J. D. Miano, and M. Umbreit, *J. Am. Chem. Soc.*, 97, 1–198 (1975).
143. P. Jallet, L. Patrice, H. Pieri, and J. Pieri, *Acad. Sci. Ser. D.* 274(4): 603–606 (1972).
144. B. L. Strehler and W. D. McElroy, Assay of ATP, in *Methods of Enzymology* (S. P. Colowick and N. O. Kaplan, eds.), Vol. 3, Academic Press, New York (1957).
145. B. L. Strehler, ATP and creatine phosphate in *Methods of Enzymatic Analysis* (H. U. Bergmeyer, ed.), Academic Press, New York (1963).
146. J. W. Patterson, P. L. Brezonik, and H. D. Putnam, Sludge activity parameters and their application to toxicity measurements in activated sludge in *Proceedings Purdue Industrial Waste Conference* (1969).
147. *Photophysiology* (A. C. Gese, ed.), Vols. II, IV, V, Academic Press, New York (1970).
148. J. Denburg and W. D. McElroy, *Arch. Biochem. Biophys.*, 141(2), 668 (1970).
149. J. B. St. John, *Anal. Biochem.*, 37(2), 409 (1970).
150. M. S. Manandhar and K. van Dyke, *Microchem. J.*, 19, 42 (1974).
151. J. A. Davison and G. H. Fynn, *Anal. Biochem.*, 58, 632 (1974).
152. B. L. Strehler, *Methods Biochem. Anal.*, 16, 99 (1968).
153. R. Nielsen and H. Rasmussen, *Acta Chem. Scand.*, 22 (1968).
154. R. Cortenbosch and E. Schram, *Arch. Int. Physiol. Biochim.*, 79, 195 (1971).
155. R. A. Johnson, J. G. Hardman, A. E. Broadus, and E. W. Sutherland, *Anal. Biochem.*, 35, 91 (1970).
156. L. H. Kilbert, Jr., M. S. Schiff, and P. P. Foa, *Horm. Metab. Res.*, 4, 242 (1972).
157. G. L. Picciolo and E. W. Chappelle, *Yale Sci.*, 48, 2 (1973).
158. S. A. Butenko, L. M. Mukhin, and E. I. Milekhina, *Zhizn Zemli, Metody Ee Obnaruzheniya* (1970).
159. S. Addauki, J. F. Sotos, and P. D. Rearick, *Anal. Biochem.*, 14, 261 (1966).
160. A. J. D'Eustachio and D. R. Johnson, E. I. du Pont de Nemours and Co. Institutional Products Div., Wilmington, Del. 19898.
161. L. Z. Gogilashvili and G. G. Sotnikov, *Dokl. Akad. Nauk SSSR, 208*, 460 (1973).
162. D. E. Brooks, *J. Reprod. Ferl., 23*, 525 (1970).
163. J. R. Clendenning, U. S. Patent# 3,637,655 (1972).
164. G. Gunn and M. P. Eidenbock, *Anal. Biochem., 50*, 89 (1972).
165. H. G. Albraum, ADP, ATP analysis, in *Modern Methods of Plant Analysis* (K. Peach and M. V. Tracey, eds.) Springer Verlag (1955).
166. D. Sofrova and L. Sylva, *Photosynthetica, 4*, 162 (1970).
167. S. Lin and H. P. Cohen, *Anal. Biochem., 24*, 531 (1968).

168. J. L. David and F. Herion, *Adv. Exp. Med. Biol., 34,* 341 (1972).

169. E. Storm and H. J. Day, *Anal. Biochem., 46,* 489 (1972).

170. O. Holm-Hansen, Determination of total microbial biomass by measurement of ATP, in *Estuarine Microbial Ecology* (L. H. Stevenson and R. R. Colwell, eds.), Univ. of South Carolina Press (1973).

171. V. T. Stack, Stabilization oxygen demand, in *Biological Methods for the Assessment of Water Quality,* ASTM STP 528 (1973).

172. W. S. Oleniacz, M. A. Pisano, and M. H. Rosenfeld, *Environ. Sci. Technol., 2,* 1030 (1968).

173. A. J. D'Eustachio and G. V. Levin, *Bact. Proc.,* 13 (1968).

174. E. W. Chappelle and G. V. Levin, *Biochem. Med., 2,* 41 (1968).

175. I. C. Kao, S. Y. Chiu, L. T. Fan, and L. E. Erickson, *J. Water Pollut. Control. Fed., 45,* 926 (1973).

176. P. I. Parkinson and E. Medley, *Liq. Scintill. Counting, 2,* 109 (1972).

177. J. J. Lemasters and C. R. Hackenbrock, *Biochem. Biophys. Res. Commun., 55,* 1262 (1973).

178. E. P. Sidorik, E. A. Baglei, and T. N. Yurkovskaya, *Tr. Mosk. Ova. Ispyt. Prir., 39,* 221 (1972).

179. D. Glick, Y. Katsumata, and D. Von Redlich, *J. Histochem. Cytochem., 22,* 395 (1974).

180. K. Van Dyke, R. Stitzel, T. McClellan, and C. Szustkiewiez, *Advan. Automat. Anal., Technicon Int. Congr., 1,* 47 (1969).

181. M. S. Evadi, B. Weiss, and E. Costa, *J. Neurochem., 18,* 183 (1971).

182. B. Weiss, R. Lehne, and S. Strada, *Anal. Biochem., 45,* 222 (1972).

183. R. W. Butcher and E. W. Sutherland, *J. Biol. Chem., 4,* 237 (1962).

184. B. M. Breckenridge, *Proc. Natl. Acad. Sci. U.S., 52,* 1580 (1964).

185. R. A. Johnson, *Methods Mol. Biol., 3,* 1 (1972).

186. R. Johnson, *Adv. Cyclic Nucleotide Res., 2,* 81 (1972).

187. B. L. Strehler, *Methods of Enzymatic Analysis* (H. U. Bergmeyer, ed.), Verlag Chem. Weinheim Bergstr. (1974).

188. *Chem. and Eng. News. 51(36),* 26 (1973).

189. M. S. P. Manandhar and K. Van Dyke, *Anal. Biochem., 60,* 122 (1974).

190. M. S. P. Manandhar and K. Van Dyke, *Anal. Biochem., 58,* 368 (1974).

191. H. Bigl and D. Biesold, *Acta Biol. Med. Ger., 27,* 245 (1971).

192. G. J. E. Balharry and D. J. D. Nicholas, *Anal. Biochem., 40,* 1 (1971).

193. P. Stanley, in *Liquid Scintillation Counting, Recent Developments,* Academic Press, New York (1974).

194. J. Lee, C. L. Murphy, G. J. Faini, and T. L. Baucom, in *Liquid Scintillation Counting, Recent Developments,* Academic Press, New York (1974).

195. E. Gerlo and E. Schram, *Arch. Int. Physiol. Biochim., 79,* 200 (1971).

196. M. J. Cormier and J. R. Totter, *Photophysiology, 4,* 315 (1968).

197. S. M. Zubkova, A. I. Zhuravlev, and V. M. Abramov, *Tr. Mosk. Ova. Ispyt. Prir., 39,* 116 (1972).

198. A. P. Jacobson and K. A. McDermott, *J. Lumin., 3,* 419 (1971).

199. J. Lee, *Biochem., 11,* 3350 (1972).

200. D. Bentley, A. Eberhard, and R. Solsky, *Biochem. Biophys. Res. Commun., 56,* 865 (1974).

201. D. K. Dunn, G. A. Michaliszyn, E. G. Bogacki, and E. A. Meighen, *Biochemistry, 12,* 4911 (1973).

202. M. Eley and M. J. Cormier, *Biochem. Biophys. Res. Commun., 32,* 454 (1968).

203. P. E. Stanley, *Anal. Biochem., 39,* 411 (1971).

204. E. W. Chappelle, G. L. Picciolo, and R. H. Altland, *Biochem. Med., 1,* 252 (1967).

205. D. Njus and T. O. Baldwin, *Anal. Biochem.*, *61*, 280 (1974).
206. S. E. Brolin, E. Borglund, L. Tegner, and G. Wettermark, *Anal. Biochem.*, *42*, 124 (1971).
207. S. E. Brolin, C. Berne, and U. Isacsson, *Anal. Biochem.*, *50* (1972).
208. E. Chappelle and G. L. Picciolo, *Methods Enzymol.*, *18*, 381 (1971).
209. D. J. D. Nicholas and G. R. Clarke, *Anal. Biochem.*, *42*, 560 (1971).
210. E. Gerlo, *Chem. Weekbl.*, *70*, L23 (1974).
211. D. C. White, B. Wardley-Smith, and G. Adey, *Life Sci.*, *12*, 453 (1973).
212. D. C. White and C. R. Dundas, *Cellular Biology and Toxicity of Anesthetics, Proceedings of a Research Symposium*, 168 (1970).
213. M. J. Halsey and E. B. Smith, *Nature*, 227, 1363 (1970).
214. D. E. White and C. P. Dundas, *Nature*, 226, 450 (1970).
215. R. W. Treick and G. J. Lancz, *Life Sci.*, *8*, 961 (1969).
216. A. M. Fish and R. I. Chumakova, *Izv. Sib. Otd. Akad. Nauk SSSR, Ser. Biol. Nauk, 32* (1969).
217. W. F. Serat, J. Kyono, and P. K. Mueller, *Atmos. Environ.*, *3*, 303 (1969).
218. R. Oshino, N. Oshino, M. Tamura, L. Kobilinsky, and B. Chance, *Biochem. Biophys. Acta, 273*, 5 (1972).
219. J. P. Henry and A. M. Michelson, *C. R. Acad. Sci., Ser. D., 270*, 1947 (1970).
220. G. Soli, *Atti. Soc. Peloritana Sci. Fis. Mat. Nature, 16*, 159 (1970).
221. K. T. Izutsu and S. P. Felton, *Clin. Chem., 18*, 77 (1972).
222. R. V. Lynch, F. I. Tsuji, and D. H. Donald, *Biochem. Biophys. Res. Commun., 46*, 1544 (1972).
223. O. Shimomura and F. H. Johnson, *Nature, 227*, 1356 (1970).
224. O. Shimomura, F. H. Johnson, and Y. Saiga, *J. Cell. Comp. Physiology, 59*, 223 (1962).
225. O. Shimomura and F. H. Johnson, *Biochem., 11*, 1602 (1972).
226. G. Loschen and B. Chance, *Nature (London), New Biol., 233*, 273 (1971).
227. K. T. Izutsu, S. P. Felton, I. A. Siegel, W. T. Yoda, and A. C. N. Chen, *Biochem. Biophys. Res. Commun., 49*, 1034 (1972).
228. O. Shimomura and F. H. Johnson, *Biochem. Biophys. Res. Commun., 53*, 490 (1973).
229. F. H. Johnson and O. Shimomura, *Nature (London), New Biol., 237*, 287 (1972).
230. J. W. Hastings, G. Mitchell, P. H. Mattingly, J. R. Blinks, and M. Van Leeuwen, *Nature (London), 222*, 1047 (1969).
231. C. Ashley and E. B. Ridgway, *Symp. Calcium Cell. Funct.*, 42 (1969).
232. T. Goto, *Kagaku To Seibutsu, 7*, 445 (1969).
233. F. H. Johnson, O. Shimomura, Y. Saiga, L. C. Gershman, G. T. Reynolds, and J. R. Waters, *J. Cell. Comp. Physiology, 60*, 85 (1962).
234. Y. Kishi and T. Goto, *Tetrah. Letters., 29*, 3445 (1966).
235. M. J. Cormier, Y. D. Karkhanis, and K. Hori, *Biochem. Biophys. Res. Commun., 38*, 962 (1970).
236. V. A. Poltorak, *Mikrobiologiya, 40*, 501 (1971).
237. E. Schram, in *Liquid Scintillation Counting, Recent Developments*, Academic Press, New York (1974).
238. E. Schram, in *The Current Status of Liquid Scintillation Counting* (E. D. Bransome, ed.) Grune and Stratton, New York (1970).
239. E. Schram, R. Cortenbosch, E. Grelo, and H. Roosens, in *Organic Scintillators and Liquid Scintillation Counting* (D. L. Horrock and C. T. Peng, eds.), Academic Press, New York (1971).
240. M. Ogata and S. Watanabe, *Igaku to Seibutsugaku, 74*, 36 (1967).
241. E. Schram and H. Roosens, *Liquid Scintill. Counting, 2*, 115 (1972).

242. W. Ernst, *Fresenius' Z. Anal. Chem., 238,* 35 (1968).
243. E. E. Kolesnikov, K. S. Kalugin, Z. I. Zhulanova, and E. F. Romantsev, *Vop. Med. Khim., 14,* 451 (1968).
244. L. Dufresne and H. J. Gitelman, *Anal. Biochem., 37,* 402 (1970).
245. R. H. Hammerstedt, *Anal. Biochem., 52,* 449 (1973).
246. P. E. Stanley and S. G. Williams, *Anal. Biochem., 29,* 381 (1969).
247. H. A. Strobel, *Chemical Instrumentation,* 2nd Ed., Addison-Wesley Publishing Co. (1973).
248. M. P. Neary, W. R. Seitz, and D. M. Hercules, *Anal. Lett., 7,* 583 (1974).
249. H. P. Mansberg, U. S. Patent No. 3,679,312.
250. B. N. Kelbaugh, G. L. Picciolo, E. W. Chappelle, and M. E. Colburn, U. S. Patent No. 3,756,920.
251. E. M. Silinsky, *Diss. Abstr. Int. B, 34,* 3449 (1974).
252. J. Slawinski, E. Grabikowski, and A. Murkowski, *Zesz. Nauk. Wyzsz. Szk. Roln. Szxzecinie,* 301 (1971).
253. G. B. Calleja and G. T. Reynolds, *Experientia, 26,* 221 (1970).
254. O. Holm-Hansen and C. R. Booth, *Lumnol. Oceanogr., 11,* 510 (1966).
255. R. D. Hamilton, O. Holm-Hansen, and J. D. H. Strickland, *Deep-Sea Res., 15,* 651 (1968).
256. M. L. B. Williams, *Can. Inst. Food Technol., J., 4,* 187 (1971).
257. P. D. Allen, *Dev. in Indust. Microbiology, 14* (1969).
258. C. C. Lee, R. F. Harris, J. D. H. Williams, J. K. Syers, and D. E. Armstrong, *Soil Sci. Soc. Am. Proc., 35,* 86 (1971).
259. C. C. Lee, R. F. Harris, J. D. H. Williams, D. E. Armstrong, and J. K. Syers, *Soil Sci. Soc. Am. Proc., 35,* 82 (1971).
260. N. H. Macleod, E. W. Chappelle, and A. M. Crawford, *Nature, 223,* 267 (1969).
261. P. D. Allen, *Biometer Flashes,* BF6, DuPont Corp. (1973).
262. C. Weber, in *Bioassay Techniques and Environmental Chemistry, 119,* Ann Arbor Science Publishers, (1973).
263. Bacteria in Cutting Oils and Hydraulic Fluids, *Biometer Flashes,* 60BF3, DuPont Corp. (1972).
264. Newsletter No. 4, Analysis and Quality Control Laboratory, Federal Water Pollution Control Administration, Jan. (1970).
265. E. W. Chappelle, U. S. Patent No. 3,575,812.
266. G. Steinlid, *Physiol. Plant Pathol., 25,* 397 (1971).
267. V. S. Faustov and B. M. Karlson, *Dokl. Akad. Nauk. SSSR, 176,* 728 (1967).
268. W. J. Nungester, L. J. Paradise, and J. A. Adair, *Proc. Soc. Exp. Biol. Med., 2,* 132 (1969).
269. E. W. Chappelle and G. V. Levin, U. S. Patent No. 3,575,811.
270. A. D. Carlson, *J. Exp. Biol., 49,* 195 (1968).
271. V. S. Danilov, *Biofizika, 16,* 346 (1971).
272. Bacterial Screening of Urine, *Biometer Flashes,* 760BF5, Du Pont Corp. (1972).
273. Monitoring Cooling Tower Water and Biocide Control, *Biometer Flashes,* 760BF2, Du Pont Corp. (1972).
274. W. Ernst, *Oecologia, 5,* 56 (1970).
275. R. A. Daumas, *Tethys., 5,* 71 (1973).

3

Environmental Studies of the Atmosphere with Gas Chromatography

Robert L. Grob

1. Introduction

Over the years much has been written about our environment. All facets have been discussed: air, water, soil, waste, as well as the various analytical techniques used in attempting to describe each of these more precisely. This chapter will examine the atmosphere which surrounds us, considering only the assay technique of gas chromatography. The extent of this survey will be to develop a systematic coverage, from sampling to various types of analyses for specific atmospheric components.

A chapter of this type is of interest for two reasons: it concerns a topic which is very contemporary and it treats a technique which has served to open many systems to analysis, including some that were not easily examined previously.

The author will try to develop an overall picture of how we obtain information from the atmosphere, so that we may gain insight into what man is doing to his huge "home" to make it more habitable and more enduring. The manner in which new environmental restrictions are continually being added (and a few old ones being relaxed) allows one to speculate as to what specific restrictions are necessary. A question to ponder is this: would this large "home" we call the environment survive if we leave it completely in the hands of Mother Nature? If the answer is yes, does this allow us to be completely unconcerned? If the answer is no, then how far should we extend our studies and safeguards? The author feels

Robert L. Grob • Department of Chemistry, Villanova University, Villanova, Pennsylvania 19085

that this is not an easy yes or no question, but that the answer lies in finding an equilibrium somewhere between the two extremes. Once this point of equilibrium is found, it is then simply a matter of adjusting the variables from time to time so that we don't shift to the "reactant" or "product" sides and become noncompatible with the system.

In this chapter, atmosphere will mean the gaseous environment from the surface of the earth to a distance seven miles above the earth's surface. This gaseous environment, minus contamination, is known as air and has the composition (major components): nitrogen, 78.08%; oxygen, 20.95%; carbon dioxide, 0.034%; argon, 0.93%; and smaller percentages of hydrogen, neon, krypton, xenon, and helium. Other components can then be classified as contaminants (minor components). In this category we find such species as hydrocarbons, carbon monoxide, sulfur compounds, nitrogen compounds, metals, metal ions, various oxidants, and peroxyacetyl nitrate. Various minor components are being added and their amounts are increasing in the atmosphere as a result of man's everyday activities. Natural hydrocarbons (those which are supplied by earth itself) have been estimated by Junge[1] to be 0.8–1.1 mg/m^3 or 1.2–1.5 parts per million (ppm). Koyama[2] stated that the annual emission rate of hydrocarbons from natural sources is 3×10^8 tons/year. In addition, hydrocarbons are emitted into our environment by industrial and vehicular sources. It has been estimated that 50% of hydrocarbon emissions comes from transportation sources, about 15% comes from residential and industrial fuel burning, and 26% comes from sources such as refuse, coal, and solvent evaporation.[3] Caplan[4] has described an "average automobile exhaust" as containing 394 ppm paraffins, 118 ppm acetylenes, 206 ppm aromatics, and 310 ppm olefins. It has been calculated[5] that in the United States alone, 2×10^{15} g/year of CO_2 are produced from fossil fuel combustion. Evaporation processes and urban combustion processes are the causes of most of the oxides in the air. Nitric oxide results predominantly from the high-temperature fixation of nitrogen and oxygen at power plants and in all our internal-combustion engines. Sulfur dioxide results from the combustion of high-sulfur-containing fuels. Automatic exhaust seems to be the main source of carbon monoxide. In contrast, anaerobic processes and natural-gas losses account for most of the methane in our atmosphere. It is for these two reasons that methane is not considered a hydrocarbon pollutant.

Nitrogen oxides plus sufficient ultraviolet (UV) radiation react with olefins, alkylbenzenes, and aldehydes to form ozone, peroxyacetyl nitrate (PAN), various aldehydes and ketones, and other products.[6]

Air quality standards have been issued by the Environmental Protection Agency (EPA).[7] These are in need of updating; other contaminants need to be added to the list. These are shown in Table 1. As illustrated in

Table 1. Air Contaminants (Air Quality Standards)

Pollutant	Primary standard control $(\mu g/m^3)^a$	Secondary standard control $(\mu g/m^3)^b$
Carbon monoxide (CO)		
24 h	260	150
8 h	10	10
1 h	40	40
Nitrogen oxides		
max. 24-h conc.	100	100
Hydrocarbons (total corrected for methane)		
max. 3-h avg.	160	160
Photochemical oxidants		
max. 1-h conc.	160	160
Sulfur dioxide (SO$_2$)		
annual avg.	80	60
24 h	365	260
3 h		1300

^{a} Levels necessary to protect public health (1 $\mu g/m^3 \cong 0.9$ ppm).
^{b} Levels necessary to ensure environmental well-being of public.

the table many components considered as contaminants and/or injurious to health have no standards.[8,9] Typical examples would be metals. Beryllium has been determined in air samples as its trifluoroacetylacetonate, using an electron capture detector.[10,11] Sixteen to forty-nine picograms per cubic meter have been measured.[12] Lead can be in the atmosphere as PbO or the uncombusted alkyl compounds; atomic absorption spectroscopy is best for the former compound whereas gas chromatography is superior for the latter. Cantute and Cartoni[13] used sample preconcentration on a Chromosorb P column and an electron capture detector (ECD) for the determination of tetraethyllead (TEL). The limit of detection was 0.1 ppm (maximum concentration of alkyl-lead compounds is about 1 mg/m^3).

2. Sampling and Sampling Techniques

Sample collection for atmospheric samples may have several goals; to sample all airborne components (gases, aerosols, particulate matter), to sample only gaseous components, or to sample only particulates. If the purpose is to sample all components then one must combine gas sampling with entrapment of particles smaller than 10 μm in diameter. It is particles of this size which are easily taken into the lungs. This type of sampling can be performed with the aid of high-volume fiberglass filters or freeze-out traps. Use of the latter will also collect water vapor whereas the former will collect essentially a dry sample. Glass fiber filters should be dried in a desiccator (not in an oven) thus preventing loss of volatile hydrocarbons or other organic compounds.

Samples from the gaseous environment require mechanical controls that are not necessary when sampling from an aqueous environment. Volumes and/or masses from the latter are easily controlled and measured. However, it is more difficult to measure a volume or weigh a mass of a gaseous sample. Pumps which allow accurately measured volumes are necessary, or column concentrators or freeze-out traps are required if we wish to measure mass.

The required amount of gaseous sample will be determined by two factors, the expected concentration of the analyte and the detector to be employed for the determination. These factors are interdependent: detectable amount of analyte will be controlled by detector employed and vice-versa. Use of a low-sensitivity detector (e.g., thermal conductivity) necessitates a large volume of air (10–15 liters). Ionization detectors would require a sample of about one liter (analyte concentration < 0.01 ppm).

Three components are necessary for air sampling if the resulting analyses are to be reliable. First, an accurate *volumetric flow meter*; this piece of equipment should be calibrated so that the readings may be corrected for any deviations from meter readings.[14] Second, some type of *sample collector*; this may be a filter, absorbing solution, or a chromatographic column. In each case, one must know the efficiency so that the true weight or volume of contaminant can be calculated. It should be pointed out that no collector is 100% efficient. The efficiency may be calculated in absolute or relative terms. Absolute efficiency can be determined by actually collecting a "known contaminant–air mixture" as the control and calculating the efficiency of collection by use of the experimental recovery data.[15] Relative efficiency can be determined by arranging several collectors in series and assuming 100% efficiency in the last of n collectors. The determination of the relative difference between actual values of the contaminant in $(n - 1)$ collectors would allow one to calculate relative efficiency. The assumptions being made are that there are no dilution effects and that there are no variations in the rate or mechanism of collection from collector to collector. The calculation of relative efficiency is utilized only when one does not have a known contaminant sample available.

Third, a *constant flow pump* is required so that the air sample is uniformly drawn through the flow device and collector. Sample volume is calculated from the product of average flow rate and time. Large fluctuations in the measured "initial" and "final" flow rates can lead to errors in calculated sample volumes. Sreedharan[16] describes a non-reactive-gas sampling pump useful for ozone-measuring instruments. The pump will deliver up to 200 cm^3 of air per minute. Leiby and Dunton[17] designed a pump for ultrapure and toxic gases. The pump is capable of circulating gases around a closed loop without leakage or contamination and handling gases in ultra-high-vacuum systems.

Several publications are also available regarding size of sampling inlet,[18] a multipurpose gas-handling apparatus,[19] and an automatic gas-sampling device.[20]

An important rule to remember is that the pump should always be downstream from the sample collector. This prevents deposition of pump lubricant into the air-measuring device. In other words, the pump should draw—not push—the air sample through the collecting device. Another consideration is how long should one sample. A rule of thumb is 1 to 3 h, unless one is tabulating hourly variations in a contaminant's concentration. A "normal" flow rate would be 0.5 ft^3/min. In some cases it might be well to consider the significance of the biological effect of the pollutant to the sampling time. The human body responds to pollutants in terms of their biological half-lives (time necessary for one-half the amount in the body to be removed). Roach[21] has postulated that variations in ambient pollutant concentrations over a time interval less than one-tenth of biological half-life are of no biological significance. This assumes rate of pollutant absorption being proportional to ambient concentration. Some representative biological half-lives are shown in Table 2.[22]

One should also consider the limitations imposed by the sampler. Inexpensive diaphragm pumps are able to maintain from 0.05 to 0.10 ft^3/min, whereas rotary vacuum pumps can sustain a dependable 1.0 ft^3/min. Too fast a flow through solvent decreases the efficiency of collection, but too slow a flow causes the same effect because of lack of surface contact of the air sample with the liquid.

Table 2. Biological Half-Lives of Representative Air Pollutants

Pollutant	Half-life
Chlorine, Cl_2	<20 min
Carbon monoxide, CO	2 h
Hydrogen sulfide, H_2S	<20 min
Lead, Pb	6 months
Nitrogen dioxide, NO_2	1 h
Sulfur dioxide, SO_2	<20 min

Table 3. Air Flow Devices

Type	Comments
Dry gas meter	0.3 to 30 ft³/min.
Spirometer (displacement bottle)	Displacement principle to measure exact volumes.
Wet-test meter	Accuracy ±0.5% of total volume.
Rate-of-flow instruments	Intermittent flow measurement. Not accurate.
Manometer	Must be calibrated with dry gas meter.
Orifice meter	Accurate to about 2% of flow measure.
Rotameter	Same as orifice meter. Subject to error because of particulate, soil, or hydrocarbon deposits.
Pitot tube	Best for high in-stack or in-duct flows.

Other considerations to account for would be:

1. Collection of the contaminant as it exists in the atmosphere. Sulfur dioxide is easily converted to SO_3 if collected by bubbling through water in presence of particulate. Removal of particulate first and bubbling through isopropanol will ensure SO_2 collection.
2. Collection vessel material. Gases are easily adsorbed on the surface of a container and lost for analysis. Collection vessels made from polyfluorocarbons are better than those made from polyolefins or polyvinyl materials. The latter materials readily adsorb SO_2 and NO_2.
3. Sample storage time. On standing some contaminants react with each other or the container walls; e.g., H_2S and SO_2 can form a redox couple giving elemental sulfur.[23]

Table 3 summarizes some pertinent information for air-flow measuring devices. One final piece of information regarding sampling in general: some monitoring of gas-phase pollutants is performed by automated methods. Little if any of this automation is performed by gas chromatography. Most effort in this area has been colorimetric and/or infrared spectroscopic methods. An elaborate setup wherein gas chromatography–mass-spectrometry and computers could be automated is a possibility. Several reviews have been written for this type of interfaced system.[24,25]

2.1. Grab Sampling

This technique of obtaining a sample is employed for convenience and/or necessity. It is used when one does not wish an elaborate sampling program or when field monitoring is not being utilized. The sampling technique may be carried out with the use of evacuated bulbs or inflatable plastic bags. The volume of the sample container is determined by calculation or liquid displacement technique.

2.1.1. Sample Bulbs

These are of two types: (1) resealable glass cylinders which must be connected to a vacuum pump and then evacuated to a predetermined partial pressure. The glass bulb is then sealed by means of a burner. (2) An evacuated flask with stopcock fittings; the stopcocks should not be glass which must be lubricated with a grease as they tend to absorb sample. Rather one should use Teflon stopcocks. Accurate volumes can be obtained by either dimensional measurements or liquid displacement measurements. Knowing the volume, we can calculate an accurate value for the amount of sample by correcting the container volume for pressure and temperature conditions (by means of ideal gas law).

2.1.2. Plastic Sampling Bags

These have found wide acceptance for air sampling. Mylar bags are the most satisfactory. They may be used for hydrocarbons, sulfur dioxide, nitrogen dioxide, and ozone. Other types of plastic materials are not recommended for the storage of samples. If storage is necessary, the bags should be protected from direct sunlight.

2.2. Solvent Trapping

One of the most common techniques for collecting gases or vapors is to use a solvent in which the sample is soluble or undergoes a chemical reaction. The main criterion necessary for this type of sample collection is that the gases or vapors come in intimate contact with the liquid. This is accomplished by providing collection equipment which increases the surface area of the liquid. Several factors should be kept in mind when sampling gases in solvents. Small bubbles furnish greater surface area and thus more contact with trapping solvent. Long contact time provides more chance of absorbing all the gases. Flow rate should be fast enough to cause agitation of solution. Finally, if collection is performed with the aid of a chemical reaction, one should employ a sufficient concentration of reagent to cause the reaction to go to completion (since otherwise the fraction collected is concentration dependent).

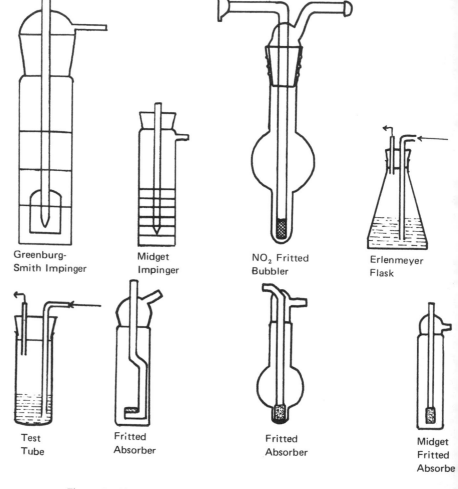

Figure 1. Absorption containers for solvent trapping of pollutants.

Two examples where gases may be collected by chemical reaction are:

(1) SO_3 in a barium chloride solution:
$$SO_3 + H_2O = H_2SO_4$$
$$H_2SO_4 + BaCl_2 = BaSO_4(s) + 2HCl$$

(2) CO_2 in a calcium hydroxide solution (lime water):
$$CO_2 + H_2O \leftrightharpoons H_2CO_3$$
$$H_2CO_3 + Ca(OH)_2 = CaCO_3(s) + 2H_2O$$

In each case an insoluble product drives the reaction to completion.

Solvent-absorption techniques may be performed by several types of bubblers. One may employ a simple bubbler made from a large test tube or a 250-cm³ Erlenmeyer flask. This type of bubbler does not ensure good contact between the gaseous sample and the absorbing liquid but is adequate for qualitative testing of pollutants. To ensure good contact between gas phase and absorbing liquid the use of fritted-glass absorbers or impingers are recommended. The fritted-glass absorbers are more efficient than the impinger type, and a 50-μm pore size is recommended for the sampling of air components. As stated above, turbulence is necessary for efficient sampling, and a flat surface in the direct path of the air stream presents this turbulence. This is the mechanism by which impingers operate. Figure 1 illustrates the various sampling absorbers mentioned above. Under optimum flow conditions these samplers can be 90–100% efficient. Impingers have the advantage of ease of cleaning. In using this technique it is important to use a solvent of low volatility. Many people do not employ fritted-glass bubblers for collection of strong oxidizing gases such as ozone because of chemical reactions which can take place at the reactive surface. The author has employed both fritted-glass absorbers and midget impingers for ozone collection and prefers the former. Little difference was found in the final results. Impingers tend to splash unless careful control of flow is monitored.

In either case one should begin with a reasonably efficient system before collecting any data. Efficiency can be checked easily with the setup shown in Figure 2. Sampling time can vary from 3 to 24 h for low levels of pollutants, to 0.25–3 h for higher levels. Using a standard pollutant gas

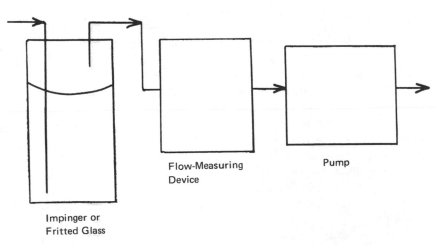

Impinger or
Fritted Glass

Flow-Measuring
Device

Pump

Figure 2. Setup for solvent trapping collection efficiency.

and careful monitoring of time, one can calculate the collection efficiency. Midget impingers use slower optimum flow rates (0.05–0.10 cfm) than do standard impingers (0.5–1.0 cfm). Another point to remember is that collection efficiencies for a system may vary with sample type.

Example

>Standard gas: 15 μg NO_2/m^3
>Sampling time: 20 min
>Flow rate: 0.2 m^3/min

$$0.2 \text{ m}^3/\text{min} \times 20 \text{ min} \times 15 \ \mu\text{g/m}^3 = 60 \ \mu\text{g NO}_2 \qquad (1)$$

If subsequent analyses showed 54 μg NO_2, then the system was 90% efficient.

Ioffe et al.[26] sampled aromatic hydrocarbons by passing 25–30 liters of air through acetic acid solution (3 cm^3) at a flow of 200 cm^3/min. Aliphatic ketones were sampled in water using 7 liters of gas sample. Both types of samples were analyzed by gas–liquid chromatography. Wartburg et al.[27] used a bubbler made from glass and polytrifluoroethylene (PTFE) and found it to be easier to use in the field and equally efficient as an all-glass bubbler. Ford et al.[28] sampled pesticides in air by use of four impingers in series (ethylene glycol solvent). Pesticides were extracted by hexane solvent, filtered through anhydrous sodium sulfate and deter-mined by electron-capture gas chromatography.

2.3. Solid Sampling Systems (Adsorption or Substrate Equilibrium Technique)

This technique is used for sampling large volumes of air, so as to concentrate an analyte present in small concentrations. Thus, it is a sample preconcentration technique. The technique requires that the air sample be drawn through a small tube packed with a suitable material. The collected material may be removed by heating or solvent stripping. Calibration for each component is necessary for this technique.

Gases are usually adsorbed on solids at temperatures below their boiling points, as defined by the gas–solid isotherm. Constant-temperature adsorption monitoring of a gas will result in an isotherm. Once the solid surface is saturated there are no more active sites available and adsorption efficiency decreases. Particle size of the adsorbent is very important; proportionately larger surface areas are of course available with smaller particles.

The literature shows a number of sampling systems to be available. Yu et al.[29] sampled and measured humid air by metering air samples directly onto Carbowax-coated Haloport F columns at 90°C. Dravnieks *et al.*[30] sampled organics in air by means of Chromosorb 102 (flow: 4 liters/min). Collected organics were released by heating and analyzed by gas chromatography. Collection efficiency was 90% and the chromatographic system was sensitive to 10^{-10}-g organic solute per liter of air. Graphitized carbon was used as a collector by Raymond and Guiochon[31] for aromatics in the air. The sampling rate was low enough so that 5–10 liters of air could be sampled without excessive condensation of water. Adsorbed sample was stripped by heating in the injector port (400°C) of the gas chromatograph. Seventeen compounds in the 1–400-μg/m^3 range were identified. Litovchenko[32] sampled hydrogen and helium in petroleum gases. Acid gases and water were removed by a column containing silica gel, and the hydrocarbons were adsorbed on a second column containing active carbon. He and H_2 were sampled directly into the chromatograph. Sensitivity was 23 ppm He (v/v) and 17 ppm H_2(v/v); error was <5%. Collected hydrocarbons were then stripped by heating and analyzed by gas chromatography. Tenax GC was used by Versino *et al.*[33] for concentrating components from air samples.

Aue and Teli[34] used a substrate equilibrium technique in which high-molecular-weight pollutants were sampled onto a column of 24% bonded silicone on Chromosorb A. In this particular system the sample was removed by hexane extraction. Mass spectrometry was used for identification of petroleum vapors, car exhaust fumes, and chlorinated pesticides. SO_2 interacted with the column and was not eluted.

Herzel and Lahmann[35] sampled pesticides on a silica gel (0.2–0.5-mm particles) packing coated with 10% hard polyethylene (3–6-cm tubes); samples were removed by acetone or benzene extraction. Pellizzari *et al.*[36] evaluated several sorbents for the concentrating of trace organics in ambient atmospheres. Polymer beads (Tenax GC, Porapak Q, Chromosorb 101 and 104), activated carbons, and liquid phases (Carbowax 400 and 600, oxydipropionitrile, didecyl phthalate, and tricresyl phosphate) were employed. Efficiencies ranged from 20 to 98%. Figure 3 shows a schematic of the monitoring system used in this study.

2.4. Freeze-Out Techniques Using Cryogenic Systems

As stated above, temperature is a variable which can be changed to aid in adsorption or desorption. One can selectively sorb various components of an air sample by using a series of collection tubes held at successively lower temperatures. Each air contaminant will be primarily sorbed in the first tube which is below its boiling point. Temperatures of some typical low-temperature baths are shown in Table 4.

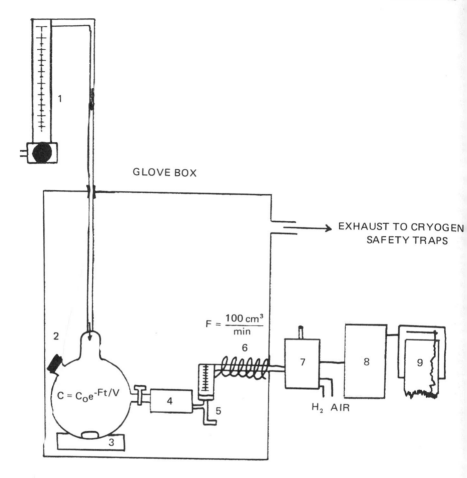

Figure 3. Monitoring system for the determination of collection efficiency of sorbents (Courtesy of *Environ. Sci. Technol.*, reference 36): (1) flow meter/regulator, (2) septum plug, (3) magnetic stirrer, (4) cartridge sampler (silica gel, polyethylene packing), (5) splitter, (6) heating tape, (7) F.I.D., (8) amplifier, (9) recorder. Abbreviations: C, concentration in chamber after time t has elapsed; C_0, initial concentration in chamber; F, purging rate (liters/hour); V, volume of chamber.

To decrease or minimize the amount of water vapor collected together with sample, the cooling trap should be preceded by a $CaCl_2$ drying tube. Additionally, slow flow rates should be employed (0.1–0.2 ft³/min) to ensure proper trapping of the cooled contaminant.

Morie and Sloan[37] used cryogenic temperatures to freeze out CO and CO_2 from cigarette smoke. Subsequent temperature programming of a Porapak Q column permitted the identification of more than 20 other components in the cigarette smoke. Snyder[38] designed an effective cryogenic trapping technique for introducing trace samples into a mass spectrometer. Liquid nitrogen was used as the thermostatting liquid.

2.5. Cryogenic Gas Chromatographic Analysis

Not only has low temperature been used for sampling but it has also been employed for the chromatographic system. Such a chromatographic system has been constructed in the author's laboratory. This unit has the capability of being programmed from $-100°C$ to $+300°C$ (liquid N_2 used as coolant). The construction details of this instrument have been published[39] and its use in air-pollution monitoring has also been demonstrated.[40]

Mignano et al.[41] used a liquid-N_2-cooled U-tube containing molecular sieve 5A to remove CO_2 and H_2O from samples and carrier stream. Neff et al.[42] employed a cross-linked polymer as a stationary phase on Chromosorb PAW, at $-60°$ to $+200°C$, for the separation of hydrocarbons, alcohols, aldehydes, and esters. Gouw and Whittemore[43] used liquid CO_2 as the oven coolant to provide cryogenic conditions for the analysis of gas samples. Variations of cryogenic sampling and/or analysis have been published by Angerer and Haag,[44] who passed their air sample through a tube cooled by liquid N_2 so as to present a thermogradient (along the column). The tube is then stripped of sample by passing air, at 220°C, from the other end and passing effluent into the chromatographic-system–mass-spectrometer. Flame ionization detection was also used but the authors claim better sensitivity and specificity with the mass spectrometer. Kaiser[45] used a similar technique to enrich volatile components in car exhaust samples and river water samples. Carbon molecular sieves were packed in the temperature-gradient tubes and concentrated compounds were removed by means of N_2 gas under temperature-gradient conditions.

Fort et al.[46] compared the concentrating of tritium as HTO by freeze-out and adsorption-collection techniques. Their conclusion was that

Table 4. Low-Temperature Baths

System	Temperature
Ice/water	0°C
Ice/salt	-16
Dry ice/acetone	-80
Liquified air	-147
Liquified oxygen	-183
Liquified nitrogen	-195

no significant difference existed provided reasonable care was used with both techniques. Cryogenic techniques have been employed to trap column effluents for further identification[47] as well as to study pyrolysis products from minireactors before sample enters the chromatograph.[48]

3. Accuracy, Precision, and Sensitivity

The accuracy of any analytical technique is influenced by many factors, each of which may affect a part or the entire system. The salient factors are sampling, subsequent treatment of the sample, and the final measurement step. We are only concerned with gas chromatographic analysis, so our measurement step is controlled by the detector system. A later section will treat detectors, but suffice it to say that the more specific and sensitive is the detector system, the more accurate the final results—provided all other variables in the system are equally controlled.

The column in the chromatographic system is primarily for the separation step; it has no effect on the concentrating of the sample (this may be performed by a precolumn external to the chromatographic column). Once a "representative" sample is on the column the detector influences the accuracy of our data the most. With the specificity of some detectors, e.g., flame photometric and electron capture, it may be possible in some cases to bypass a column completely.

Even though we can control the sampling, the column, and our detector, the accuracy obtainable can still be influenced by sample composition changes that occur during the workup. An example of this situation is the analysis of SO_2 in the presence of SO_3 or vice versa. SO_2 slowly oxidizes to SO_3 from the oxygen present and measurements for the SO_2 concentration may be high in relation to what was actually present at time of sampling. The presence of a reducing agent (e.g., isopropanol) would prevent this oxidation from occurring.

Another factor which can influence accuracy is the dynamic range of our system; we cannot expect to work outside this range with accuracy. The precision of gas chromatographic data is primarily a function of the detector sensitivity. However, sample size and the readout system used will also influence precision.

One may determine how accurate and/or precise a particular chromatographic analysis is by utilizing a sample of known concentration (e.g., a certified sample from the National Bureau of Standards or some reliable commercial laboratory). Repetitive measurement of this sample, under rigidly controlled conditions, will enable one to calculate accuracy and precision. Accuracy is best determined from the arithmetic mean ($\bar{x}$) or the median (M). The latter is obtained by arranging all the data in either ascending or descending numerical order and choosing the middle value (if odd number of pieces of data are available) or the average of the two middle pieces of data (if an even number of pieces are available). The mean is obtained by totalling all the individual results and dividing by the number of pieces of data in the group:

$$\bar{x} = \sum x_i/N \tag{2}$$

x_i is a piece of data or result of an analysis and N is the number of pieces of data. If a result appears more than once in a set of data, the mean may be calculated by

$$\bar{x} = \sum nx/N \tag{3}$$

Here, n is the number of times a result repeats itself and x is the repeated result.

Precision can be expressed either as a range (w) or standard deviation (s). The range, as a measure of precision, can be misleading when outlaying values are present at the high or low end of the tabulation. An outlaying piece of data can often be accounted for on the basis of nonconformity (a known mistake or deviation from the procedure) to the conditions of sampling, sample treatment, or the measurement step. If extreme values cannot be accounted for on the basis of nonconformity, then they should be checked for rejection by statistical methods (Q Test,[49] or validity test using t-tables[50]).

$$s = \left[\frac{\sum (x_i - \bar{x})^2}{N - 1} \right]^{1/2} \tag{4}$$

By instrument sensitivity we mean the smallest concentration of a contaminant that is necessary to produce a signal which is readily distinguished from noise. Some representative sensitivities for air pollutant gas chromatographic methods are shown in Table 5.

Table 5. Sensitivities of Gas Chromatographic Methods

Pollutant	Sampling technique	Sensitivity, ppm
Hydrocarbons, flame ionization detection	Grab sampling	1
	Collection bottle	1
	Gas pipet	1
Pesticides (Cl-containing), electron capture detection	Collection bottle	0.001
	Gas pipet	0.001
Peroxyacetyl nitrate, electron capture detection	Collection bottle	0.001
	Gas pipet	0.001

4. Calibration Techniques

If the data obtained from an environmental sample are to have meaning, comparison must be made to some standard. It follows then that the equipment used in each step of the analysis of an environmental sample must be calibrated. To cite an example that is often overlooked we can consider detector response. It makes no difference if one is using a thermal conductivity detector (TCD), a flame ionization detector (FID), or an electron capture detector (ECD). Each will be affected by the operating parameters and the type of sample being separated. If the data are to be meaningful the detector must be calibrated for each component that will be present in the samples. The response of the TCD will be more affected by sample type than will that of the FID; however, contrary to the belief of many workers in gas chromatography, the FID will give a different response for each type of compound passing through the detector.

Most complications arise when the concentration levels to be monitored are low (part per million or less), or when the materials (samples) to be analyzed exhibit chemisorption. High-concentration calibration gases retain their calibration values better than low-concentration calibration gases. Known-composition gas mixtures may be purchased from many commercial laboratories. Most of these are at the ppm to the percentage range. Lower concentrations of standards should be prepared in the laboratory, immediately prior to use.

The use of standards and/or calibration techniques requires an understanding of the system and the instrumentation required for the analysis. Some measurement systems have a time requirement and thus a large quantity of standard is required. Knowing the volume flow and the time response permits one to calculate the minimum volume for calibration. As a rule of thumb, one should prepare somewhere between 5 to 10 times the volume of standard required for one measurement. If one is utilizing a continuous sampler requiring 120 cm^3/min of sample flow and the response time is 8 min, it would be necessary to use about 1000 cm^3 for one measurement. Thus, it would be prudent to prepare at least 5000 cm^3 of the standard gas so that additional checks can be performed. More than this volume of standard would be economically unsound, because static volumes are not stable for long periods of time (more than several days). Long storage periods result in changes in concentration because of sorption effects, permeation through container walls, and photochemical effects. Thus, before the standards are even prepared one should account for the inertness of the container (metal or porous polymers should be avoided), purity of purchased standard gas, purity of diluent gas and materials, and construction design of all transfer parts in the system.

4.1. Methods Available

The analytical chemist and/or the analyst have a number of calibration methods which can be used when working with environmental samples. Several of these will be discussed briefly and pertinent references cited. No single method is the best or is workable under all conditions; the scientist must consider all factors before choosing any one technique.

Some of the more commonly employed calibration techniques are: dynamic calibration gas generator,[51,52,53] diffusion cells,[54,55] dilution,[56-60] permeation tubes,[61-64] exponential dilution flask,[65,66] rotating syringe technique,[67] and plastic bag calibration techniques.[57,68,69]

4.2. Total Hydrocarbons Corrected for Methane

The method recommended by the EPA uses a gas chromatograph equipped with a flame ionization detector and two injection valves. The total hydrocarbon sample is injected onto a column and the effluent is directed into a FID. The signal generated is proportional to total hydrocarbon (THC). Another sample is then injected onto a stripper column, which removes everything but methane from the air sample. The methane value (from the FID signal) is then subtracted from the THC value and the difference calculated as ppm of THC as carbon, which is then expressed as methane (i.e., 16/12 times ppm carbon). This EPA method is based upon two assumptions; namely, that the signal from the FID is caused only by hydrocarbons, and that one ethane molecule will yield twice the response of a methane molecule, a propane molecule three times, etc. Using these assumptions the hydrocarbon content from one vicinity may differ significantly from that of another vicinity, although both might be reporting the same THC corrected for methane.

Williams and Eaton[70] have described a method for preparing pure air as a diluent gas or a support gas for the FID used in the above THC assay. The method employs the use of 0.5% Pd on aluminum pellets. Various other publications have appeared for specific type samples and/or systems. Celegin et al.[71] devised a method for injecting small amounts of CO into samples for the preparation of known amounts of carboxyhemoglobin. Back et al.[72] describe a technique for removing small amounts (1 nmol–10 μmol) of gases from sealed glass tubes. Paule[73] presented a unique technique for storing small volumes of gases as bubbles over Hg in a capillary tube. When needed, the stored gas is mixed with a known volume of diluent gas. Standards of CO were prepared by Smith et al.[74] by drawing known amounts into evacuated stainless steel tanks and pressurizing with CO-free air. Authors caution that at least four hours should elapse before use to ensure proper mixing. Daines[75] used a gravimetric technique to prepare mixtures of up to 14 different gases. Lower concentrations were prepared by double or triple dilutions. Accuracy for methane (1:99 v/v) was ±0.01% and CO (200 ppm) was ±1 ppm. Most gases could be prepared within ±0.25%. A pyrolysis technique for preparing gaseous pollutants in known concentrations was described by Tsang[76] in which known compounds were pyrolyzed in dilute solutions and the decomposition products studied. Pyrolysis products were examined by gas chromatography and detected by a FID.

4.3. Individual Hydrocarbon Component Calibrations

To make a total hydrocarbon analysis corrected for methane (as described above) more meaningful, a calibration curve can be prepared for each hydrocarbon. After identification, the number of milligrams per cubic meter is obtained for each component, using the calibration curve for that component. The conversion to ppm carbon is then accomplished by using the gas law. In this manner concentration values are corrected for a nonlinear response to carbon number.[40] When one employs this method of calibration in place of the EPA total hydrocarbon corrected for methane, lower values are obtained for the hydrocarbon analyses. These lower values are more accurate and give a more correct picture of the amount of contamination in the atmosphere.

4.4. Dynamic-Calibration Gas Technique

This technique is useful when one needs a continuous source of a calibrated gas or when the gases under study are unstable. The technique is straightforward, and concentrations are readily calculated from the vapor pressure of the liquids or the flow rates of the gases employed. Figure 4 depicts a representative system for this type of calibration gas. Braman[51] has found that the equipment used in this technique is slow to equilibrate (at low concentrations) and changing from one sample type to another is time consuming. Mixtures can be prepared in which the concentration of one component ranges from several ppm to 50%. The method is not useful for compounds which are liquids at >0°C.

The concentration of each contaminant gas can be varied by varying its flow rate in the diluent stream. The calculation is based upon the partial flow rates of each gas:

$$C_{cg} = \frac{F_{cg}}{F_{cg} \times F_{dg}} \times 10^6$$

where C_{cg} is the volume concentration of contaminant gas in ppm, and F_{cg} and F_{dg} are the flow rates of contaminant and diluent gases, respectively.

Angely et al.[77] have discussed the preparation of standard samples for detector calibration. Averett[78] constructed an all-glass apparatus suitable for preparing corrosive-gas standards for gas chromatography; pressures were monitored by a Hg manometer. Brockway et al.[79] described a method for the simultaneous calibration of gas analyzers and meters. It was used to monitor animal-respiration products. The sample stream was simultaneously calibrated by the addition of pure test gases. A pressure device made of ceramic has been made by Axelrod et al.[80] to prepare gas contaminants at the 100-ppm level. The ceramic disk is connected directly to a high-pressure tank of the contaminant gas, and the flow rate through the ceramic frit can be precisely controlled by regulation of the pressure. The nitrogen-fixing capacity of legumes was studied by a dynamic system using serum bottle stoppers and rubber balloons. With this setup,[81] gas mixtures were easily prepared and dispensed. A pneumatically operated system permitted the preparation of binary gas mixtures even when one of the gases was present at the ppm level.

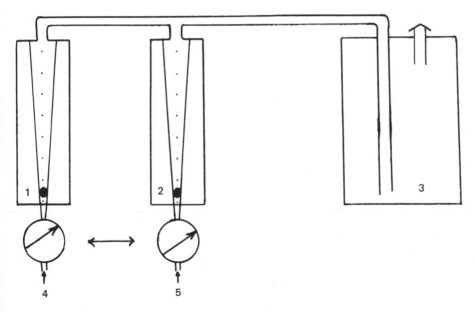

Figure 4. Dynamic calibration gas technique system: (1) rotameter, (2) rotameter, (3) calibration gas mixing chamber, (4) contaminant gas supply in, (5) diluent gas supply in.

Admixtures of O_2 in He were made with a coefficient of variation of 1.1% (in the range of 42–4870 ppm). The technique[82] was also used for mixtures of He in O_2 and Ar in H_2; admixtures were analyzed by gas chromatography.

4.5. Permeation Tubes

The design of permeation tubes is based on the fact that nearly all plastic materials will retain a liquid while allowing its vapor to dissolve in the walls and diffuse at a calculable rate. The rate at which these gases diffuse is an exponential function of absolute temperature and a direct function of thickness of the tube wall and the surface area of the tube. Figure 5 is an illustration of a permeation tube with an explanation of its operation. Rate of diffusion is also dependent upon the molecular weight of any carrier (diluent gas) which passes over outside wall and its moisture (if the permeating gas will react with water).

The diffusion rate of gases from permeation tubes is approximately described by Fick's law of diffusion:

$$D = dsa \left[\frac{P_i - P_o}{W} \right] \tag{5}$$

D is volume of gas diffusing; d, the gas diffusion constant; s, solubility of gas in tube material; a, area of tube; W, tube wall thickness; P_i, inside pressure of tube; P_o, outside pressure of tube. Since solubility (s) and diffusion rate (d) are temperature dependent, a constant temperature assures that the permeation rate will be constant during calibration. Because the rate of permeation is not instantaneous, one must allow the tube and the system to come to equilibrium before actually performing the calibration.

Permeation rates may be calculated by measuring the weight loss (using a microbalance) with respect to time. These weight losses are usually expressed as nanograms per centimeter of tube length per minute. A detailed study of calibration using permeation tubes is described by O'Keefe and Ortman,[61] using Du Pont fluorinated ethylene–propylene resin (FEP Teflon, Du Pont). A number of gases were studied by this technique, e.g., SO_2, NO_2, C_3H_8, CO_2, C_6H_6, etc. Both inside diameter (i.d.) and wall thickness of the tubes were varied.

Knowing the rate of permeation of a particular gas at a particular temperature, one can calculate the length of tube necessary for a particular calibration. For example, to calibrate for H_2S at 40°C, the permeation rate for H_2S (at 40°C) is 280 ng/cm·min. If H_2S monitor operates at 100 cm³/min in an atmosphere of 0.10 ppm in H_2S, the length may be calculated by dividing the required rate R_R of permeation (ng/min)

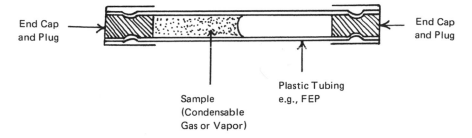

Figure 5. Permeation tube (cross-sectional diagram). Condensable gas or vapor is sealed as a liquid in the plastic tube at its saturation vapor pressure. After initial equilibration period, vapor permeates through the tube wall at constant rate. Temperature controlled to within $\pm0.1°C$ to maintain 1% accuracy. Tube thermostatted in a chamber permitting diluent gas to fully flush the chamber.

by the rate, R_M, of tube of similar material in ng/cm·min, i.e.,

$$L = R_R/R_M \qquad (6)$$

Converting ppm to ng/cm³:

$$0.10 \text{ ppm} \times 34.04 \text{ (mol wt } H_2S)/2.404 = 3.404/2.404 = 1.42 \text{ ng/cm}^3$$

$$1.42 \text{ ng/cm}^3 \times 100 \text{ cm}^3/\text{min} = 142 \text{ ng/min}$$

$$L = (142 \text{ ng/min})/(280 \text{ ng/cm·min})$$

$$= 0.51 \text{ cm}$$

If one wishes a wider range of concentrations, then dilution of the permeating gases can be used. In the calibration of gas chromatographic systems, this can be accommodated by gas-sampling valves. One outstanding feature of permeation tubes is that permeation rates remain relatively constant over the life of the tubes.

A gasometric microtechnique employing a compensated Warburg syringe manometer has been used for calibration of permeation tubes.[83] Rates for SO_2, NO_2, HF, and several hydrocarbons were checked for air pollution studies. With proper care and technique it is possible to maintain rates within $\pm1\%$ at 95% confidence interval. Studies of the preparation and storage of permeation tubes have shown that the tubes remain constant (gravimetric monitoring) for at least 3 months.[84] Lifetimes of permeation tubes have been extended up to one year by attachment of glass or stainless steel reservoirs to one end.[85] Lucero[86] derived expressions to describe sample gas emission rates during the steady-state, saturation, and depletion stages of the tubes. DeMaio[87] described a technique for sealing the liquefied test gas in a tube made of FEP Teflon

(Du Pont) and then allowing the gas to permeate into a diluent gas (air). Gases checked were butane, NH_3, NO_2, HF, $COCl_2$, H_2S, and SO_2. Tube lives were about 6 months. A rapid and sensitive method of calibration for permeation tubes was described by Purdue and Thompson[88] in which the permeation tube was hung within another tube which was connected to one arm of a vacuum recording microbalance. Permeation rate was determined from the slope of a weight-vs.-time graph.

4.6. Diffusion Cell and Exponential Dilution Flask Technique

One of the first reported diffusion cells was that of McKelvey and Hoelscher,[54] which allows easy preparation of low concentrations of volatile materials in air. The cell consisted of two flasks joined in barbell design by a diffusion tube. The lower tube contained the liquid and saturated vapor. The assumption was that this saturated vapor would then diffuse to the top flask, where it would mix with a flowing diluent gas. The rate of diffusion, γ, could then be calculated:

$$\gamma = 2.303 \frac{DPMA}{RTL} \log \frac{P}{P - p} \qquad (7)$$

D is the diffusion coefficient of vapor; P, total pressure in chambers (atm); M, molecular weight of diffusing liquid vapor; A, cross-sectional area of connecting diffusion tube (cm^2); R, gas constant (liter-atm/mol°K); T, absolute temperature (°K); L, length of diffusion tube (cm); and p, vapor pressure of diffusion liquid vapor at T(atm). As with permeation tubes, diffusion cells must be used at a constant temperature (constant-temperature bath). Experimental diffusion rates are usually found by weight loss. A disadvantage of the double-diffusion tube is that it must be calibrated each time it is used. Altshuller and Cohen[55] reported that observed diffusion rates differ widely from rates calculated using the standard diffusion rate equation. The single-chamber diffusion cell gave better agreement with literature data for diffusion coefficients; it is recommended for contaminant concentration of the range of 10–10,000 ppm.

A unique way of preparing standards for calibration is the use of an exponential dilution flask.[89] Utilizing this technique, one can prepare a dynamic range of concentrations of the contaminant gas. Diluent gas enters the top of the flask, is mixed with an initial known concentration of gas (by means of a magnetic stirrer and paddle wheel assembly), and then exits at the bottom of flask. One must know the volume of the mixing chamber. Knowing also the gas flow rate of the diluent gas, one can calculate the concentration C of contaminant gas at any point in time:

$$C = C_0 e^{-tQ/V} \qquad (8)$$

C_0 is the initial gas composition; V, the flask volume; and Q, the flow rate.

This technique is well adapted for studying the dynamic range of a gas chromatographic detector in a simple experiment.

Several workers have combined permeation tubes and the exponential dilution flask.[90,91] The calibration curve begins to lose linearity as the contaminant gas approaches very low concentrations. This may be attributed to adsorption of the contaminant gas on the walls of the container. A final variation of this type of calibration is the syringe-dilution technique[91] where successive dilutions of the air sample are made with a glass injection syringe.

5. Detectors

There are a number of detectors available for use in gas chromatography.[92-94] The most widely used detectors for atmospheric sample analysis are the thermal conductivity detector (TCD), the flame ionization detector (FID), and the electron capture detector (ECD). Of the three, only the FID destroys the sample during its measuring process. However, with the employment of gas stream splitters, one can divert a predetermined percentage or ratio of the column effluent away from the detector and thus trap the peak component for further study and/or identification. Another means of identification and quantification which is receiving attention is the interfacing of gas chromatographs and mass spectrometers.[95-97] For the purpose of classifying instrumentation it is difficult to decide whether the mass spectrometer is an auxillary piece of equipment (detector) for the gas chromatograph or whether the gas chromatograph is a sample system for the mass spectrometer. In any case, the mass spectrometer is very useful for qualitative identification of peak components.

I propose to define a few terms used with detectors and to state their applicability and sensitivity; details may be found in the references 92–94. Detectors as a whole may be placed in two categories. As shown in Table 6, these categories are defined according to whether the response of the detector is concentration dependent (I) or mass-flow-rate dependent (II).

Detector noise refers to fluctuations from baseline with time when no sample flows through the detector. It is expressed as electrical units or as a percentage of full-scale deflection. Sensitivity is the ability of a detector to respond to compounds entering its environment. Detectors which are concentration dependent have their sensitivity expressed as

$$\text{Sensitivity} = \frac{\text{peak area} \times \text{flow rate}}{\text{sample weight}} \tag{9}$$

$$= \frac{\text{mV} \times \text{cm}^3}{\text{mg}} \tag{10}$$

Detectors which are mass-flow-rate dependent have their sensitivity expressed as

$$\text{Sensitivity} = \frac{\text{peak area}}{\text{sample weight}} \qquad (11)$$

$$= C/g \qquad (11a)$$

Detectability or lower limit of detection (LLD) refers to the smallest amount of sample which will cause a measurable signal over the noise signal. The measurable signal is usually taken as twice the noise signal (referred to as a signal-to-noise of 2:1). Specificity of a detector refers to a ratio of the detector response of a contaminant (interfering substance) to the detector response of the desired component. Linearity is the range over which a detector will maintain a constant sensitivity to increasing concentration of a specific component. A final term which is commonly used to describe a detector is its response time. A generally accepted definition here is the time necessary for the signal from the detector to reach 63% of its true value. Caution should be employed here because the system may limit the response time of the detector; e.g., one may employ a recorder which has a response time slower than that of the detector.

Table 6 summarizes some of the more widely used detectors and their characteristics. Since no one detector is good for all types of components, the analytical chemist must choose his detector on the basis of the type of compound to be quantified and the concentration range in which he expects most of his samples to fall.

Table 6. Commonly Used Detectors in Gas Chromatography

Type	Components detectable	Linearity	Category[a]	Sensitivity, g
Thermal conductivity (TCD)	All components	10^4	I	$10^{-6}–10^{-7}$
Flame ionization (FID)	Organic compounds	$5 \times 10^6–7 \times 10^7$	II	10^{-10}
Electron capture (ECD)	Halogenated and oxygenated compounds	$10^1–10^2$	I	$10^{-12}–10^{-13}$
Flame photometric (FPD)	Phosphorus and sulfur compounds	$5 \times 10^2(S)$ $1 \times 10^3(P)$ log/log scale	II	10^{-11}
Thermionic (TID)	Phosphorus-, sulfur- and nitrogen-containing compounds	10^3	II	10^{-12}

[a] I, concentration dependent; II, mass-flow-rate dependent.

6. Atmospheric Pollution

This section will concern atmospheric pollution and the resulting analyses which do not specifically fit in other sections. A large portion of the organic pollutants which enter our atmosphere are in the category of hydrocarbons. Of this large amount of hydrocarbons about 50% are traceable to transportation sources, about 15% are accounted for from all types of industrial and residential fuel burning and industrial-process evaporation, and another 26% result from a variety of sources such as coal, waste and refuse, and evaporation processes.[3] This last grouping includes oil refineries, coke plants, pharmaceutical processes, and the textile, foods, plastics, and paint industries. Hydrocarbons are of interest because the lower members of the group are reactive in producing products which are photoirritants, but the heavier hydrocarbons produce products that are more irritating on a hydrocarbon-per-mole basis. It should be borne in mind that most internal-combustion engines do not burn fuel with 100% efficiency to form water, carbon dioxide, and various oxides of nitrogen and sulfur. In fact, Merrion[98] has shown that chromatograms of diesel exhaust resemble the chromatogram obtained with the uncombusted fuel. This escape of unburned fuel is due to channeling of the uncombusted fuel surrounding the flame and quenching of the flame by the combustion channels at the walls.

Many reactive hydrocarbons (e.g., ethylene, 2-methyl-2-butene, cis-2-butene, butene-1, etc.) are present in higher concentrations in the early part of the day and exhibit marked decreases by afternoon. This change in concentration can be rationalized on the basis of photochemical reactions taking place in the atmosphere[99]:

$$NO_2 + UV \text{ radiation } (3400 \text{ A}) = NO + O \qquad (12)$$

$$O + O_2 = O_3 \quad \text{and} \quad O + \text{Hydrocarbons(HC)} = HCO\cdot \qquad (13)$$

$$HCO\cdot + O_2 = RCO_3\cdot \xrightarrow{O_2} RCO_2\cdot + O_3$$

$$\xrightarrow{NO_2} R-C-OONO_2 \qquad (14)$$

peroxyacetyl

nitrate

(PAN)

7. Industrial and Manufacturing Pollution

Capillary columns with gas–liquid chromatography (GLC) and mass spectrometry (MS) have been used to investigate ultratraces of organic compounds in the C_6–C_{20} range.[100,101] Two columns (column I for more volatile components was 80 m × 0.33 mm, Ucon LB550; column II for less volatile compounds, 120 m × 0.32 mm, HB1500 substrate) were used to separate the volatile from the less volatile components. The separated sample showed 460 peaks (108 identified). Beckman Instruments, Inc. was issued a patent on this work for a system capable of measuring ambient air pollutants using columns packed with molecular sieves and a hydrogen flame ionization detector.[102] The C_3–C_4 olefins (used in production of polymers) isolated from cracked petroleum gases have been separated and determined[103] by GLC. Davies and Bowen[104] developed a portable hydrocarbon-leak detector; principles of the device were the mixing of sample with H_2 and passing the mixture effluent into a FID. Range of detection was 0–1000 ppm. Enrichment and selective detection techniques are many times necessary to determine trace amounts of hydrocarbons in air samples. Kaiser[105] devised a formula for relating percentage of ethylene in a mixture to the operating characteristics of the detector; 1.6 parts ethylene in 10^9 parts of air were easily detected. Villalobos and Chapman[106] described a gas chromatographic system which was capable of automatically monitoring air pollutants. Total hydrocarbons were measured directly by flame ionization detection. A stripper porous-polymer column removed the compounds heavier than ethane. A molecular sieve column separated CO and methane. Full-scale sensitivity was 1 ppm per individual component.

A number of publications describe the measurement of low-molecular-weight hydrocarbons in air samples. Some of these methods have been completely automated,[107,108] others have used chemically bonded phases for the separations[109] and still others have employed temperature-programming and flame ionization detection.[110] Rasmussen et al.[111] reviewed the chromatographic techniques used for analysis of hydrocarbon pollutants and recommended the establishment of a standard reference gas chromatographic method for atmospheric hydrocarbon analysis. Higher-molecular-weight aliphatic hydrocarbons (up to C_{36}) have been extracted from particulate samples.[112–114] The extraction step would also carry the polynuclear hydrocarbons into solution. Initial separation of these two general classes can be effected by liquid–solid chromatography. Identification and quantification of a wide range of molecular weights requires the use of standards in which both gas and liquid phases are present; this presents a tricky sampling technique. Malan et al.[115] have described a unique method for handling this type of sample. The method depends upon the expansion of the mixture into a known volume and a pressure of 2 Torr. C_1–C_5 may be easily analyzed by this method. Odorants from whiskey fermentation units have been adsorbed on carbon,[116] stripped off and then analyzed by GC; sulfur compounds in paper-mill-stack gas were similarly determined using bromine titration cell detection.[117]

8. Low-Molecular-Weight Hydrocarbons

Mixtures of O_2, N_2, CO, CO_2, CH_4, and C_2H_6 have been analyzed after using either a molecular sieve 13X (80/100 mesh) column or a 10% squalane on 150/200 mesh Porapak Q column at 210°C using H_2 as the carrier gas.[118] Subbarao[119] has described a unique way of analyzing binary and ternary gas mixtures without the use of standard gas mixtures. Each gas, in pure form, is vented into a sample loop. The partial pressure of the gas at constant total pressure and the peak area are then measured. These data are then utilized with unknown gas mixtures to determine the composition of the unknown mixture. A multiple column setup of a primary column (Porapak T, or Porapak T followed by Porapak S) connected in series with a pair of parallel columns (Porapak S and molecular sieve 13X) has been used with thermal conductivity detection to analyze gas mixtures of H_2, O_2, N_2, CO, CO_2, CH_4, C_2H_6, C_2H_4, and C_2H_2.[120] Columns were operated at 35°–100°C, with He as carrier gas. Petroleum-polluted air samples have been analyzed for nonaromatic hydrocarbons and some aromatics using FID with N_2 or He as carrier gas.[121] Heptane, benzene, or toluene have been used as reference compounds. Repeatability and reproducibility were within 5–10%, respectively.

Gasified carbon products (products from high-temperature reaction of carbon-containing materials with hydrogen) have been separated and determined by use of a FID[122]; a Spherochrome column was used. Total carbon compounds were determined using the same column with a 20-cm length of hydrogenation catalyst (10% metallic Ni on crushed firebrick). Combustion gases from burning residual fuel oil have been determined by a gas chromatographic method.[123] Two columns and a column-switching valve were employed. Column 1 was Porapak Q and column 2 was molecular sieve 5A. A 0.25-cm³ sample was employed and detection was by hot wire. Component ranges (in percentages) were: CO_2, 2–16; C_2H_4 and H_2, 0.1–20; O_2, 0.1–5; N_2, 40–80; H_2O, 1–20; CH_4, 0.1–10; CO, 0.1–20 and C_2H_6, 0.1–5. Reproducibility for each component was 0.1%.

Free carbon and carbide-carbon in metallic beryllium can be determined by gas chromatography.[124] Free carbon is burned to CO_2 at 900° ± 50°C and passed through a silica gel column with He as carrier. Carbide-carbon is converted to methane with boiling HCl (1:1).

9. Polynuclear Aromatic Hydrocarbons (PAH)

Polynuclear aromatic hydrocarbon compounds contribute to emphysema and lung cancer; they are usually associated with soots, tars, and oils, and are usually separated from the source sample by either benzene or cyclohexane extraction. Once separated as a group, they may be determined in several ways: initial separation by thin-layer chromatography (TLC) followed by spectrofluorometry; separation by TLC and then analysis by UV spectrophotometry; and gas chromatography (GC). A number of GC techniques are available and have appeared in the literature. The compounds which make up the general classification of PAHs and related heterocompounds include benz(a)anthracene, benz(a)pyrene, benz(e)pyrene, chrysene, perylene, anthanthrene, naphtho(2,3)(a)pyrene, dibenzo(e,l)pyrene, dibenzo(a,i)pyrene, indeno(1,2,3-cd)pyrene, dibenz(a,h)acridine, 7H-benz(de)anthracene-7-one, and phenalen-1-one.

Halot[125] has written a review describing the sources and nature of hydrocarbons in the atmosphere, including the toxicity and carcinogenicity of PAHs. The review concludes with a discussion of methods for their detection, separation and determination. A GC method by Bhatia[126] uses a 20-ft-×-0.125-in. column packed with 60/80 mesh glass beads coated with OV-7; temperature programming from 170°–260°C, and FID was employed.

Problems associated with an initial separation by TLC, followed by GC, have been reviewed by Chatot et al.[127,128] The initial separation by TLC was achieved with alumina-cellulose acetate, and the spots were extracted with benzene. The gas chromatographic separation used 4.5% SE-52 on Chromosorb G; flame ionization was used to detect the effluent components. Other authors[129–132] have also utilized the combined techniques of TLC and GC for the separation and analysis of polycyclic aromatic hydrocarbons.

Novotny et al.[133] used gel fractionation, high-performance liquid chromatography, nondestructive spectral methods and GC for the determination of PAHs. Lao et al.[134] coupled mass spectrometry with GC to identify seventy major ring compounds (2–7 rings per molecule). The 12-ft column employed was packed with 6% Dexsil 300 on Chromosorb W (80/100 mesh). Liquid–liquid extraction of the PAHs followed by capillary GC has also been successfully employed.[135] Grant et al.[136] separated and determined PAHs and dihydric phenols by use of Bentone-34 on glass beads. Burchfield[137] employed an unusual technique for the collection of the PAHs. High-volume filters collected the particulate matter, and the polynuclear arenes were removed by a stream of N_2 at 300°C. The N_2 stream was then trapped on a cold column of Dexsil 300. Heating of the trapping column transferred the arenes to an analytical Dexsil 300 column where they were subsequently determined. Janini *et al.*[138] used high-temperature nematogenic liquid crystals as the stationary phases for the separation of PAHs. These phases were able to separate the geometric isomers of the three- to five-ring PAHs. Theoretical plate count and retention time of chrysene were reproducible to 3%, over a period of 150 h of continuous operation at 260°C on column packings of 2.5%(w/w) N,N'-bis(p-butoxybenzylidene)-α,α'-bi-p-toluidine on 100/120 mesh Chromosorb W.

10. Nonhydrocarbon Compounds

This section provides a catchall listing for sample components which do not strictly fit into other headings. Several representative types of compounds will be mentioned, and the reader will find similar references in other sections as well. Lewis[139] reviewed studies which have been carried out for additives and contaminants in air, water, and foods. Angerer *et al.*[140] have discussed the various methods for the determination of ultramicro constituents in air samples. Techniques employed were mass spectrometry with gas chromatography and a thermal-gradient tube. The air sample is dried by passage through the thermal-gradient tube packed with Dexsil 300. The gas–liquid chromatographic system employed a column 4 m × 1.9 mm packed with 20% SE-30 on Chromosorb G-AW-DMCS. Helium was the carrier gas and a FID was used for quantification. The column was programmed from $-150°$ to 200°C at an increment of 7.5°C per minute. Sixty compounds in a 2-liter air sample have been detected. Okita[141] drew air samples through filters (100 liters/min) impregnated with 3% $Hg(CN)_2$, 5% $HgCl_2$ or H_2SO_4. Amines, thiols, and organic sulfides were trapped and subsequently released by treatment with HCl or NaOH. The released compounds were trapped in a CO_2–acetone bath and then separated by GLC. Carrier gas was N_2 and detection was by flame ionization. Air samples containing unsaturated compounds which react with O_3 have also been determined.[142] Excess ethylene is added to samples to deplete the O_3 concentration. Compounds such as hexanal, hex-2-en-1-al, limonene, allyl isothiocyanate, and pyridine were easily determined in the range of 0.1–10 μmol/liter.

A general review has been published in Russian[143] of air pollution associated with the chemical and related industries. Various topics were discussed; e.g., industrial pollution of the atmosphere, photoelectric gas analyzers, electrochemical gas analyzers, ionization methods, kinetic methods, and infrared absorption. The separation of permanent gases and light hydrocarbons by subambient temperature programming on a Porapak Q column has been discussed by Sarkar and Haselden.[144] The separation was programmed from $-65°$ to $200°C$. The carrier gas was helium and thermal conductivity was used for the detection of components.

11. Halogenated Compounds

Atmospheric halogen-containing components fall into two categories, inorganic and organic. Inorganic halides are predominantly alkali halide salts. These enter the atmosphere from industrial sources where minerals are processed, e.g., limestone kilning, iron refining. They may also result from such industries as cement manufacture, fertilizer manufacture, ceramic and glass processing, and aluminum refining. In addition, some enter the ambient air from coal and oil combustion. Inorganic halides are usually determined by a method other than GC.

Gas chromatography has been used to monitor directional sources of pollution. In studies of this type a compound which is used as a tracer must not itself be a pollutant, must have low or negligible toxicity, and must not react by photolysis or photooxidation; also, it must also be easily sampled and result in a dependable and rapid analysis technique. Tracers can be used for following horizontal as well as vertical pollution pathways. Most of the work to date has been utilized for vertical tracing studies.

Two compounds used for tracer studies are sulfur hexafluoride (SF_6) and dichlorodifluoromethane (Cl_2CF_2). It is possible to collect and analyze these compounds at distances up to several miles from their point of release.[145] Quantities as low as a few parts per billion have been determined for SF_6, and 50–100 parts per thousand for CCl_2F_2. The most widely used detector is the electron capture detector (ECD).[146] Many compounds of similar structure have been studied and their sensitivities reported. Some representative compounds and their respective sensitivities (in ppm) are: C_6H_6, 1000; SF_6, 580; CF_3Br, 40; $CFCl_3$, 370; CF_2Br_2, 500; and BrF_2CCF_2Br, 230.

Sulfur hexafluoride has been detected 100 miles from its point of release.[146] One of the major problems in the gas chromatographic determination of SF_6 has been its separation from oxygen. This separation was greatly improved by Dietz and Cote[147] who used a nitric-oxide-treated molecular sieve column; a lower limit of detection was 4×10^{-13} cm^3/cm^3.

11.1. Fluorinated Compounds

A comparative study of methods for determining fluorides in air was presented by Monteriolo and Pepe.[148] Inorganic fluorides (e.g., F_2, MoF_6,

UF_6, SbF_5, and SbF_3) were separated with a nickel metal column packed with PTFE; a sensitive ECD was employed for quantification.[149] Samples were first passed through drying tubes to remove the moisture and then into the column for separation and determination. Rauws et al.[150] studied chlorofluorocarbons in air and body fluids. Air samples could be determined directly, or diluted with N_2. Body fluid samples (0.5–1.0 cm^3) were set aside in contact with N_2 (in the dark) to allow equilibration of chlorofluorocarbons between liquid and gas phases. 100 μliters of air sample or headspace sample was then injected into the chromatograph for analysis, using Porapak Q (100/120 mesh) at 150°C with N_2 as carrier gas and electron capture detection. Limits of detection (ng/cm^3 of blood) were calculated to be: Cl_3CF, 0.21; Cl_2CF_2, 1.35; 1,1,2-trichloro-1,2,2-trifluoro-ethane, 4.2; and 1,2-dichloro-1,1,2,2-tetrafluoroethane, 14.

Million et al.[151] have described an automatic gas chromatographic system for the analysis of corrosive fluorine gases. The sampling technique allowed pressures of 1 to several Torr. The analytical technique has a halocarbon sensitivity of 20 ppm and an analytical cycle of 1.5 min.

A unique application, employing a fluoride derivative, is the determination of beryllium in particulates.[152] A 400-liter air sample is passed through a filter (20 liters/min). Sample and filter are ashed and the residue fused with KHF_2. The fusion residue is heated in H_2SO_4 and HCl until all solvent is evaporated. Metals in sample residue are converted to trifluoroacetylacetonates and separated on a Chromosorb G support (80/100 mesh) coated with silicone nitrile XE-54 at 150°C. Nitrogen is the carrier gas (50 cm^3/min). The method is capable of determining Be down to 10 ng/m^3 of air.

11.2. Vinyl Chloride

Pellizzari et al.[153] used cartridges filled with Tenax to collect organic vapors in the ambient air over several large cities. Once collected, the vapors were thermally desorbed and a gas–liquid chromatographic system coupled with a mass spectrometer was used for the analysis and identification. An on-line computer recorded the data and generated mass fragmentograms. Most of the hydrocarbons from auto exhaust were resolved and twenty-one halogenated hydrocarbons, including vinyl chloride and trichloroethylene, were detected. In addition, many organic oxygen-, sulfur-, nitrogen-, and silicon-containing compounds were also detected.

11.3. Alkyl Halides

A variety of halogenated compounds have been determined in atmospheric samples. The inhalation anesthetic Halothane ($F_3CCHBrCl$)

is a slightly water-soluble nonflammable liquid which is now of interest in air pollution analysis. Gelbicov-hunckova *et al.*[154] determined it in the atmosphere of an operating theater. Halothane was concentrated from the sample and separated by GLC with a FID. The concentration range varied between 10–50 ppm.

Chlorinated hydrocarbons have been determined indirectly, by dechlorination with sodium bis(2-methoxy-ethoxy) aluminum hydride at 160°C.[155] The hydrocarbons resulting from this reaction are separated on a 3% Dexsil 300 GC on Chromosorb W. Identification and determination was by infrared and NMR spectroscopy. Aerosol irritant projectors have been separated and determined by gas chromatography and mass spectrometry (GC/MS). The separation of dichlorodifluoromethane, trichlorofluoromethane, acetone, chloroacetone, bromoacetone, 1,1-dichloroacetone, and 1,1-dibromoacetone was effected by either Porapak Q or Carbowax 20M columns with helium as the carrier gas.[156]

One-half to one liter of air sample was bubbled through a washing bottle containing tetrachloroethylene for the concentration of the chlorinated hydrocarbons.[157] Polyoxypropylene glycol MG-550 was the gas-chromatographic liquid substrate coated on Celite 545. The detection limit for 1,1,1-trichloroethane is 125 mg per square meter of air. Bis(chloromethyl)ether at the ppb level, in air samples, was separated on a Porapak Q column.[158] Detection was by high-resolution MS at m/e of 78.995. The method is also capable of concentrating and determining other organic contaminants.

Mann *et al.*[159] determined hexachlorobenzene and hexachlorobuta-1,3-diene in air. The air sample was passed through a Chromosorb 101 column using a portable air pump. OV-17 and QF-1 were the combined liquid chromatographic phases. It was possible to determine 28 ng (hexachlorobenzene)/m^3 in a 360-liter sample of air. Lower-molecular-weight compounds (e.g., $CHCl_3$, CCl_4, C_2HCl_3, and C_2Cl_4) have been separated and determined in air samples using electron capture detection.[160] Activated charcoal traps were employed for concentrating the sample, and the chlorohydrocarbons were desorbed by a heated N_2 stream. If one is interested in the same low-molecular-weight compounds in water samples, it is only necessary to pass N_2 through the water sample and directly into the concentrator column (activated charcoal). The desorption step and determination of the halogenated compounds are the same as for the air sample.

Free chlorine in air can be determined by use of a gas-density balance detector with N_2 carrier gas.[161] CO_2 is used as the reference gas for quantification. The ratio of the peak areas of Cl_2 and CO_2 (allowing for molecular-weight differences) are needed to calculate the amount of chlorine in the air sample.

The presence of 3,3'-dichloro-4,4'-diaminodiphenylmethane was determined in air[162] by collection on Gas Chrom S followed by elution with acetone. Aliquots of the acetone solution were analyzed on a Anakrom ABS column coated with Dexsil 300 GC. Flame ionization detection was used for quantification.

11.4. Other Halogenated Compounds and Polychlorinated Biphenyls

Volatile metal chlorides (e.g., $SiCl_4$, $GeCl_4$, $SnCl_4$, and $TiCl_4$) can be separated on a number of gas chromatographic columns.[163] The order of separation is according to their boiling points: $SiCl_4$ (57.6°), $GeCl_4$ (86.5°), $SnCl_4$ (114.1°), $TiCl_4$ (136.4°). A relationship of this type is typical for a series of compounds which are nonassociated under the conditions studied.

Mustard gas, bis(2-chloroethyl)sulfide, may be sampled from air by solvent (diethyl succinate) entrapment.[164] The air is sampled for 24 h at a rate of 100 cm^3/min. The entrapping solution is injected directly onto a 6 ft $\times$ 0.125 in., 4% FFAP on Chromosorb W-AW-DMCS (60/80 mesh) column at 155°C using N_2 as the carrier gas at 55 cm^3/min. A linear calibration curve, over the range of 1–120 μg/cm^3, results. The minimum detectable amount was 0.2 ng, resulting in an accuracy of ±3%.

Several papers have appeared for the gas chromatographic analysis of chlorinated acetyl chloride and phosgene.[165,166] Dahlberg and Kihlman[165] determined the mono-, di-, and tri-chloroacetyl chlorides. Isopropyl alcohol was the absorbing solvent. A 3-μliter liquid sample was subjected to gas chromatographic analysis with a 20% silicone oil DC-200 on Chromosorb B; electron capture detection was used for the detector. Sensitivity for the di- and trichloro- derivatives was 1 ppm or better. The monochloro derivative has a sensitivity of 1 ppm.

Chloromethyl methyl ether and bis(chloromethyl)ether have been determined in air samples. Evans et al.[167] used a Porapak Q column to concentrate the sample; the separation was performed on a polyethylene glycol Celite column. Monitoring of the ether compound was done at m/e 78.9950 using a mass spectrometer. Solomon and Kallos[168] collected air samples by means of impingers containing methanolic solution of the sodium salt of 2,4,6-trichlorophenol. Sodium methoxide was added and the derivatives analyzed on QF-1 + OV-17 coated glass beads at 140°C. A nickel-63 ECD was used for the detection and quantification.

Hydrogen chloride was analyzed for trace impurities of organic substances.[169] A 4-m-×-4.5-mm column packed with 20% dinonyl phthalate on INZ-600 brick was used, with N_2 carrier gas. 1–2 $\times$ 10^{-3} volume percent of methane, ethane, propane, isobutane, and methyl chloride was obtained from the reaction of sulfuric acid and hydrogen chloride solution. DiCorcia *et al.*[170] found that carbon blacks (graphitized with H_2 at 6000°C) were excellent supports for gases such as the halogen hydrides and boron trifluoride. Hydrogen fluoride and HCl could be separated from air sample on a 2-m-×-3-mm column of Graphon coated with a 1:1 mixture of polyfluoroether and benzophenone (40% w/w). The benzophenone acts as the adsorbate while the fluorinated substrate aids in the dispersion of the benzophenone on the carbon support. Boron trifluoride is usually difficult to separate by GC because of its strong Lewis acid properties.

Polychlorinated biphenyls (PCBs) have been on the market for over 40 years, yet it was only in the last 9–10 years that they were recognized as environmental hazards. Jensen[171] first identified PCBs as an ecology problem when he discovered them in the bodies of pike fish in Sweden. He also found them in other fish, as well as in eagle feathers. The PCBs have also been detected in human adipose tissue,[172] human milk,[173] and such foods as margarine, vegetable oils, and fish.[174]

Since the PCBs are extremely stable, fat soluble, and persistent in the environment, it has been theorized that they have four pathways into the environment: (1) containers that contaminate feeds and foodstuffs, (2) industrial smoke, (3) manufacturing wastes (including transformer oils) flushed into our waterways, and (4) insecticidal formulations in which PCBs are added to increase kill ratios.

PCBs are similar in structure to the DDT homologues and thus may be carried through the usual pesticide workups and be mistaken for DDT. When doing pesticide analyses by GC, one should confirm the analysis by additional techniques, because the ECD is nonspecific in function and will respond to either type of compound. One can separate the pesticide eluate on a polar substrate (e.g., DEGS or DEGA) and then convert to derivatives and separate again.[175] Another approach is to use SCOT columns and electrolytic conductivity or microcoulometric detection.[176] This technique in conjunction with mass spectrometry can confirm a PCB peak and furnish the number of chlorine atoms per molecule.

Qualitative and quantitative analyses of PCBs using flame ionization detection has been described.[177] In this study retention indices (RI) on six different liquid phases have been calculated for the mono-, di-, and trichlorobiphenyls. The authors present a method for predicting relative molar responses of PCBs in a hydrogen flame ionization detector.

12. Carbon Monoxide and Carbon Dioxide

Carbon monoxide's toxicity is well known and documented. One of the most important air pollutants in urban areas, it results from the incomplete combustion of industrial processes, household heating, motor vehicle exhausts, and tobacco. This places it in a category all to itself, because no other pollutant is present in such high concentrations and possesses such high toxic potential. At levels in excess of 100–200 ppm the toxicity is acute. Long-term effects below these levels are not as well documented.

The Intersociety Committee for Methods of Ambient Air Sampling and Analysis is considering a method for preparing CO standards. The method will involve taking known volumes of cylinder CO, volumetrically diluting with CO-free air and oxidizing the CO to CO_2 by use of a CuO-MnO_2 catalyst. The CO_2 produced will be absorbed by Ascarite (NaOH) and determined gravimetrically. As 1.0 mg of CO_2 at 25°C and 1 atmosphere of pressure is equivalent to 555.54 ppm CO,

$$CO(ppm) = \frac{CO_2 \text{ (mg)} \times 555.54}{\text{sample volume (liters)}} \qquad (15)$$

The chief gas chromatographic methods for CO involve either conversion to CH_4 or direct determination. Stevens and O'Keefe[178] devised an automatic gas chromatographic system capable of determining 0.01 to 200 ppm CO on a continuous basis. A nickel catalyst converts the CO to CH_4, which in turn is detected by an FID. Dubois and Monk-man[179] published a sampling technique utilizing frontal analysis in which the air is sampled until an equilibrium is established between the CO in the column and the CO in the air. The same authors[180] also established a method for the continuous determination of CO by a frontal analysis technique. Four columns are used (1 for sample and 3 for standards) to sample the air for 6 min at 75 cm^3/min. N_2 and O_2 are flushed from columns with H_2. The CO is then eluted with H_2 and the combined gases passed over a Ni-catalyst at 250°C, which converts the CO to CH_4. The CH_4 can be quantified with an FID. Fast flows of H_2 (>200 cm^3/min) will strip the column of compounds having higher retention volumes than CO. The technique is sensitive to 0.1 ppm and samples may be taken every 6 min.

Tesarik and Krejci[181] were able to determine CO below the 1 ppm level. In this method, oxygen, nitrogen, and water are removed before the sample enters the chromatograph. Carbon monoxide was determined in the usual way (converted to CH_4 and assayed via FID). Zielinski and Mayer[182] concentrated the CO on a 100-cm copper column packed with active carbon at liquid N_2 temperatures. Hydrogen was used as the mobile-phase gas, and the CO peak (by thermal conductivity detection) was compared to a standard CO sample peak.

Pyroelectric katharometer detection has also been used for CO (low concentrations) in the air.[183] This method is capable of determining from

0.5 ppm up to 0.1% (v/v). An activated charcoal G-5 column packing was employed. Precision of the determination at the ppm level was about 2%, which agrees with those data obtained by GLC and a He ionization detector.

A group in Sweden have invented a submicroliter sampling device for quantitative collection of gases.[184] The device consists of a 10-μliter Hamilton syringe in which the end plate of the plunger is pressed by a spring against a metal disk. The sampling device is to be used for the preparation of known concentrations of carboxyhemoglobin by injection of measured volumes of CO. Coefficient of variation of peak area was 2.1% for a 0.36-μliter volume (10 injections). Peak area is in proportion to volume in the range of 0.091–0.546 μliters.

Steffer *et al.*[185] invented a portable gas chromatograph and electrochemical detector (ELCD) for CO. The ELCD is connected in series with a TCD and the CO is electrochemically oxidized at a Pt electrode. Ambient air samples are separated on a 5A molecular sieve column and the order of elution is H_2, O_2, N_2, CO. The instrument should be adaptable to other gases which are electrochemically active, e.g., NO, NO_2, C_2H_5OH, and H_2S.

Schaefer and Douglas have described the response of the FID to CO and CO_2.[186] In order to obtain a response which was greater than that of the TCD, they increased the background current by continuous addition of trace amounts of hydrocarbon to the flame or by operating with a hot burner. Carbon monoxide and methane can be monitored on one analyzer in the low ppm range by the use of an event programmer.[187] The programmer shuts off the sample flow before the sample is injected, which allows the sample to be equilibrated at the same pressure. The sample is then injected into the pressurized system, unwanted material is back-flushed, and a peak measurement is taken. Periodically an analyzed calibration mixture is run to check the system.

The determination of CO in blood and tissue taken at postmortem was performed by GLC[188] and the results compared with the spectrophotometric and differential protein-precipitation techniques. A 5-ft-×-0.25-in. stainless steel column containing molecular sieve 5A was employed. Helium was the carrier gas and a katharometer detector was used for the quantification. The sensitivity could be greatly enhanced by introducing heated nickel catalyst at the end of the column, converting the CO to CH_4, and performing the detection and quantification with a FID.

The determination of carbon dioxide in the ambient air has been considered of secondary importance in regard to air contamination. The concentration of CO_2 in "uncontaminated" air is usually in the range 300–350 ppm. At this level it is not considered to be dangerous, and moderate fluctuations around this range are of no concern.

Bethea and Meador[189] did a review on the analysis of reactive gases in the air. Retention times of air, CO_2, N_2O, NO, NO_2, H_2S, SO_2, Cl_2, HCl, and NH_3 were studied on 21 different columns. One outcome of this work was the achievement of lower detection limits for practical separations. Subgroups of the above gases could be separated on some columns, but a three-column system was necessary to separate all ten gases.

Traces of CS_2, CO_2, and SO_2 in the air can be determined after being trapped on glass particles at $-183°C$. The collection tube [190] is then heated to 20°C, and H_2 carrier gas (80 cm^3/min) transports the gases into a gas chromatograph where they are separated on a silanized Chromosorb B column.

A Porapak R column enabled Bailey et al.[190] to separate many common gases; hydrogen was used as the carrier gas. As little as 0.7 μliter SO_2/cm^3 and 0.2 μliter/cm^3 of other gases were easily detected. A carbon molecular sieve column with helium carrier has been used[191] to separate NO, N_2O, CO, H_2S, SO_2, C_2H_2, CH_4, C_2H_6, and CO_2. Temperature programming from 25° to 150°C is necessary for the entire separation.

Radioactive carbon in the form of CO and CO_2 has been determined by several techniques. In one instance[192] a proportional counter and a shielded anticoincidence counter are employed. Error is 9%. Complete apparatus diagrams for each of the methods are presented. Isotope fractionation during a chromatographic separation has been noted by Gunter and Gleason.[193] While investigating volcanic and hydrothermal gases for radioisotope measurement they noted severe fractionation of the effluent gas. In order to quantify the data it was found necessary to collect all the effluent. Elution of the CO_2 from a Porapak column indicated an enrichment of ^{13}C in the leading edge of the peak whereas elution of CO_2 from silica gel resulted in enrichment of the leading edge with ^{13}C and the back edge with ^{18}O.

Amberlyst 15 resin (42/60 mesh) in the H^+ form was treated with aqueous $Ni(NO_3)_2 \cdot 6H_2O$ and then packed into a column. In the Ni^{2+} form the resin was able to separate CO_2 but not CO; however, when reduced by H_2 gas to metallic Ni it was possible to separate CO_2 and CO.[194] A similar system used CuO on a silica column for the separation of H_2, air, CO, CH_4, and CO_2. Calibration curves for each gas were rectilinear when peak area was plotted against log concentration.[195]

The use of a helium-plasma detector was employed to monitor CO, CO_2, N_2O, and SO_2 from either a molecular sieve or Porapak Q column.[196] Adjustment of the operating conditions to suit the determination of an individual gas causes loss of peak resolution for the other gases. Calibration curves were rectilinear up to 1000 ppm. Although this detector is sensitive and selective it is not applicable to quantitative analysis of each gas in one run. Castello and Munari[197] employed a metastable helium detector to determine gases, in atmospheric samples, without preconcentration. A number of column packings such as molecular sieve, silica gel, activated charcoal, and Porapak Q were used in the study. The following gases with their minimum effective concentration (in parts per 10^9) were detected; H_2, C_2H_6, C_2H_4 (1.5); C_2HF_3 (10); C_2F_4 (7); SO_2 (2); N_2, CO (0.5); O_2, CH_4, NO_2, CO_2 (0.5); C_2H_2, NO, H_2S (1.0). Bros *et al.*[198] also applied a helium detector for the determination of small amounts of CO, CO_2, CH_4, COS, and H_2S. The laboratory-constructed detector contained two parallel electrodes and incorporated a 0.6-Ci tritium source of β-particles. Detector voltage, gas flow, and the effects of temperature on the operating conditions are discussed in the reference.

13. Nitrogen-Containing Compounds

Two of the most important oxides of nitrogen are nitric oxide (NO) and nitrogen dioxide (NO_2), the former being found in greater quantities in air. It has been estimated that transportation emissions account for about 32% and fossil fuel combustion for 48.5% of the nitrogen oxides in the atmosphere. The nitrogen oxides cycle is of great importance for air pollution chemistry. It begins with the dioxide absorbing ultraviolet radiation:

$$NO_2 + UV \leftrightharpoons NO + (O) \qquad (16)$$

$$O_3 \longleftarrow \rceil\, O_2$$

Data have shown that 0.04–0.16 ppm of NO_x usually occurs with 0.3–1.4 ppm of nonmethane hydrocarbons.[199]

The presence of NO in air results in the formation of peroxyacetylnitrate (PAN) and peroxybenzylnitrate (PBN). The lower hydrocarbons are more reactive than the higher hydrocarbons as sources of these photoirritants. The concentration of PAN in air is dependent on the oxidant level and the diurnal variation of sunlight. PAN is a fairly easily determined contaminant because it requires no special sampling or sample preparation.

Goedert and Guiochon[200] carried out a reproducibility study for the response of a katharometer using the nitrogen in an air sample as the monitored species. Under precisely controlled system parameters the nitrogen content of air was determined with an error of 10^{-3}% at the 95% confidence level.

13.1. Nitric Oxides

Paule[201] described a technique for encapsulating, measuring, and storing small volumes of pollutant standards. The gas sample is stored as a microbubble over mercury in a capillary tube (for up to six months). At the time of analysis, it is mixed with a large volume of diluent gas. The paper illustrates the encapsulation apparatus and presents recovery data for NO and SO_2.

Subambient analysis of NO ($-70°C$) was achieved[202] by means of a stainless steel column (12 ft × 0.125 in.) packed with Chromosorb 102; the carrier gas was helium, and the detection was by thermal conductivity. The technique was capable of determining 100–3000 ppm of NO. At the lower level the error was 5%. When the oxides of nitrogen were analyzed in exhaust samples, gas chromatography was used strictly as a separation technique and the analysis was performed by either chemiluminescence or infrared spectroscopy.[203] One may employ either a 7-m column of Chromosorb 104 ($-50°$ to $+70°C$) or a 10-m column containing molecular sieve 13× ($90°C$). Infrared spectroscopy results were higher than chemiluminescence data.

The ratio of NO to NO_2 emitted from auto exhausts is usually 99:1; the nitric oxide forms directly from the oxygen and nitrogen of the air at these elevated temperatures.[204] The NO is then slowly oxidized to NO_2. Once the nitrogen oxides are in the atmosphere they react with many of the organic compounds present. They are most reactive with the hydrocarbons; the products which may form include peroxyacetylnitrate, acrolein, formaldehyde, ketones, acids, and ozonated olefins.

The determination of nitric oxide by GC is complicated by the equilibrium between it and nitrogen dioxide:

$$2NO + O_2 = 2NO_2 \tag{17}$$

The dioxide can in turn dimerize to dinitrogen tetraoxide:

$$2NO_2 = N_2O_4 \tag{18}$$

13.2. Nitrous Oxide (N_2O)

LaHue et al.[205] determined atmospheric N_2O by first passing the air sample over $CaSO_4$ and Ascarite, which removed H_2O and CO_2. The N_2O was collected on activated molecular sieve 5A. Helium saturated with water vapor was employed to transfer the sorbed N_2O molecules from the adsorbent to an activated silica gel column, held at solid CO_2-isopropanol temperature. Finally the N_2O is transferred to a Porapak Q column at $150°C$ and detected by thermal conductivity. At concentration of 300 parts per 10^9 of air, the accuracy of determination is ± 5 parts per 10^9 and reproducibility is good. The outstanding advantage of this technique is that the sampling can be performed anywhere. In addition the sample can be stored (in the sealed molecular sieve column) for at least one month before analysis. Hahn[206] collected air samples in a similar manner aboard ship. Fifteen to twenty liters of air are sampled. He adapted the technique to water samples also by heating and transferring the volatile materials into the molecular sieve column. The concentration of N_2O is calculated from the peak area and a calibration factor. The standard deviation for measurement of N_2O in air was $\pm 1.4\%$ and in water was $\pm 1.7\%$. LaHue *et al.*[207] published additional work on their technique, in which the standard deviation was $\pm 1\%$ and the total analysis time was 30 min. Nitrous oxide has also been determined in air samples with the use of a ^{63}Ni electron capture detector.[208] Porapak Q was the packing utilized and N_2 was the carrier gas. The detector was operated at 395°C in the pulse sampling mode, which made it about 1000 times more sensitive than at 20°C. Under these conditions it was possible to detect ppb of N_2O in air.

Nitrous oxide is found in unpolluted air as well as seawater, in concentrations of about 0.3 ppm. It is usually formed by the decomposition of nitrogen-containing inorganic and organic substances, and is found in tobacco smoke (40 $\mu g/g$ of tobacco). N_2O is used quite extensively in dental practice and some forms of surgery.

The higher concentrations of nitrous oxide found in the atmosphere (250–300 ppb) are usually attributed to soil bacteria and/or photochemical reactions.

One technique of determining nitrous oxide in air is to concentrate it on a column of silica gel at -70°C and then desorb it at room temperature. A thermistor probe is used for detection after separation on a 10.5-m-×-6-mm column of 19% propylene carbonate and 16.5% glutaronitrile on Sterchamol.[209,210] The detection limit for a concentrated 10-liter air sample is 0.05 ppm.

13.3. Nitrogen Dioxide

The biological half-life of this contaminant is one hour. A sensitive method for the detection of NO_2 is to react it with styrene, at 22°C, in air and collect the products (5–8 products) which may be separated by GLC on a 5% SE-30 column at 60° or 100°C using N_2 as the carrier and a ECD.[211] The sensitivity of the method is 2 ng and H_2O, O_2, Cl_2, NO, and SO_2 do not interfere, even when present in excess.

NO_2 has a characteristic odor and can be detected at concentrations as low as 5 ppm. When the concentration reaches 10–20 ppm it becomes irritating to the eyes, nose, and upper respiratory system. Above 50 ppm it is dangerous even for short exposures. Deaths attributed to nitrogen dioxide result from pulmonary edema, not from nitrite effects.[212]

Nitrogen dioxide will react with Porapak Q and Chromosorb 102. Nitration of the rings takes place and NO and water are released as products; these have been identified by infrared spectroscopy.[213] The author has noted the same results in his laboratory. The use of glass columns permitted us to visually observe the reaction *in situ*.

13.4. Mixed Nitrogen Gases

The most commonly used packing for the separation of nitrogen oxides has been molecular sieve 5A. Silica gel is also used, but less frequently. Lawson and McAdie[214] determined the oxides of nitrogen, with and without and preconcentration of sample, using an electron capture detector for quantification. They were able to detect concentrations as low as 10 ppm. Shaw[215] has written a highly recommended review on formation, incidence measurement, and control of nitrogen-oxides in flue gas. Blades[216] studied the response of the FID to nitrogen compounds and phosphine. Ion formation was monitored with molecular emission in a flame photometric system; mechanisms for the formation of cyano-group and ions from NH_3 are discussed. A rapid method for analyzing nitric oxide reduction products is that of Landau and Petersen.[217] A two-column system with katharometer detection enables hydrogen, nitrogen, oxygen, nitric oxide, ammonia, and nitrous oxide to be determined. Total time for the analysis is 11 min. A method for the determination of ammonia in dilute aqueous alkali and in nitrogen–ethylene environments was devised by Sims.[218] The results are comparable with those of other methods. Ammonia and water can be separated on a Chromosorb 102 column treated with Theed.[219] This was one-column used for the quantitative determination of NH_3 oxidation products. Other components determined were O_2, N_2, N_2O, NO_2.

13.5. Other Compounds Containing Nitrogen

Karasek and Denney[220] detected 2,4,6-trinitrotoluene in the air by plasma chromatography. The TNT compound formed characteristic negative- and positive-ion molecule pairs, which were detected by their mobility spectra. Ten nanograms could be detected at a signal-to-noise ratio of greater than 10^3. Aliphatic amines volatilized from a cattle feedyard could easily be identified by GLC after being collected in H_2SO_4 traps.[221] Ten compounds, including methyl-, dimethyl-, propyl-, isopropyl-, butyl-, and amylamines, were easily identified. The identification was confirmed by the gas chromatographic separation of their pentachloro-benzyl derivatives.

Thin-layer and gas chromatography were combined to determine aza-heterocyclic compounds in atmospheric dust.[222] One to two grams of dust were collected on glass fiber filters and extracted with benzene. Initial separation was on silanized silica gel by TLC. The band of interest was dissolved in benzene and then separated by GLC on 40-m capillary columns coated with Versamind 900. Eighteen such compounds were identified, and recovery of model compounds added to the dust was about 90%. Neiser *et al.*[223] determined pyridine and its homologues in the atmosphere using a Carbowax 1500 column, argon carrier gas, and a FID. With this 3-m column at 180°C, up to 12 components were separated and identified.

Using an Apiezon L-Chromosorb G trapping column (at liquid oxygen temperatures), nitriles in 20–50 liters of air were separated and identified from a 5% dinonyl phthalate column 8 m × 4 mm. The effect of other constituents in the air was also investigated.[224] Detection was by flame ionization. Zamir-ul Haq *et al.*[225] identified and quantified some *N*-heterocyclics in the condensate from marijuana smoke. Carbazole, indole, and skatole were determined on a glass column packed with Silar 5CP on Gas Chrom Q, with helium as carrier gas and a FID. Mass spectrometry was used as a confirming technique.

Steam-volatile *N*-nitrosamines in food were detectable at the 1-μg-per-kg level using a 5.5-m-×-3-mm column of 15% FFAP on Chromosorb W and a Coulson conductivity detector.[226] For additions of nitrosamines at the level of 10 μg/kg, recoveries were between 70–90%. Bryce and Telling[227] did a similar study of nitrosamines in luncheon meats at the 10-, 2-, and 1-part-per-10^9 level. Recoveries varied between 27 and 133%. In this study the column effluent was monitored by a mass spectrometer, rather than the usual GC detector.

13.6. Peroxyacetyl Nitrates (PAN)

Peroxyacetyl nitrates (PANs) may be viewed as a series of organic nitrogen compounds which result from the action of sunlight on polluted air containing small concentrations of nitrogen oxides and organic compounds. They have the general formula

$$R\!-\!\underset{\displaystyle \underset{O}{\|}}{C}\!-\!OONO_2$$

Five analogues are important: acetyl, propionyl, butyryl, isobutyryl, and benzoyl; the latter is considered the most irritating to the eyes. PANs may be determined in the presence of hydrocarbons by use of an electron capture detector and a 3-ft column packed with 5% Carbowax 400 on Chromosorb W-HMDS (100/120 mesh).[228] Using nitrogen carrier gas and the ECD allows a sensitivity of 1 ppb or less with an overall accuracy of ±5%. Smith *et al.*[229] described an automatic gas chromatographic system for the analysis of PAN; this system was an extension of that described in reference 228. Automatic sampling system samples were taken every 15 min; the column was operated at 25°C, and N_2 at a flow of 40 cm^3/min was used as carrier.

Bellar and Slater[230] reported a quantitative method for PAN and methylnitrate. Polyethylene glycol 400 or 1540 on Gas Chrom Z offered adequate resolution. An electron capture detector was used in the pulse mode.

14. Oxygenated Materials

14.1. Ozone

When we think of oxygen and air pollution, the first component to come to mind is ozone. Ozone is a very unstable but highly reactive gas which is formed in nature at high altitudes, as well as by technological processes and industrial equipment. In nature O_3 is formed in the upper atmosphere by the photodissociative action of UV radiation on O_2 molecules. It is in the stratosphere that we find the highest concentrations of this gas (11 ppm), while at ground level the concentration is usually in the range of 1–3 parts per hundred million. The actual concentration depends on weather conditions and the height above sea level.[231]

Under conditions of photochemical smog we find higher concentrations of ozone at ground level (0.6 ppm). Direct manmade sources of ozone include arc-welding devices, high-voltage electrical equipment (X-ray equipment, spectrographs), neon signs, ultraviolet quartz lamps, and office copy equipment. It is also a product of the high-voltage ozonizers used in sewage treatment and water purification, as well as those being used to control molds and bacteria in cold storage plants.[232,233]

Donohue and Jones[234] developed a gas–solid chromatographic technique for the analysis of the gaseous products produced from the electrolysis of wet hydrogen fluoride (H_2, O_2, OF_2, and O_3). A silica gel column, programmed from −75° to −10°C, was employed for the separation. It is important for a system of this type that all materials that come in contact with the sample be made from fluorocarbon plastic. Helium was used both to sweep the electrolysis products onto the column, and as the carrier gas.

Pate *et al.* have studied the gas-phase reaction of ozone with a series of aromatic hydrocarbons.[235] Compounds studied were benzene, toluene, the xylenes, 1,2,3-, 1,2,4-, and 1,3,5-trimethylbenzenes. The reaction products were analyzed using a 50-cm-×-0.3-cm Teflon column packed with Carbowax 400 on 60/80 mesh firebrick. An electron capture and flame ionization detector were both employed. Product identification was substantiated by means of a 40-m long-path cell and infrared spectroscopy.

14.2. Other Oxygen Compounds

A quantitative GC-MS method for the analysis of phenols in wood-smoke concentrates, smoked foods, and beverages has been reported by Kornreich and Issenberg.[236] Separations were carried out on a 150-m capillary column coated with OV-17. The phenols were identified after derivatization to their trimethylsilyl ethers.

Selke *et al.*[237] reported a technique for the identification of the volatile components from Tristearin in air. The volatile oxidation products were collected on a gas chromatographic column cooled to −60°C. The OV-17 column was then temperature programmed to 250°C and 86 peaks, most of which were aldehydes and ketones, were identified by mass spectrometry.

Carbonyl compounds have been determined in a variety of samples. Halvarson[238] volatilized the carbonyls from 5–10-g samples of autooxidizing fats by steam distillation. The volatiles were trapped on a PTFE column containing Celite impregnated with 2,4-dinitrophenylhydrazone. The derivatives were separated on a Porapak S or Porasil C column and quantified by flame ionization. Papa and Turner[239] determined carbonyls in a similar manner as they were collected from auto exhaust.

Volatile carbonyl compounds in edible fats were also converted to their hydrazone derivatives and group separated on thin-layer plates.[240] The groups were then separated by GLC using Dow Corning silicone oil 200. Carbonyl compounds from auto exhaust were converted to oxime derivatives and separated on either a Carbowax 20M or UCON-50-HB-660 coated glass beads; nitrogen was used as the carrier gas and detection was by flame ionization. Smythe and Karasek determined low-molecular-weight carbonyls from diesel engine exhaust, after conversion to their hydrazone derivatives.[242] The column was 15% Dexsil 300 on Anakrom A (90/100 mesh). The low-molecular-weight compounds identified were formaldehyde (20 to 30 mg per 1000 liters exhaust gas), acrylaldehyde (0.5 to 0.8 mg), and crotonaldehyde (0.02 to 3.2 mg).

15. Sulfur-Containing Compounds

We begin this section with an encouraging piece of information: in the 1975–76 Annual Report of the Environmental Protection Agency to Congress, Russell E. Train (EPA Administrator) reported that the U.S. air is a little cleaner, especially with regard to sulfur dioxide. Levels declined from a value of 38 $\mu g/m^3$ in 1970 to 26 $\mu g/m^3$ in 1974. These values were from 258 sampling sites and represent an overall decrease of 9% per year.[243]

Sulfur dioxide and trioxide are the only major sulfur oxides present in the atmosphere. About 33% of the sulfur entering our atmosphere is man-generated, and most of this is in the form of SO_2; the smelting of ores is probably the largest source. Sulfur compounds affect the quality and the properties of our air by causing smog formation, eye irritation, offensive odors, and hydrocarbon fixation. Sulfur dioxide affects the respiratory tract in humans, usually resulting in gastritis or stomach ulcers. If the SO_2 becomes oxidized to SO_3 then absorbs water, it results in the formation of sulfuric acid mist which can be very detrimental to textile fibers, paints, metals, and building materials.

Sulfur dioxide and sulfur trioxide have been determined by gas chromatographic techniques, although this is not the method used by the EPA. Mixtures of SO_2, H_2S, and CH_3SH have been separated on a 30% Triton X-100 on Chromosorb by Adams and Koppe[244]; the order of elution was H_2S, CH_3SH, and SO_2, with helium as carrier and a thermal conductivity detector. Air mixtures containing SO_2 were separated, and the SO_2 was quantified on Chromosorb W coated with diisodecyl phthalate.[245] The air mixtures ranged from 0.1–50 mole percent of SO_2 and the resulting peak areas were reproducible to within ±1.0%. Levchuk et al.[246] followed the methane reduction of sulfur dioxide with the use of two columns: Column I was silica gel, which separated SO_2, CS_2, H_2S, COS, and CO_2. Column II was molecular sieve CaA; it separated CH_4, CO, H_2, O_2, and N_2. Both columns utilized thermal conductivity detection; argon was the carrier. Water and elemental sulfur were not estimated by this method.

Sulfur dioxide in nitrogen was determined by means of an AT-cut piezoelectric quartz crystal, coated with Carbowax 400, triethanolamine, Amine No. 220, or squalene.[247] Stevens and O'Keefe claim instrumentation with great specificity and sensitivity; SO_2 is separated on a column (34 ft × 0.085 in.) packed with 5-ringed polyphenylether (containing H_3PO_4) on PTFE. A flame photometric detector with a 394-nm optical filter was used for the detection and quantification. The sensitivity claimed is 2 parts in 10^9.[248] A review of the various instruments using electrochemical, spectrometric, and chromatographic methods for the determination of SO_2 in ambient and stationary sources has been presented by Hollowell *et al.*[249] This article tabulates the principal features of the commercially available monitors. The determination of sulfur dioxide by means of a dry adsorbent has been reported by Bourbon *et al.*[250] A metered flow of air is passed through a cellulose-based filter impregnated with zinc acetate in glycerol; the SO_2 is converted to zinc sulfate and may be stored for at least 20 days before measurement. Although this is not a chromatographic method, it does present a means of storing a sample when it is not possible to analyze it immediately.

The determination of SO_3 is less easy than SO_2, because of SO_3's reactivity and tendency to polymerize. Bond *et al.*[251] determined SO_3 indirectly by reacting it with oxalic acid, which was the packing in the first column. The reactions taking place are

$$(COOH)_2 \cdot 2H_2O = 2H_2O + HCOOH + CO_2 \tag{19}$$

$$HCOOH + SO_3 = CO + H_2SO_4 \tag{20}$$

The products from these reactions were passed into a second column containing 40% phenyl oxitol on Sil-O-Cel C22, and then through one arm of a katharometer (CO and CO_2 were not resolved). This composite peak then entered a longer column (20 ft) of 40% oxitol on Sil-O-Cel, and the effluent passed into the second arm of the katharometer. The CO and CO_2 peaks were thereby resolved as negative peaks. The method has been used to determine SO_3 at concentrations of about 0.3%.

15.1. Carbonyl Sulfide

This compound is a product of the destructive distillation of coal, petroleum purification processes, coking operations, and viscose plants. It is always found when carbon monoxide, carbon disulfide, and sulfur dioxide are present at high temperatures. In the presence of chlorine it forms phosgene and sulfur dichloride, and in the presence of ammonia it will yield urea and hydrogen sulfide.

Mixtures of COS, SO_2, Cl_2, and H_2S have been separated by Berezkina and Mel'nikova,[252] using a 1-m column of 0.25-to-0.5-mm silica gel particles. The separation of COS, H_2S, CS_2, and SO_2 was investigated by Staszewski *et al.*[253]; a complete separation of all four components was not possible on three different columns. The liquid substrates investigated were squalene (I), dinonyl phthalate (II), and polyethylene glycol 400 (III), all on Teflon 6. Column I did not resolve H_2S and SO_2, column II only partially resolved H_2S and COS, and column III partially resolved the same pair but retained SO_2 for 30 min. Hall[254] has described a quantitative gas–solid chromatographic method for the determination of COS in CO_2. A 4-ft-×-4-mm glass column packed with 40/60 mesh activated silica gel was used at 25°C with a Lovelock beta-ionization detector. The COS peak was linear over the range 0.5 to 2700 ppm, with a minimum detectable concentration of 0.3 ppm.

15.2. Carbon Disulfide and Hydrogen Sulfide

These two compounds have offensive odors, and maximum safe concentrations have been established. Hydrogen sulfide is one of the more dangerous industrial chemicals of today because of its toxicity and explosive nature when mixed with either air or sulfur dioxide. Its maximum safe concentration is 13 ppm; when inhaled it attacks the nerve centers of the body. Carbon disulfide has a maximum permissible concentration of 20 ppm for an 8-h day. It is absorbed mainly through the lungs and then enters the bloodstream, where it is transported throughout the body. Very little CS_2 is absorbed through the skin. Most of the CS_2 which enters the body is metabolized and eliminated in the urine in the form of inorganic sulfates. The source of most of these sulfur compounds is natural gas and petroleum deposits. When sulfur reacts with natural gas, about 33% of it ends up as hydrogen sulfide.

Osin et al.[255] developed a method for low concentrations of CS_2 in air. Although this was not a gas chromatographic method, it could be adapted to GC for a reference method. The air sample is drawn through a tube which contains silica gel, piperazine and ethanolic copper acetate solution. A reference tube contains silica gel, auramine, safranine, and methylene blue. When the color of the indicating tube matches the reference tube, the determination can be made. By passing air through a third tube and directing the effluent into a FID one can compare gas chromatographic data with colorimetric data.

Cook and Ross[256] have presented a method for the determination of H_2S after separation from water and air. Two columns are employed; 0.25% Triton 305 on Porapak Q, and 5% Carbowax on Teflon 6. The calibration graph is linear from 0.2 to 12 μg of H_2S, with a detection limit of 50 ppm. The method is usable for analysis of stack gases from chemical recovery plants. Bollman and Mortimore[257] were able to determine CO_2, H_2S, SO_2, C_2H_6, and C_3H_8 after separation on a molecular sieve column. Helium was the carrier gas, with thermal conductivity detection. The procedure was applied to the analysis of flue gases from copper sulfide ore processors.

The Melpar flame photometric detector was employed for the detection of hydrogen sulfide and sulfur dioxide by Greer and Byda-lek.[258] It was found that the detector lost sensitivity at concentrations greater than 70 ng. This loss of sensitivity was due to self-absorption rather than detector fatigue. In using a system such as this, log detector response should be plotted versus log sulfur mass. When the slope of this line changes it is an indication that self-absorption is taking place.

15.3. Miscellaneous Sulfur-Containing Compounds

The use of an ionization detector with a ^{63}Ni source has been demonstrated for the determination of hydrogen sulfide, sulfur dioxide, and carbonyl sulfide in air samples. The detector provides improved detectability for all three gases.[259] Two columns were used for the complete determination; a 40-cm silica gel column and a 1-m 5% polyoxyethylene glycol on Celite.

Manfred et al.[260] determined traces of impurities (COS, CS$_2$, thiophen, thiols, organic sulfides, and disulfides, as well as saturated, unsaturated, and aromatic hydrocarbons) in coal fuel gas by GLC. A beta-ray argon ionization detector was used. Errors were as high as ±15–20% for each 100 m^3 of flue gas sampled. Stevens et al.[261] determined reactive sulfur gases in air at the ppb level. An automated system was employed with a flame photometric detector. Sulfur dioxide, H$_2$S, methanediol, and dimethyl sulfide were easily determined down to 0.02 ppm. A 36-ft column packed with PTFE powder coated with a mixture of poly(phenyl ether) and H$_3$PO$_4$ using N$_2$ as the carrier gas was found to be adequate for the separation. The detection limits were approximately 5 ng.

Gelman filters were used to trap mixtures of thiols, dimethyl sulfide, and methylamines. The filters had been impregnated with Hg(CN)$_2$, HgCl$_2$, or H$_2$SO$_4$. The trapped compounds were regenerated and trapped in an organic solvent held at dry-ice–acetone temperature. Separation was achieved with nitrogen as carrier, and a flame ionization detector. Named compounds could be detected at the parts-per-10^9 level.[262]

An automated method for the determination of sulfur compounds at the ppb level has been presented by Bruner et al.[263] The compounds are separated on a 40/60 mesh Graphon carbon column with a flame photometric detector. Calibration was accomplished by means of the conventional exponential-dilution-flask technique. Sensitivity in parts per 10^9 were: thiols, 30; H$_2$S, 20; and SO$_2$, 10, which compares favorably with other methods. A similar system was also discussed by Pecsar and Hartman.[264]

Robertus and Schaer developed a portable continuous chromatographic coulometric sulfur-emission analyzer.[265] It uses a Br-titration cell for the continuous determination of methanediol, hydrogen sulfide, and sulfur dioxide, the limits of determination being 0.1, 0.2, and 1 ppm, respectively. This instrumental system has been used for analyzing stack gases at a paper mill.

Gibson et al.[164] and Casselman et al.[266] published a method for the analysis of mustard gas. The air sample was trapped in decalin at $-15°C$ at a flow rate of 100 cm^3/min. The analysis used a 2-ft-×-0.125-in. column packed with 2% FFAP on Chromosorb W, and Ar-methane as carrier with a ^{63}Ni electron capture detector. In the range of 1–8 $\mu g/cm^3$, the method was accurate to ±2%, with a limit of detection of 0.2 $\mu g/cm^3$.

16. Human Volatile Material

In addition to the various contaminants that we expel into the atmosphere from our autos, pleasure devices, and industrial processes, we as individual human machines can pollute the air by several means. Every time that we breathe we pollute the immediate atmosphere. In addition, our body fluids, whether they be excreted or exposed to the atmosphere by other means (e.g., volatiles in blood), add to the ever-increasing amount of pollutants. This can be more of a problem when the person(s) is in an enclosed area (e.g., room, confined working area, mines or even medical–surgical rooms).

Table 7. Gas Chromatographic Methods for Human Volatile Materials

Volatile material	Matrix	Reference
Ethyl alcohol	Expired air	267, 268
Acetone	Breath, plasma	269
Carbon monoxide	Expired air, blood	270
Dichloromethane	Breath, blood, urine	271
Tetrachloroethylene	Expired air	272
Volatile selenium-75 metabolites	Expired air	273
Volatile compounds	Breath and urine	274
Ethyl alcohol	Expired air, blood, urine	275, 279
Nicotine	Urine, cigarette smoke	276
Sedatives, pesticides, amphetamines, $CHCl_3$	Biological fluids	277
Cannabis constituents	Smoke, body fluids	278
Volatile fatty acids	Urine	280
Volatile metabolites from sulfonylureas or insulin	Urine	281
Volatiles	Human effluents	282
Volatile acids	Cooked mutton	283
Volatile phenols	Urine	284
Mercaptans, H_2S	Breath	285

Table 8. Components Determined in Tobacco Smoke

Sample	Components found	Column used	Detector	Reference
Cigarette smoke	Polycyclic hydrocarbons	SE-30	ECD	286
Cigarette smoke	Acrylaldehyde and HCN	Capillary	FID	287
Cigarette smoke	Tar and nicotine	Porapak Q	Spectrometric	288
Cigarette smoke	Naphthylamines	Silicone QF-1 DC-200	ECD	289
Cigarette smoke	9-Methylcarbazoles	OV-225	Liquid scintillation counter	290
Cigarette smoke	133 Volatile components	Capillary	GC/MS	291
Cigarette smoke	Hydrocarbons, amines phenols, cyclic ketones			292
Cigarette smoke	Tar and nicotine	Porapak	Spectrometric	293
Cigarette smoke	Hydrocarbons C_{10}–C_{33}	Polysev	GC/MS	294
Cigarette smoke	1-Alkylindoles	XE-60	FID	295
Cigarette smoke	Water	Porapak Q	TCD	296
Cigarette smoke	Volatiles (215 compounds)	Capillary	FID	297
Cigarette smoke	Volatile sulfur compounds	1,2,3-tris(2-cyanoethoxy)propane on AW Chromosorb W	Flame photometric	298
Cigarette smoke	Sulfur-containing compounds	Porapak Q, Chromosorb 104	S-specific flame photometric	299
Cigarette smoke	Fluorenes	OV-17, OV-1	MS	300
Cigarette smoke	Carbon monoxide	Molecular sieve 13×	FID	301
Cigarette smoke	Carbon monoxide and carbon dioxide	Porapak Q	TCD	302

Sample	Compounds	Column	Detector	Ref.
Cigarette smoke	Acetic and higher acids	Chromosorb 101	FID	303
Cigarette smoke	26 Components	Propylene glycol LB-550×, Poly-glycol 400 + -oxydipropionitrile	FID	304
Cigarette smoke	Semivolatile compounds	OV-1	FID	305
Tobacco and tobacco smoke	Ammonia	Polyethylimine on Porapak Q	Hot wire	306
Tobacco and tobacco smoke	Free phytosterols	OV-101	FID	307
Tobacco smoke	Volatile phenols	UC-W98 silicone	FID	308
Tobacco smoke	Nitrobenzenes	OV-225	FID	309
Tobacco smoke	1,1- and 1,2-trans-dimethylcyclo-propanes, 1-chloro-5-methyl-hexane, hexa-1,3,5-triene and 2-methylacetene	SF-96 open tubing	MS	310
Tobacco and tobacco products	Vanillin, ethylvanillin, coumarin, dihydrocoumarin	QF-1, OV-17	FID	311
Tobacco and tobacco smoke	Polyhydric alcohol humectants	SE-30	FID	312
Tobacco and tobacco smoke	Thiodan and thiodan sulfate	QF-1, Dow 11	ECD	313
Tobacco smoke	N-Dimethylnitrosamine	Carbowax 1540 Porapak Q	MS	314
Tobacco smoke	Carbon monoxide	Molecular sieve 5A	TCD	315
Tobacco and cigarette smoke	Hydrazines	OV-17 + QF-1	ECD	316
Tobacco smoke	Chlorinated hydrocarbon pesti-cides	Silicone QF-1 + DC-200	MS	317
Cigarette and cigar smoke	Carbon monoxide + carbon dioxide	Silica Gel	TCD	318
Marijuana smoke	Marijuana	SF-96 Silicone oil	MS	319

The extent to which these types of pollutants affect our atmosphere depends upon the individual situation. In a confined area, such as a room in one's home or a medical–surgical room, they would be atmospheric pollutants. Although some may classify these as medical problems rather than pollution problems, pollution problems in general are in fact also medical problems. The author considers them both and accordingly lists some representative methods in Table 7.

17. Cigarette, Cigar, and Pipe Smoke

Smoking, especially cigarette smoking, is the cause of hundreds of thousands of deaths annually in this country alone, lung cancer and emphysema being the two end results of heavy smoking. It is no wonder that so much research is being carried out in the analysis of smoke products. We will not attempt to cover all of this work, but enough for the reader to have have an appreciation for various studies completed.

Table 8 summarizes the various compounds which have been determined in tobacco smoke.

Kaburaki et al.[320] carried out some studies for the identification of low-boiling amines in tobacco smoke. Several groups of columns were investigated for the separations. In all cases helium was the carrier. Nicotine interfered with the GLC separation and had to be removed by precipitation by tungstosilicate. A profiling procedure to survey compounds which had biological significance was reported by Guerin et al.[321] in which they used sulfur- and nitrogen-selective detectors. Grob[322] has discussed the difficulties encountered with high-resolution gas chromatographic analysis of cigarette smoke using capillary columns. The difficulties encountered were: (a) inability to separate both acidic and basic components, (b) nonelimination of the holdup of solvent vapor by slow release from the injection septum, and (c) poor coupling of the mass spectrometer to the column. A review has been written which discusses the application of the Janak method to the gas chromatographic analysis of cigarette smoke.[323] The article also discusses the method's application to the analysis of industrial gases.

A technique whereby fresh tobacco smoke is first group-separated on a packed column of 2,2'-oxydipropionitrile and then the effluent fractions analyzed on a glass capillary column coated with SF-96 connected to a mass spectrometer has been described by Blomberg and Widmark.[324] Maleic hydrazide has been determined in cigarette smoke and particulate matter after extraction and formation of the trimethylsilyl derivative. A column of OV-11 on Chromosorb W was used for the separation.[325]

18. Exhaust Samples

This is the area of atmospheric analysis which has received the lion's share of analytical investigations. Exhaust gases from all types of vehicles are major source of atmospheric pollution in many parts of the world. The author has had personal experience in this area in his role as a research analytical chemist for one of the large petroleum companies. This was followed by active participation as a consultant for a municipal air pollution department and Consumers' Union in an air pollution survey of the Ohio Valley.[326]

The results of the work in this area have led to changes in the exhaust systems of new cars and use of catalytic converters. I will not attempt to cover this topic in depth because this would be in direct competition with other in-depth treatments, e.g., the publication of the Intersociety Committee is composed of thirteen societies, namely; ACGIH, ACS, AIChE, AIHA, AOAC, APCA, APHA, APWA, ASCE, ASME, HPS, ISA, and SAE. Two chapters in *Advances in Chromatography* published by Marcel Dekker, Inc.[328,329] and the publication of the ASTM Committee D-22[330] provide additional references.

Tailpipe emissions are composed of nitrogen, carbon dioxide, water vapor, oxygen, carbon monoxide, hydrogen, unburned hydrocarbons, oxygenated hydrocarbons, and oxides of nitrogen. Table 9 illustrates the composition of dry automotive exhaust.[329] The chemical reactions which take place in the combustion process depend on such variables as reaction time, nature, and concentration levels of the hydrocarbons in the fuel, as well as the nitrogen oxide concentrations. The hydrocarbons, oxygenates, NO_x, and carbon monoxide are the important components of the exhaust which are of concern to environmentalists because of their toxicity and consequences to human, animal, and plant life. Several selected methods for some of the components in exhaust samples are tabulated in Table 10.

Table 9. Composition of Automotive Exhaust[a]

Pollutant	Amount
Nitrogen	82%
Carbon dioxide	12%
Carbon monoxide	3%
Oxygen	2%
Hydrogen	1%
Hydrocarbons	1000 ppm
Oxides of nitrogen	1500 ppm
Oxygenated hydrocarbons	75 ppm

[a] Several selected methods for some of the components in exhaust samples are tabulated in Table 10.

Table 10. Selected Methods for Exhaust Samples

Component system	Column used	Detector	Reference
Low-molecular-weight organic compounds	1,2,3-tris(2-Cyanoethoxy)propane/Chromosorb W, Porapak Q	FID	331
Hydrocarbons and carbon monoxide	Molecular sieve 5A, Porapak Q	TCD	332
Hydrocarbons and oxygen derivatives	1,2,3-tris(2-Cyanoethoxy)propane/Chromosorb P	TCD	333
Hydrocarbons	Packed column and open tubular columns	FID	334
Polycyclic aromatic hydrocarbons	OV-101 or OV-17/capillary, OV-17/Gas Chrom Q	FID	335
Polycyclic aromatic hydrocarbons	OV-101 or OV-17/capillary	FID	336
Hydrocarbons	$HgSO_4/H_2SO_4$, $PdSO_4/H_2SO_4$	FID	337
Hydrocarbons and oxygenated derivatives	Molecular sieve 5A, 2-(2-methoxyethoxy)ethyl ether	FID	338
Polycyclic aromatic hydrocarbons	OV-101 or OV-17	FID	339
Total hydrocarbons, CO, and CO_2	Silica gel, Apiezon L/Chromosorb W	FID	340
Hydrocarbons	OV-17/Gas Chrom Q, OV-1/Gas Chrom Q, Alumina GSC-121	FID	341
Hydrocarbons	Molecular sieve 5A, Porapak Q	FID	342
Organic compounds	OV-101 + SF-96 capillary columns	MS	343
CO, CO_2, SO_2, COS, O_2, N_2	Porapak R, molecular sieve 5A	TCD	344
Boron hydrides (industrial effluents)	Kel F-3/Teflon 6	ECD, MCD, FID, MS	345
Mine air for CO_2, O_2, CH_4, and N_2	Molecular sieve 5A	Ar detector, ^{90}Sr cross-sectional ionization	346
Organic chelating agents in steam propulsion systems	SE-30/Chromosorb W	FID	347
O_2, Ar, CO, CH_4, CO_2, N_2O, H_2, N_2, C_1–C_6 HC	Porapak Q and R	FID	348

19. Soil and Plant Atmospheres

Discussions of analysis of plants or plant environment immediately bring to mind topics such as pesticides, herbicides, and fungicides. Of course, there is no intention of attempting to cover such a wide area in a section such as this. The reader can easily find many articles, books, and reviews written on these topics. Grob[349] has written a chapter on gas chromatographic analysis in soil chemistry which should be consulted for more detailed information. The nomenclature and cross-referencing of pesticides may be found in two excellent publications.[350,351] My aim here is to point out techniques of determining those components which may find their way into plants and the soil and would therefore constitute an environmental problem. In some cases the soil or plant may itself give off some volatile material or degradation product. In cases such as these the analytical chemist may either be faced with a gas or solid sampling problem. If one is concerned with the plant or soil per se, than a larger problem of sample cleanup presents itself. Here the chemist will have to use not only GC, but TLC, PC, and various forms of liquid chromatography. The reader is referred to two publications which treat the determination of fumigants and fumigant residues[352] and chlorinated insecticides and their congeners.[353]

Table 11 lists some of the representative techniques for the analysis of plant and/or soil environments. More details for any of the references may be found in the original publications. One may consult two reviews which have been published on the chromatographic and biological aspects of DDT and its metabolites[354] and the chromatographic analysis of pesticide residues.[355]

20. Summary

The author has presented a survey of the use of gas chromatographic analyses as applied to atmospheric analysis. As stated in the introduction, the aim was not to give a complete coverage of everything which has been published, but to present selected topics in the area. The reader can add to this coverage by consulting other publications on this topic. *Analytical Chemistry* issues reviews on an annual basis; the even-numbered years cover technique reviews, whereas the odd-numbered years cover application reviews. *Environmental Science and Technology* is a monthly journal of the American Chemical Society which is devoted to the economics, laws, and feasibility of the techniques used in air, water, and land contamination analysis. The *Journal of Environmental Science and Health* (formerly *Environmental Letters*) has now expanded to three parts. Part A (monthly)

Table 11. **Plant and/or Soil Environmental Samples**

Sample type and analyte(s)	Column	Detector	Reference
Plant tissue for ethylene	Porapak Q	FID	356
Plant environment for ethylene and carbon dioxide	Porapak Q	FID	357
Plant tissue environment for ethylene	Porapak Q	FID	358
Plant environment for insecticides	OV-17 + OF-1/Gas Chrom Q, SE-30 + QF-1/Chromosorb W	ECD	359
Propoxur (pesticide) in plant environment	OV-1/Gas Chrom Q	ECD	360
Pesticides in plant environment (dichlorvos, parathion, diazinon, DDT, dieldrin)	SE-30/Chromosorb W, OV-101/Chromosorb W	ECD	361
Soil atmosphere gases (NO, NO_2, CO_2, N_2O, O_2, N_2)	Porapak Q, molecular sieve 5A, Carbowax 1550/silanized glass beads	TCD	362
Volcanic gases	Porapak Q, molecular sieve 5A	TCD	363
Soil atmosphere for bromomethane	Dow-Corning high vac. grease/Chromosorb W	FID	364
Soil atmosphere for O_2, N_2, Ar, CO_2, N_2O, C_1–C_4 hydrocarbons	Molecular sieve 5A, Porapak Q	FID	365
Pesticides in soil environments	OV-101, OV-17, OF-1, SE-30	ECD, FID	366
Pesticides in soil and plant environment	OV-101, OV-17, OF-1	FID, MCD	367
Ekalux on Plants	OV-25, OV-210/Gas Chrom Q	CsBr thermionic	368
Organochlorine pesticides on vegetation	DC-200	ECD	369
Chloro-s-triazines in soil	Carbowax 20M/Chromosorb W	ECD and electrolytic conductivity	370
Dichlorvos residues on plants	Chromosorb 101	FPD	371
Toxaphene insecticide in plant atmosphere	OV-17 + QF-1, SE-30 + QF-1	ECD	372

deals with environmental science and engineering, Part B (quarterly) deals with pesticides, food contaminants, and agricultural wastes, and Part C (semiannually) deals with environmental health sciences. Consultation of many of the reference-book series will yield separate volumes or chapters on various aspects of environmental analyses. Finally, scanning of the *Federal Register* and the publications of the Environmental Protection Agency (EPA)[373] will aid in keeping abreast of what is being studied and what has been studied in regard to our environment. Air quality publications since 1970 are now in one volume.[374] There are books which may be consulted on such topics as source sampling and analysis,[375] the monitoring of air pollutants,[376] detection of aerosols and their evaluation,[377] odor control and its measurement,[378] trace metals in the atmosphere,[379] diffusion of pollutants,[380] and detailed discussion and description of sampling and monitoring instruments.[381,382] There has been published an article on the gas chromatographic correlation techniques for trace analysis.[383]

21. References

1. C. F. Junge, Air chemistry and radioactivity, *Int'l. Geophysical Series* (J. Van Meighem, ed.), Vol. 4, p. 382, Academic Press, New York (1963).
2. T. Koyama, Gaseous metabolism in lake sediments and paddy soils and production of atmospheric methane and hydrogen, *J. Geophys. Res., 63*(13), 3971–3983 (1963).
3. P. O. Warner, *Analysis of Air Pollutants,* p. 62, Wiley-Interscience, New York (1976).
4. J. D. Caplan, Smog Chemistry Points the Way to Rational Vehicle Emission Control, SAE Preprint 650641, Chicago, Ill. (1965).
5. A. P. Altshuller, in *Advances in Chromatography,* Vol. 5 (J. C. Giddings and R. A. Keller, eds.), p. 230, Dekker, New York (1968).
6. A. P. Altshuller and J. J. Bufalini, *Photochem. Photobiol., 4,* 97 (1965).
7. Federal Register, *36*(84), 8195 (1971).
8. J. A. Cardina, *Rubb. Chem. Technol., 46*(1), 232 (1973).
9. W. D. Washington and C. R. Midkiff, *J. Assoc. Anal. Chem., 56*(5), 1239 (1973).
10. M. S. Black and R. E. Sievers, *Anal. Chem., 45*(9), 1773 (1973).
11. *Health Lab. Sci., 8*(2), 101 (1971).
12. W. D. Ross and R. E. Sievers, *Environ. Sci. Technol., 6,* 155 (1972).
13. V. Cantute and G. P. Cartoni, *J. Chromatogr., 32,* 641 (1968).
14. H. J. Burkhardt, M. F. Pool, and C. Elliger, *Anal. Biochem., 43*(2), 601 (1971).
15. P. O. Warner, *Analysis of Air Pollutants,* p. 196, Wiley-Interscience, New York (1976).
16. C. R. Sreedharan, *J. Phys. E, 4*(8), 614 (1971).
17. C. C. Leiby and E. C. Dunton, *Rev. Sci. Instrum., 43*(8), 1202 (1972).
18. V. M. Yamada and R. J. Charlson, *Environ. Sci. Technol., 3*(5), 483 (1969).
19. A. O. Niedermayer, *Anal. Chem., 42*(2), 310 (1970).
20. D. C. Weber and E. J. Spanier, *Anal. Chem., 42*(4), 546 (1970).
21. S. A. Roach, *Am. Ind. Hyg. Assoc. Q., 27,* 1 (1966).
22. B. E. Saltzman, *J. Air Pollut. Control Assoc., 20,* 660 (1970).
23. E. S. Gould, *Inorganic Reactions and Structure,* rev. ed., p. 292, Holt, Rinehart and Winston, Inc., New York (1962).

24. R. C. Leo, R. S. Thomas, H. Oja, and L. DuBois, *Anal. Chem.*, *45*(6), 908 (1973).
25. R. A. Young, *Pollut. Eng.*, *4*(9), 32 (1972).
26. B. V. Ioffe, A. G. Vitenberg, and V. N. Borisov, *Zh. Analit. Khim.*, *27*(9), 1811 (1972).
27. A. F. Wartburg, J. B. Pate, and J. P. Lodge, *Environ. Sci. Technol.*, *3*(8), 767 (1969).
28. J. H. Ford, C. A. McDaniel, F. C. White, R. E. Vest, and R. E. Roberts, *J. Chromatogr. Sci.*, *13*(6), 291 (1975).
29. J. Yu, C. P. Hedlin, and G. H. Green, *J. Chromatogr. Sci.*, *8*(8), 480 (1970).
30. A. Dravnieks, A. Kowszynski, K. Boguslaw, J. Whitefield, A. O'Donnel, and T. Burgwald, *Environ. Sci. Technol.*, *5*(12), 1220 (1971).
31. A. Raymond and G. Guiochon, *Analusis*, *2*(5), 357 (1973).
32. A. Y. Litovchenko, *Ref. Zh., Khim.*, *8*, 19GD (1972).
33. B. Versino, M. DeGroot, and F. Geiss, *Chromatographia*, *7*(6), 302 (1974).
34. W. A. Aue and P. M. Teli, *J. Chromatogr.*, *62*(1), 15 (1971).
35. F. Herzel and E. Lahmann, *Z. Anal. Chem.*, *264*(4), 304 (1973).
36. E. D. Pellizzari, J. E. Bunch, and B. H. Carpenter, *Environ. Sci. Technol.*, *9*(6), 552 (1975).
37. G. P. Morie and C. H. Sloan, *Beitr. Tabakforsch.*, *6*(4), 178 (1972).
38. R. E. Snyder, *J. Chromatogr. Sci.*, *9*(10), 638 (1971).
39. J. A. Giannovario, R. J. Gondek, and R. L. Grob, *J. Chromatogr.*, *89*, 1 (1974).
40. J. A. Giannovario, R. L. Grob, and P. W. Rulon, *J. Chromatogr.*, *121*, 285 (1976).
41. M. J. Mignano, P. R. Rony, D. Grenoble, and J. E. Purcel, *J. Chromatogr. Sci.*, *10*(10), 637 (1972).
42. W. E. Neff, E. H. Pryde, E. Selke, and J. Cowan, *J. Chromatogr. Sci.*, *10*(8), 512 (1972).
43. T. H. Gouw and I. M. Whittemore, *Chromatographia*, *2*(4), 176 (1969).
44. J. Angerer and A. Haag, *Z. Klin. Chem. Klin. Biochem.*, *12*(7), 321 (1974).
45. R. E. Kaiser, *Anal. Chem.*, *45*(6), 965 (1973).
46. C. W. Fort, V. E. Andrews, and A. Goldman, Report of the Atomic Energy Commission, U.S. NERC-LV-539-12, 20 pp., (1972).
47. C. H. Van Dyke, *J. Chromatogr.*, *60*(2), 248 (1971).
48. R. G. Schaefer and G. Schomburg, *Chromatographia*, *4*(11), 508 (1971).
49. R. B. Dean and W. J. Dixon, *Anal. Chem.*, *23*, 636 (1951).
50. J. F. Fritz and G. W. Schenk, *Quantitative Analytical Chemistry*, Allyn and Bacon, Boston (1974).
51. R. S. Braman, in: Chromatographic Analysis of the Environment (R. L. Grob, ed.), p. 82, Marcel Dekker, New York (1975).
52. J. E. Lovelock, in: *Gas Chromatography 1960* (R. P. W. Scott, ed.), Butterworths, London (1960).
53. D. H. Desty, C. J. Geach, and A. Goldup, in: *Gas Chromatography 1960* (R. P. W. Scott, ed.), Butterworths, London (1960).
54. J. M. McKelvey and H. E. Hoelscher, *Anal. Chem.*, *29*, 123 (1957).
55. A. P. Altshuller and I. R. Cohen, *Anal. Chem.*, *32*, 802 (1960).
56. B. E. Saltzman and C. A. Clemons, *Anal. Chem.*, *38*, 800 (1966).
57. B. E. Saltzman, C. A. Clemons, and A. E. Coleman, *Anal. Chem.*, *38*, 753 (1966).
58. F. H. Huyten, G. W. A. Rijimders, and W. V. Beersum, in: *Gas Chromatography 1962* (M. van Swaay, ed.), Butterworths, London (1963).
59. E. R. Stephens and M. H. Price, *J. Air Pollut. Control Assoc.*, *15*, 320 (1965).
60. G. F. Collins, F. F. Barlett, A. Turk, S. M. Edmonds, and H. L. Mark, *J. Air Pollut. Control Assoc.*, *15*, 109 (1965).
61. A. E. O'Keefe and G. C. Ortman, *Anal. Chem.*, *38*, 760 (1966).
62. A. L. Lynch, R. F. Stalzer, and D. T. Lefferts, *Am. Ind. Hyg. Assoc. Q.*, *28*, 79 (1968).

63. Methods of Air Sampling and Analysis, Intersociety Committee Publication, American Public Health Association, Washington, D.C., p. 22, 24, 29 (1972).

64. Analysis of Atmospheric Inorganics, U.S. Dept. of HEW, EPA, NAPCO, Cincinnati, Ohio, 1976, Section VII-2, p. 1.

65. J. E. Lovelock, *Anal. Chem., 33*, 162 (1961).

66. F. Bruner, C. Canulli, and M. Possanzini, *Anal. Chem., 45*, 1790 (1973).

67. R. S. Braman and E. S. Gordon, *IEEE Trans. Instrumental and Measurement, IM-14*, 11–19 (1965).

68. T. A. Bellar, M. F. Brown, and J. E. Sigsby, Jr., *Anal. Chem., 35*, 1924 (1963).

69. A. P. Altshuller and C. A. Clemons, *Anal. Chem., 34*, 466 (1962).

70. F. W. Williams and H. G. Eaton, *Anal. Chem., 46*(1), 179 (1974).

71. I. M. Celegin, R. Hansson, and G. Sundstroem, *Scand. J. Clin. Lab. Invest., 27*(4), 367 (1971).

72. R. A. Back, J. N. Friswell, J. C. Boden, and J. M. Parsons, *J. Chromatogr. Sci., 7*(11), 708 (1969).

73. R. C. Paule, *Anal. Chem., 44*(8), 1537 (1972).

74. R. G. Smith, R. J. Bryan, M. Feldstein, B. Levadie, F. A. Miller, E. R. Stephens, and N. G. White, *Health Lab. Sci., 7*, 72 (1970).

75. M. E. Daines, *Chemy Ind. (N.Y.), 31*, 1047 (1969).

76. W. Tsang, *J. Res. Nat. Bur. Stand.*, A, *78*(2), 157 (1974).

77. L. Angely, E. Levart, G. Guiochon, and G. Peslerbe, *Anal. Chem., 41*(11), 1446 (1969).

78. W. R. Averett, *J. Chromatogr. Sci., 8*(9), 552 (1970).

79. J. M. Brockway, A. W. Boyne, and J. G. Gordon, *J. Appl. Physiol., 31*(2), 296 (1971).

80. H. D. Axelrod, R. J. Teck, J. J. Lodge, and R. H. Allen, *Anal. Chem., 43*(3), 496 (1971).

81. W. D. Bennett, *Lab. Pract., 20*(7), 583 (1971).

82. L. Angely, G. Guiochon, E. Levart, and G. Peslerbe, *Analusis, 1*(2), 103 (1972).

83. B. E. Saltzman, C. R. Feldman, and A. E. O'Keefe, *Environ. Sci. Technol., 3*(12), 1275 (1969).

84. F. P. Scaringelli, A. E. O'Keefe, E. Rosenberg, and J. P. Bell, *Anal. Chem., 42*(8), 871 (1970).

85. B. E. Saltzman, W. R. Burg, and G. Ramaswamy, *Environ. Sci. Technol., 5*(11), 1121 (1971).

86. D. P. Lucero, *Anal. Chem., 4*(13), 1744 (1971).

87. L. DeMaio, *Instrum. Technol. 19*(5), 37 (1972).

88. L. J. Purdue and R. J. Thompson, *Anal. Chem., 44*(6), 1034 (1972).

89. L. J. Lorenz, R. A. Culp, and R. T. Dixon, *Anal. Chem., 42*(9), 1119 (1970).

90. F. Bruner, C. Canulli, and M. Possanzini, *Anal. Chem., 45*(9), 1790 (1973).

91. O. F. Folmer, *Analytica Chim. Acta, 56*(3), 440 (1971).

92. D. J. David, *Gas Chromatographic Detectors*, Wiley-Interscience, New York (1974).

93. H. M. McNair and E. J. Bonelli, *Basic Gas Chromatography*, Varian Aerograph, Walnut Creek, Ca. (1967).

94. Sevcik, *Detectors in Gas Chromatography*, Elsevier, New York (1976).

95. E. O. Oswald, P. W. Albro, and J. D. McKinney, *J. Chromatogr., 98*(2), 363 (1974).

96. Y. Talmi and A. W. Andren, *Anal. Chem., 46*(14), 2122 (1974).

97. G. Lucien, *Chim. Analyt., 52*(10), 1089 (1970).

98. D. F. Merrion, Effect of Design Revisions on Two Stroke Cycle Diesel Engine Exhaust, SAE Trans. (1968).

99. A. P. Altshuller, S. L. Kopczynski, W. A. Lonneman, and F. D. Sutterfield, *Environ. Sci. Technol., 4*(6), 503 (1970).

100. K. Grob and G. Gorb, *J. Chromatogr., 62*(1), 1 (1971).

101. K. Grob, *Mitt. Geb. Lebensmittelunters. Hyg.*, *63*(1), 23 (1972).
102. Beckman Instruments, Inc., British Patent 1,326,520, 1/3/72.
103. M. Jaworski and H. Szewczyk, *Kh. Analit. Khim.*, *25*(6), 1215 (1970).
104. A. J. Davies and A. W. Bowen, *Erdoel Kohle Erdgas Petrochem. Brennst. Chem.*, *23*(6), 363 (1970).
105. R. E. Kaiser, *Chromatographia, 11*(5), 281 (1972).
106. R. Villalobos and R. L. Chapman, *Anal. Instrum.*, *9*(D-6), 1 (1971).
107. D. Siegel, F. Muller, and K. Neuschwander, *Chromatographia, 7*(8), 399 (1974).
108. R. Jeltes and E. Burghardt, *Atmos. Environ.*, *6*(10), 703 (1972).
109. H. H. Westburg, R. A. Rasmussen, and M. Holdren, *Anal. Chem.*, *49*(12), 1852 (1974).
110. D. Brocco, V. Cantuti, and F. Moscatelli, *Riv. Combust.*, *25*(4), 164 (1971).
111. R. A. Rasmussen, H. H. Westburg, and M. Holdren, *J. Chromatogr. Sci.*, *12*(2), 80 (1974).
112. T. T. Hauser and J. N. Pattison, *Environ. Sci. Technol.*, *6*(6), 549 (1972).
113. D. Brocco, V. DiPalo, and M. Possanzini, *J. Chromatogr.*, *86*(1), 234 (1973).
114. W. Solarski and E. Zielinski, *Chim. Anal.*, *16*(5), 1067 (1971).
115. E. Malan, N. H. Louben, A. S. J. Boshoff and O. Eps., *Chromatographia, 4*(10), 475 (1971).
116. R. V. Carter and B. Linsky, *Atmos. Environ.*, *8*, 57 (1974).
117. R. J. Robertus and M. J. Schaer, *Environ. Sci. Technol.*, *7*(9), 849 (1973).
118. E. Davidson, *Chromatographia, 3*(1), 43 (1970).
119. B. V. Subbarao, *Indian J. Technol.*, *10*(7), 282 (1972).
120. D. R. Deans, M. T. Huckle, and R. M. Peterson, *Chromatographia, 4*(7), 279 (1971).
121. J. J. Himmelreich, *Erdoel Kohl Erdgas Petrochem. Brennst. Chem.*, *23*(6), 366 (1970).
122. O. L. Gorshunov and L. G. Leitina, *Zav. Lab.*, *37*(2), 159 (1971).
123. J. S. Archer, *J. Inst. Fuel, 43*, 56 (1970).
124. T. L. Evseeva, E. P. Cherstvenkova, V. A. Nikol'skii, and V. I. Denisova, *Zav. Lab.*, *39*(4), 397 (1973).
125. D. Halot, *Talanta, 17*(8), 729 (1970).
126. K. Bhatia, *Anal. Chem.*, *43*(4), 609 (1971).
127. G. Chatot, M. Jequier, M. Jay, and R. Fontages, *J. Chromatogr.*, *45*(3–4), 415 (1969).
128. G. Chatot, R. Dangy-Caye, and R. Fontages, *J. Chromatogr.*, *72*(1), 202 (1972).
129. D. Brocco, V. DiPalo and M. Possanzini, *J. Chromatogr.*, *86*(1), 234 (1973).
130. G. Chatot, M. Castegnaro, J. L. Roche, and R. Fontages, *Chromatographia, 3*(11), 507 (1970).
131. R. E. Schaad, *Chromatogr. Rev.*, *13*(1), 61 (1970).
132. L. Zoccolillo, A. Liberti, and D. Rocco, *Atmos. Environ.*, *4*(10), 715 (1972).
133. M. Novotny, M. L. Lee, and K. D. Bartle, *J. Chromatogr. Sci.*, *12*(10), 606 (1974).
134. R. C. Lao, R. S. Thomas, H. Oja, and L. Dubois, *Anal. Chem.*, *45*(6), 908 (1973).
135. G. Grimmer, *Erdoel Kohle Erdgas Petrochem. Brennst. Chem.*, *25*(6), 339 (1972).
136. D. W. Grant, R. B. Meiris, and M. G. Hollis, *J. Chromatogr.*, *99*, 721 (1974).
137. H. P. Burchfield, E. E. Green, R. J. Wheeler, and S. M. Billadeau, *J. Chromatogr.*, *99*, 694 (1974).
138. G. M. Janini, G. M. Muschik, and W. L. Zielinski, *Anal. Chem.*, *48*(6), 809 (1976).
139. D. T. Lewis, *Lab. Pract.*, *20*(1), 24 (1971).
140. J. Angerer, A. Haag, D. Szadkokski, and G. Lehnert, *Z. Klin. Chem. Klin. Biochem.*, *11*(3), 133 (1973).
141. T. Okita, *Atmos. Environ.*, *4*(1), 93 (1972).
142. M. Gottauf, *Z. Anal. Chem.*, *246*(1), 31 (1969).
143. O. V. Chebotarev, G. I. Kryukov, S. L. Simkina, V. Z. Al'perin, R. N. Saifi, E. T.

Gainullina, R. N. Nurmukhametov, A. Y. Bonn, A. M. Drobiz, I. A. Pushkin, M. K. Yarmak, I. A. Lyamin, A. A. Druzhinin, I. V. Korablev, L. T. Lositskii, V. M. Matsnev, A. G. Melamed, V. A. Rylov, S. K. Bordakov, *Kh. Vses. Khim. Obshch, 15*(5), 482 (1970).

144. M. K. Sarkar and G. G. Haselden, *J. Chromatogr., 104*(2), 425 (1975).

145. G. F. Collins, F. E. Bartlett, A. Turk, S. M. Edmonds, and H. L. Mark, *J. Air Pollut. Control Assoc., 15,* 109 (1965).

146. R. L. Braman, in: *Gas Chromatographic Analysis of the Environment,* p. 119, (R. L. Grob, ed.), Marcel Dekker, New York (1975).

147. R. N. Dietz, E. A. Cote, *Environ. Sci. Technol., 7,* 338 (1973).

148. S. C. Monteriolo and A. Pepe, *Pure Appl. Chem., 24*(4), 631 (1970).

149. O. Pitak, *Chromatographia, 2*(7), 304 (1969).

150. A. G. Rauws, M. Olling, and A. E. Wibovo, *J. Pharm. Pharmac., 25*(9), 718 (1973).

151. J. G. Million, W. S. Pappas, and W. C. Weber, *J. Chromatogr. Sci., 7*(3), 182 (1969).

152. Y. S. Drugov, G. V. Murav'eva, K. M. Grinberg, B. N. Nesterenko, and D. N. Sokolov, *Zav. Lab., 38*(11), 1305 (1972).

153. E. D. Pellizzari, J. E. Bunch, R. E. Berkley, and J. McRae, *Anal. Chem., 48*(6), 803 (1976).

154. J. Gelbicov-hunckova, J. Novak, and J. Janak, *J. Chromatogr., 64*(1), 15 (1972).

155. H. Panzel and K. Ballschmiter, *Z. Anal. Chem., 271,* 182 (1974).

156. A. A. Casselman, R. A. B. Bannard, and R. F. Pottie, *J. Chromatogr., 80*(2), 155 (1973).

157. L. Grupinski, *Staub. Reinhalt. Luft,* (English edition), *31*(10), 3 (1971).

158. L. Collier, *Environ. Sci. Technol., 6*(10), 930 (1972).

159. J. B. Mann, H. F. Enos, J. Gonzalez, and J. F. Thompson, *Environ. Sci. Technol., 8*(6), 584 (1974).

160. A. J. Murray, J. H. Riley, *Analytica Chim. Acta, 65*(2), 261 (1973).

161. A. A. Aratskova, A. A. Balaukhin, D. A. Kalmanovskaya, G. A. Karpov, and D. N. Ivanov, *Zav. Lab., 37*(8), 911 (1971).

162. S. K. Yasuda, *J. Chromatogr., 104*(2), 283 (1975).

163. I. Tohyama and K. Otozai, *Z. Anal. Chem., 271,* 117 (1974).

164. N. C. C. Gibson, A. A. Casselman, and R. A. B. Bannard, *J. Chromatogr., 92*(1), 162 (1974).

165. J. A. Dahlberg and I. B. Kihlman, *Acta Chem. Scand., 24*(2), 644 (1970).

166. R. Jeltes, E. Burghardt, and J. Breman, *Br. J. Ind. Med., 28*(1), 96 (1971).

167. K. P. Evans, A. Mathias, N. Mellor, R. Silvester, and A. E. Williams, *Anal. Chem., 47*(6), 821 (1975).

168. R. A. Solomon and G. J. Kallos, *Anal. Chem., 47*(6), 955 (1975).

169. V. Y. Dudorov and N. K. Agiulov, *Zh. Anal. Khim., 25,* 162 (1970).

170. A. DiCorcia, P. Ciccioli, and F. Bruner, *J. Chromatogr., 62,* 128 (1971).

171. S. Jensen, *New Scientist, 32,* 612 (1966).

172. F. J. Biros, A. C. Walker, and A. Medbury, *Bull. Environ. Contam. Toxicol. 5,* 317 (1970).

173. R. Risebrough and V. Brodine, Environment, *12,* 16 (1970).

174. G. Westoo, K. Noren, and M. Anderson, *Var Foeda, 2–3,* 10 (1970).

175. L. M. Reynolds, *Residue Rev., 34,* 27 (1971).

176. D. L. Stalling, in: *Workshop on Pesticide Residue Analysis, 2nd International Congress of Pesticide Chemistry* (A. S. Tahori, ed.), p. 413, Gordon and Breach, London (1972).

177. P. W. Albro and L. Fishbein, *J. Chromatogr., 69,* 273 (1972).

178. R. K. Stevens and A. E. O'Keefe, *Anal. Chem., 42*(2), 143A (1970).

179. L. Dubois and J. L. Monkman, *Microchim. Acta,* (2), 313 (1970).

180. L. Dubois and J. L. Monkman, *Anal. Chem.*, *44*(1), 74 (1972).
181. K. Tesarik and M. Krejci, *J. Chromatogr.*, *91*, 539 (1974).
182. E. Zielinski and K. Mayer, *Chemia Analit. (Warsaw)*, *18*(4), 745 (1973).
183. V. G. Guglya, V. E. Shepelev and A. A. Khukovitskii, *Zh. Anal. Khim.*, *28*(6), 1223 (1973).
184. M. Celegin, R. Hansson, and G. Sundstroem, *Scand. J. Clin. Lab. Invest.*, *27*(4), 367 (1971).
185. J. R. Stetter, R. D. Rutt and K. F. Blurton, *Anal. Chem.*, *48*(6), 924 (1976).
186. B. A. Schaefer and D. M. Douglas, *J. Chromatogr. Sci.*, *9*, 612 (1971).
187. A. A. Poli, International Conference of the Instrument Society of America, Chicago, Oct. 4–7, p. 1, (1971).
188. D. J. Blackmore, *Analyst*, *95*, 439 (1970).
189. R. M. Bethea and M. C. Meador, *J. Chromatogr. Sci.*, *7*(11), 655 (1969).
190. J. H. W. Bailey, N. E. Brown, and C. V. Phillips, *Analyst (London)* *96*, 447 (1971).
191. A. Zlatkis, H. R. Kaufman, and D. E. Durbin, *J. Chromatogr. Sci.*, *8*(7), 416 (1970).
192. M. Seliga, P. Povinec, and M. Chudy, *Colin Czech. Chem. Commun.*, *35*(4), 1278 (1970).
193. B. D. Gunter and J. D. Gleason, *J. Chromatogr. Sci.*, *9*(3), 191 (1971).
194. O. Kunio and K. Tomihito, *J. Chromatogr.*, *55*(2), 319 (1971).
195. K. Kishimoto and K. Kinoshita, *Japan Analyst*, *20*(7), 866 (1972).
196. R. M. Dagnall, D. J. Johnson, and T. S. West, *Spectrosc. Lett.*, *6*(2), 87 (1973).
197. G. Castello and S. Munari, *Chimica Ind. (Milano)*, *51*(5), 469 (1969).
198. E. Bros, J. Lasa, and M. Kilarska, *Chem. Anal. (Warsaw)*, *19*(5), 1003 (1974).
199. Air Quality Criteria for Nitrogen Oxides, Section 4, p. 19, Environmental Protection Agency, APCO, Washington, D.C. (1971).
200. M. Goedert and G. Guiochon, *J. Chromatogr. Sci.*, *7*(6), 323 (1969).
201. R. C. Paule, *Anal. Chem.*, *44*(8), 1537 (1972).
202. C. W. Quinlan and J. R. Kittrell, *J. Chromatogr. Sci.*, *10*(11), 691 (1972).
203. L. Charpenet and R. Ponsonnet, *Analusis*, *2*(5), 383 (1973).
204. R. M. Campau and J. C. Neerman, Automotive Engineering Congress, Detroit, Mich., Jan., 1966, Society of Automotive Engineers, New York.
205. M. D. LaHue, H. D. Axelrod, and J. P. Lodge, *Anal. Chem.* *43*(8), 1113 (1971).
206. J. Hahn, *Anal. Chem.*, *44*(11), 1889 (1972).
207. M. D. LaHue, H. D. Axelrod, and J. P. Lodge, *J. Chromatogr. Sci.*, *11*(11), 585 (1973).
208. W. E. Wentworth and R. R. Freeman, *J. Chromatogr.*, *79*, 322 (1973).
209. V. W. Leithe and A. Hofer, *Allgem. Prakt. Chem.*, *19*, 78 (1968).
210. W. Leithe, *The Analysis of Air Pollutants*, p. 176, Ann Arbor Press, Ann Arbor, Mich. (1971).
211. O. Grubner and A. S. Goldin, *Anal. Chem.*, *45*(6), 944 (1973).
212. T. R. Carson, M. S. Rosenholtz, F. T. Wilinski, and N. H. Weeks, *Am. Ind. Hyg. Assoc. J.*, *23*, 457 (1962).
213. J. M. Trowell, *J. Chromatogr. Sci.*, *9*, 253 (1971).
214. A. Lawson and H. G. McAdie, *J. Chromatogr. Sci.*, *8*, 731 (1970).
215. J. T. Shaw, *J. Inst. Fuel*, *46*, 170 (1973).
216. A. T. Blades, *J. Chromatogr. Sci.*, *10*(11), 693 (1972).
217. J. I. Landau and E. E. Petersen, *J. Chromatogr. Sci.*, *12*(6), 362 (1974).
218. E. W. Sims, *J. Chromatogr. Sci.*, *12*(4), 172 (1974).
219. E. Moretti, G. Leofanti, D. Andreazza, and N. Giordano, *J. Chromatogr. Sci.*, *12*(2), 64 (1974).
220. F. W. Karasek and D. W. Denney, *J. Chromatogr.*, *93*(1), 141 (1974).

221. A. R. Mosier, C. E. Andre, and F. G. Viets, *Environ. Sci. Technol., 7*(7), 642 (1973).
222. D. Brocco, A. Cimmino, and M. Possasanzini, *J. Chromatogr., 84*(2), 371 (1973).
223. J. Neiser, V. Masek, and J. Pospisil, *Chemichy Prum., 22*(8), 408 (1972). *Anal. Abstr., 9,* 3312, 4887 (1972).
224. M. E. Dimitriev and N. A. Kitrosskii, *Gig. Sanit., 8,* 63 (1970).
225. M. Zamir-ul Haq, S. J. Rose, L. R. Deiderich, and A. P. Patel, *Anal. Chem., 46*(12), 1781 (1974).
226. N. T. Crosby, J. K. Foreman, J. F. Palframan, and R. Sawyer, *Nature (London), 238,* 342 (1972).
227. T. A. Bryce and G. M. Telling, *J. Agric. Food Chem., 20*(4), 910 (1972).
228. E. F. Darley, K. A. Kettner, and E. R. Stephens, *Anal. Chem., 35,* 589 (1963).
229. R. G. Smith, R. J. Bryan, M. Feldstein, B. Lavadie, F. A. Miller, and E. R. Stephens, *Health Lab. Sci., 8,* 48 (1971).
230. T. A. Bellar and R. W. Slater, 150th American Chemical Society Meeting, Atlantic City, N.J., Sept., 1965.
231. A. Ehmert, *J. Atmos. Terr. Phys., 2,* 189 (1952).
232. R. Erlich, *Frontier, 22,* 16 (1960).
233. R. Nagy, *Advan. Chem. Ser., 21,* 57 (1958).
234. J. A. Donohue and F. S. Jones, *Anal. Chem., 38,* 1858 (1966).
235. C. T. Pate, R. Atkinson, and J. N. Pitts, Jr., *J. Environ. Sci. Health., A11*(1), 1 (1976).
236. M. R. Kornreich and P. Issenberg, *J. Agric. Food Chem., 20*(6), 1109 (1972).
237. E. Selke, W. K. Rohwedder, and H. J. Dutton, *J. Am. Oil Chem. Soc., 52*(7), 232 (1975).
238. H. Halvarson, *J. Chromatogr., 76*(1), 125 (1973).
239. L. J. Papaand, L. P. Turner, *J. Chromatogr. Sci., 10*(12), 744 (1972).
240. D. C. Johnson and E. G. Hammond, *J. Am. Oil Chem. Soc., 48*(11), 653 (1971).
241. J. W. Vogh, *Anal. Chem., 43*(12), 1618 (1971).
242. R. J. Smythe and F. W. Karasek, *J. Chromatogr., 86*(1), 228 (1973).
243. Chemical and Engineering News, May 24, 1976, p. 4.
244. D. F. Adams and R. F. Koppe, *Tappi, 42,* 601 (1959).
245. R. Patrick, T. Schrodt, and R. Kermonde, *J. Chromatogr. Sci., 9,* 381 (1971).
246. N. N. Levchuk, S. L. Zal'tsman and N. G. Vilesov, *Khim. Promst. Ukr., 6,* 40 (1969), *Chem. Abstr., 72,* 50687m (1970).
247. M. Janghorban and H. Freund, *Anal. Chem., 45*(2), 325 (1973).
248. R. K. Stevens and A. E. O'Keefe, *Anal. Chem., 42*(2), 143A (1970).
249. C. D. Hollowell, G. Y. Gee, and R. D. McLaughlin, *Anal. Chem., 45*(1), 63A (1973).
250. P. Bourbon, R. Malbose, and M. J. Bel, *Trib. CEBEDEAU, 25,* 182 (1972).
251. R. L. Bond, W. J. Mullin, and F. J. Pinchin, *Chem. Ind.,* 1902 (1963).
252. L. G. Berezkina, S. V. Mel'nikova, and N. A. Eleterova, *Khim. Prom., 42,* 619 (1966). *Chem. Abstr., 65,* 16044 (1966).
253. R. Staszewski, J. Janak, and T. Pomopowski, *Chem. Anal. (Warsaw), 8,* 897 (1963).
254. H. L. Hall, *Anal. Chem., 34,* 61 (1962).
255. I. A. Osin, V. I. Zhukov, E. V. Utenkova, V. I. Vostrikov, and M. I. Bukowskii, *Zav. Lab., 40*(10), 1199 (1974).
256. W. G. Cook and R. A. Ross, *Anal. Chem., 44*(3), 641 (1972).
257. D. H. Bollman and D. M. Mortimore, *J. Chromatogr. Sci., 10*(8), 523 (1972).
258. D. G. Greer and T. J. Bydalek, *Environ. Sci. Technol., 7*(2), 153 (1973).
259. E. Zielinski, *Chem. Anal. (Warsaw), 14*(3), 521 (1969).
260. R. Manfred, O. Rene, and K. Hartwig, *Chem. Tech. Berl., 22*(8), 481 (1970).
261. R. K. Stevens, J. D. Mulik, and A. E. O'Keefe, *Anal. Chem., 43*(7), 837 (1971).
262. T. Okita, *Atmos. Environ., 4*(11), 93 (1974).

263. F. Bruner, A. Liberti, M. Possanzini, and I. Allegrini, *Anal. Chem.*, *44*(12), 2070 (1972).

264. R. E. Pecsar and C. H. Hartman, *J. Chromatogr. Sci.*, *11*(9), 492 (1973).

265. R. J. Robertus and M. J. Schaer, *Environ. Sci. Technol.*, *7*(9), 849 (1973).

266. A. A. Casselman, N. C. C. Gibson, and R. A. B. Bannard, *J. Chromatogr.*, *78*(2), 317 (1973).

267. K. Hartman, *Wass. Luft Ter.*, *15*(6), 211 (1971).

268. H. C. Dodd, *Anal. Chem.*, *43*(12), 1724 (1971).

269. M. D. Trotter, M. J. Sulway, and E. Trotter, *Clinica Chim. Acta*, *35*(1), 137 (1971).

270. K. Grieder and H. Buser, *Beitr. Tabakforsch.*, *6*(1), 36 (1971).

271. G. D. Divincenzo, F. J. Yanno, and B. D. Astil, *Am. Ind. Hyg. Assoc. J.*, *32*(6), 387 (1971).

272. J. Golacka and W. Bolanowska, *Chem. Anal. (Warsaw)*, *17*(5–6), 1375 (1972).

273. V. Vlasalova, J. Benes, and J. Parizek, *Radiochem. Radioanal. Lett.*, *10*(5), 251 (1972).

274. R. Teranishi, T. R. Mon, A. B. Robinson, P. Carey, and L. Pauling, *Anal. Chem.*, *44*(1), 18 (1972).

275. N. C. Jain and R. H. Cravey, *J. Chromatogr. Sci.*, *12*(5), 214 (1974).

276. J. P. Cano, J. Catalin, R. Badre, C. Dumas, A. Viala, and R. Guillerme, *Ann. Pharm. Fr.*, *28*(9–10), 581 (1970).

277. H. Bradenberger, *Pharm. Acta Helv.*, *45*(7), 394 (1970).

278. F. W. H. M. Merkus, M. G. J. Jasters-van Wouw, and J. F. C. Roovers-Bollen, *Pharm. Weekbl.*, *107*(7), 98 (1972).

279. N. C. Jain and R. H. Cravey, *J. Chromatogr. Sci.*, *10*(5), 257 (1972).

280. B. F. Gibbs, K. Itiaba, J. C. Crawhall, B. A. Cooper, and O. A. Mamer, *J. Chromatogr.*, *81*(1), 65 (1973).

281. A. Zlatkis, W. Bertsch, H. A. Lichtenstein, A. Tishbee, F. Shunbo, H. M. Liebich, A. M. Coscia, and N. Fleischer, *Anal. Chem.*, *45*(4), 763 (1973).

282. R. I. Ellen, R. L. Farrand, F. W. Oberst, C. L. Crouse, N. B. Billups, W. S. Koon, N. P. Musselman, and F. R. Sidell, *J. Chromatogr.*, *100*(1), 133, 152 (1974).

283. E. Wong, B. C. Johnson, and L. N. Nixon, *Chem. Ind. (London)*, *1*, 40 (1975).

284. M. Duran, D. Ketting, P. K. deBree, C. van der Heiden, and S. K. Wadman, *Clin. Chim Acta*, *45*(4), 341 (1973).

285. A. R. Blanchette and A. D. Cooper, *Anal. Chem.*, *48*(4), 729 (1976).

286. H. J. Davis, *Talanta*, *16*(5), 621 (1969).

287. A. Artho and R. Koch, *Mitt. Geb. Lebensmittelunters. Hyg.*, *60*(5), 379 (1969).

288. H. C. Pillsbury, C. C. Bright, K. J. O'Connor, and F. W. Irish, *J. Assoc. Off. Anal. Chem.*, *52*(3), 458 (1969).

289. Y. Masuda and D. Hoffman, *Anal. Chem.* *41*(4), 650 (1969).

290. D. Hoffman, G. Rathkamp, and S. Nesnow, *Anal. Chem.*, *41*(10), 1256 (1969).

291. K. Grob and J. A. Voellmin, *Beitr. Tabakforsch*, *5*(2), 52 (1969).

292. J. F. Graham, *Beitr. Tabakforsch*, *5*(5), 220 (1970).

293. C. L. Ogg and E. F. Schultz, *J. Assoc. Off. Anal. Chem.*, *53*(4), 659 (1970).

294. E. Gelpi and J. Oro, *J. Chromatogr. Sci.*, *8*(4), 210 (1970).

295. D. Hoffman and G. Rathkamp, *Anal. Chem.*, *42*(3), 366 (1970).

296. D. C. Watson, R. W. Hale and H. R. Randolph, *J. Chromatogr. Sci.*, *8*(3), 143 (1970).

297. G. Neurath, M. Duenger, and I. Kuestermann, *Beitr. Tabakforsch*, *6*(1), 12 (1971).

298. P. J. Groenen and L. J. Gemert, *J. Chromatogr.*, *57*(2), 239 (1971).

299. M. R. Guerin, *Anal. Lett.*, *4*(11), 751 (1971).

300. D. Hoffman and G. Rathkamp, *Anal. Chem.*, *44*(6), 899 (1972).

301. H. J. Klimisch and K. Meissner, *Beitr. Tabakforsch.*, *6*(5), 216 (1972).

302. G. P. Morie and C. H. Sloan, *Beitr. Tabakforsch.*, *6*(4), 178 (1972).

303. G. P. Morie, *Beitr. Tabakforsch.*, *6*(4), 173 (1972).
304. H. Binder, *J. Chromatogr.*, *82*(2), 420 (1973).
305. M. S. Baggett, G. P. Morie, M. W. Simmons, and J. S. Lewis, *J. Chromatogr.*, *97*(1), 79 (1974).
306. C. W. Ayers, *Talanta*, *16*(7), 1085 (1969).
307. C. Grunwald, *Anal. Biochem.*, *34*(1), 16 (1970).
308. I. Kushnir, P. A. Barr, and O. T. Chortyk, *Anal. Chem.*, *42*(13), 1619 (1970).
309. D. Hoffman and G. Rathkamp, *Anal. Chem.*, *42*(13), 1643 (1970).
310. K. D. Bartle and M. Novotny, *Beitr. Tabakforsch.*, *5*(5), 215 (1970).
311. E. Nesemann and F. Seehofer, *Beitr. Tabakforsch.*, *5*(6), 290 (1970).
312. N. Carugno, S. Rossi, and G. Lionetti, *Beitr. Tabakforsch.*, *6*(2), 79 (1971).
313. H. Hengy and J. Thirton, *Beitr. Tabakforsch.*, *6*(2), 57 (1971).
314. J. W. Rhodes and D. E. Johnson, *Nature (London)*, *236*, 307 (1972).
315. A. J. Kruszynski and A. Henriksen, *Beitr. Tabakforsch.*, *5*(1), 9, (1969).
316. Y. Y. Liu, I. Schmeltz, and D. Hoffman, *Anal. Chem.*, *46*(7), 885 (1974).
317. D. Hoffman, G. Rathkamp, and E. L. Wynder, *Beitr. Tabakforsch.*, *5*(3), 140 (1969).
318. K. D. Brunnemann and D. Hoffman, *J. Chromatogr. Sci.*, *12*(2), 70 (1974).
319. M. Novotny and M. I. Lee, *Experientia*, *29*(8), 1038 (1973).
320. Y. Kaburaki, Y. Mikami, Y. Okabayashi, and Y. Saida, *Japan Analyst*, *18*(9), 1100 (1969).
321. M. R. Guerin, G. Olerich, and A. D. Horton, *J. Chromatogr. Sci.*, *12*(7), 385 (1974).
322. K. Grob, *Chemy Ind.*, (6), 248 (1973).
323. G. Jakob, *G.I.T. Fachz. Lab.*, *16*, 1139, 1144 (1972).
324. L. Blomberg and G. Widmark, *J. Chromatogr.*, *106*(1), 59 (1975).
325. A. F. Haeberer and O. T. Chortyk, *J. Agric. Food Chem.*, *22*(6), 1135 (1974).
326. Consumers Report, *25*, 400 (1960).
327. Intersociety Committee, Methods of Air Sampling and Analysis, American Public Health Association, Washington, D.C., 1972.
328. A. P. Altshuller in *Advances in Chromatography*, Vol. 5, p. 229 (J. C. Giddings and R. A. Keller, eds.), Marcel Dekker, New York (1968).
329. B. Dimitriades, C. F. Ellis, and D. E. Seizinger, in: *Advances in Chromatography*, Vol. 8, p. 327 (J. C. Giddings and R. A. Keller, eds.) (1969).
330. ASTM Committee D-22, 1974, Annual Book of ASTM Standards, Part 26, American Society for Testing and Materials, Philadelphia, Pa. (1974).
331. T. R. Bellar and J. E. Sigsby, *Environ. Sci. Technol.*, *4*(2), 150 (1950).
332. W. B. Innes and A. J. Andreatch, *Environ. Sci. Technol.*, *4*(2), 143 (1970).
333. K. L. Malik and M. L. Khuraha, *Z. Anal. Chem.*, *253*(2), 125 (1971).
334. B. Dimitriades and D. E. Seizinger, *Environ. Sci. Technol.*, *5*(3), 223 (1971).
335. G. Grimmer, A. Hildebrandt, and H. Boehnke, *Erdoel Kohle Erdgas Petrochem. Brennst. Chem.*, *25*(9), 531 (1972).
336. G. Grimmer and H. Boehnke, *Z. Anal. Chem.*, *261*(4–5), 310 (1972).
337. B. Dimitriades, C. J. Rable, and C. A. Wilson, Report of the Investigators of the U.S. Bureau of Mines, RI7700, 21pp, (1972).
338. J. A. Beech, *J. Chromatogr.*, *72*(1), 176 (1972).
339. G. Grimmer, A. Hildebrandt, and H. Boehnke, *Erdoel Kohle Erdgas Petrochem. Brennst. Chem.*, *25*(8), 442 (1972).
340. P. Konopczynski and E. Prasek, *Chemia Analit. (Warsaw)*, *18*(2), 397 (1973).
341. H. Siegert, H. H. Oelert, and J. Zajontz, *Chromatographia*, *7*(10), 599 (1974).
342. M. Krejci and K. Tesarik, *J. Chromatogr.*, *91*, 525 (1974).
343. F. W. Karasek, R. J. Smythe, and R. J. Laub, *J. Chromatogr.*, *101*(1), 125 (1974).
344. J. B. W. Bailey, N. E. Brown, and C. V. Phillips, *Analyst (London)*, *96*, 447 (1971).

345. E. J. Sowinski and I. H. Suffet, *Anal. Chem.*, *46*(9), 1218 (1974).
346. B. R. Pusali and A. C. Dugar, *J. Phys. E*, *5*(9), 862 (1972).
347. P. J. Shiegoski, and D. L. Venezky, *J. Chromatogr. Sci.*, *12*(6), 359 (1974).
348. A. Lamb, K. Larson, and E. L. Tallefson, *J. Air Pollut. Contr. Assoc.*, *23*(3), 200 (1974).
349. R. L. Grob, in: *Chromatographic Analysis of the Environment*, p. 245 (R. L. Grob, ed.), Marcel Dekker, New York (1975).
350. D. E. H. Frear, *Pesticide Index*, 4th ed., College Science Publishers, State College, Pa. (1969).
351. *Dictionary of Pesticides 71, Farm Chemicals*, (H. Shepard, ed.), Meister Publishing Co., Willoughby, Ohio (1971).
352. B. Berck, *J. Chromatogr. Sci.*, *13*(6), 256 (1975).
353. W. L. Oller and M. F. Cranmer, *J. Chromatogr. Sci.*, *13*(6), 296 (1975).
354. L. Fishbein, *J. Chromatogr.*, *98*(1), 177 (1974).
355. J. Sherma, *C.R.C. Crit. Rev. Anal. Chem.*, *3*(3), 299 (1973).
356. K. A. Ramsteiner, B. A. Karlhuber, and W. D. Hormann, *Anal. Chem.*, *46*(11), 1621 (1974).
357. J. Becka and P. Ocelka, *Collect. Czech. Chem. Commun.*, *38*(8), 2242 (1973).
358. E. M. Beyer and P. M. Morgan, *Plant Physiol.*, *46*(2), 352 (1970).
359. K. Beyerman and W. Eckrich, *Z. Anal. Chem.*, *265*(1), 4 (1973).
360. C. W. Miller, M. T. Shafik, and F. J. Biros, *Bull. Environ. Contam. Toxic.*, *8*(6), 285 (1972).
361. J. W. Miles, L. E. Fetzer, and G. W. Pearce, *Environ. Sci. Technol.*, *4*(5), 420 (1970).
362. O. van Cleemput, *J. Chromatogr.*, *45*(2), 315 (1969).
363. K. Conrad, F. Conillard, M. Kraft, and J. G. Bartaire, *C.R. Hebd. Scanc. Acad. Sci., Paris, D.*, *272*(7), 928 (1971).
364. M. J. Kolbezen and F. J. Abu-El-Haj, *Pestic. Sci.*, *3*(1), 73 (1972).
365. K. A. Smith and R. J. Dowdell, *J. Chromatogr. Sci.*, *11*(12), 655 (1973).
366. C. W. Stanley, J. E. Barney, M. R. Helton, and A. R. Yobs, *Environ. Sci. Technol.*, *5*(5), 430 (1971).
367. J. D. Tessari and D. L. Spencer, *J. Assoc. Off. Anal. Chem.*, *54*(6), k376 (1971).
368. R. Maes, R. Drost, and H. Sauer, *Bull. Environ. Contam. Toxicol.*, *11*(2), 121 (1974).
369. W. Debska and S. Domeracki, *Herba Pol.*, *19*(1), 15 (1973); *Anal. Abstr.*, *26*(2), 1181 (1974).
370. H. Y. Young and A. Chu, *J. Agric. Food Chem.*, *21*(4), 711 (1973).
371. W. E. Dale, J. W. Miles, and D. B. Weathers, *J. Agric. Food. Chem.*, *21*(5), 858 (1973).
372. T. F. Bideman and C. E. Olney, *Nature*, *257*(5526), 475 (1975).
373. Air Pollution Technical Information Center (APTIC), Air Pollution Abstracts, U.S. Government Printing Office, Washington, D.C.
374. Air Quality Abstracts, Pollution Abstracts, Inc., LaJolla, Ca. (1974).
375. D. L. Brenchley, C. D. Turley, and R. F. Yarmac, Industrial Source Sampling, Ann Arbor Science Publishers, Ann Arbor, Md. (1973).
376. M. Sittig, *Pollution Detection and Monitoring Handbook*, Noyes Data Corp., Park Ridge, N.J. (1974).
377. T. Mercer, *Aerosol Technology in Hazard Evaluation*, Academic Press, New York, N.Y. (1973).
378. F. L. Cross, Jr., *Air Pollution Odor Control Primer*, Technomic Publishers, Westport, Conn. (1973).
379. J. A. Hodgeson, W. A. McClenny, and P. L. Hanst, *Science*, *182*, 248 (1973).
380. *Air Pollution and Atmospheric Diffusion*, (M. E. Beryand, ed.), Halsteas Press, New York (1973).

381. Air Sampling Instruments Committee ACGIH, *Air Sampling Instruments,* 4th ed., American Conference of Governmental Industrial Hygienists, Cincinnati, Ohio (1972).

382. Environmental Instrumentation Group, *Instrumentation for Environmental Monitoring Air,* 1st ed., Lawrence Berkeley Laboratory, University of California, Berkeley, Ca. (May, 1972); Updates: Feb. 1973, Dec. 1973.

383. G. C. Moss and K. R. Godfrey, *Meas. Control,* 5, 351 (1972).

4

Photodiode Arrays for Spectrochemical Measurements

Gary Horlick and Edward G. Codding

A rather large number of spectrochemical studies and analyses could be greatly facilitated by the simultaneous measurement of spectral information over a range of wavelengths or spatial positions. Aside from the obvious time and signal-to-noise-ratio advantages of simultaneous measurement, simultaneous multielement analyses based on atomic spectrochemical methods could more readily be implemented, complete spectra of transient events could easily be acquired, and time resolution of spectral events would be facilitated. Often these types of spectrochemical measurements are required by scientists in order to elucidate the complex problems and analyses encountered in industrial, clinical, environmental, and basic research fields. The study of many problems in these areas has been hampered by a lack of convenient, versatile, and inexpensive multichannel spectrochemical measurement systems. New detector subsystems based on modern electronic image sensors are now helping to overcome this obstacle. One particularly useful image sensor is the self-scanning silicon photodiode array. In this chapter the development and nature of these devices will be outlined, their operational characteristics for spectrochemical measurements presented, and their present range of spectrochemical applications reviewed.

Gary Horlick • Department of Chemistry, University of Alberta, Edmonton, Alberta, Canada T6G 2E1   **Edward G. Codding** • Department of Chemistry, University of Calgary, Calgary, Alberta, Canada T2N 1N4

1. Introduction: Electronic Image Sensors

To date, the dominant technique for the spatial dispersion of spectra is based on the diffraction grating. The typical dispersive system forms the spectrum as an optical spatial array in the focal plane of an instrument such as a monochromator or a spectrograph. The two main detection systems used to measure the spectral intensities in this image have been the photographic emulsion (plate) and the photomultiplier (PM) tube in conjunction with an exit slit. The photographic plate, although it is capable of integrating and recording thousands of lines in a single exposure as a permanent record, has a nonlinear response, limited dynamic range, and tedious readout. Thus, even though the PM-tube–exit-slit combination is limited to the measurement of one spectral-resolution element at a time, its wide linear dynamic range, sensitivity, and the fact that it transduces light intensity directly to an electronic signal have made it the detection system of choice for the majority of spectro-chemical measurements. However, a very large number of spectrochemi-cal measurements would be greatly facilitated if effective detection systems were available that combined the desirable characteristics of the PM tube with the one major advantage of the photographic plate, that of simulta-neous multichannel detection.

The classic direct reading spectrometer, in which an exit-slit–PM-tube is mounted at each point in the exit focal plane of a dispersive instrument at which measurements are to be made, is a reasonably powerful multi-channel measurement system. But even at best, this method allows only a very small fraction of the spectrochemical information available in a spectrum to be measured. This limitation is now being overcome by the utilization of electronic image sensors as spectrochemical detectors[1,2,3] which, over a moderate wavelength range, provide a continuous multi-channel measurement system.

The image orthicon, one of the first electronic image sensors, was applied at an early stage to spectrochemical measurements.[4,5] Newer versions of the orthicon[6] and more recently the silicon vidicon,[7-10] secondary electron conduction (SEC) image tubes,[11] and self-scanning linear silicon photodiode arrays[12] have all been applied to a variety of multichannel spectrochemical measurements.

Electronic image sensors[3,13] can be broken down into three basic elements: a radiation sensor that converts an optical image focused on the sensor to a charge pattern, a charge-storage element that stores the image pattern long enough so that it can be read out, and a readout system to convert the charge pattern to a video electrical signal. In the classic image orthicon these elements are distinct; the radiation sensor is a photocathode, the charge-storage element (target) is frequently a thin film of MgO, and readout is accomplished using a scanning electron beam with the signal measured in the return beam mode. The secondary electron conduction (SEC) image tube has a somewhat similar construction except that the charge-storage element is a multilayer target of Al_2O_3–Al–KCl, and the scanning electron beam readout operates in the so-called direct-beam mode. Vidicon is a generic name for electronic image tubes that use a single element which combines the radiation-sensor and charge-storage functions. Readout is still accomplished using a scanning electron beam in the direct-beam mode. These tubes include the Plumbicons and the silicon diode array camera tube or silicon vidicon.

The most recent developments in electronic image sensors are devices that combine all the basic functional elements (radiation sensor, charge storage, and readout) into a single integrated circuit. The two main devices of this type are self-scanning silicon photodiode arrays and charge-coupled devices (CCDs) both of which are available in linear and area array configurations.

It is interesting to briefly trace the development of the silicon vidicon and the self-scanning photodiode arrays. Both these devices use the same system for radiation sensor and charge storage, an array of reversed-biased PN junction photodiodes. The diodes operate in a so-called charge-storage mode, and it was the development of this mode of operation for the silicon photodiode that made the development of these image sensors possible.[14,15] The silicon vidicon uses conventional scanning electron beam readout methods,[13,15,16] while the self-scanning silicon photodiode array uses on-chip shift-register-controlled electronic scanning[13,17,18] for readout.

Silicon photodiode arrays have recently been reviewed by Fry.[19] He covers the nature of silicon as a photoconductor, charge-storage operation, scanning (readout) methods including the silicon target vidicon, the serially multiplexed array (i.e., the self-scanning arrays described in this chapter), capacitive access, the charge-injection device, the "bucket brigade" array and the charge-coupled device, array characteristics, and, finally, applications.

In the next two sections self-scanning photodiode arrays will be described, clocking and measurement circuits will be presented, and the operational characteristics of photodiode arrays with reference to spectrochemical measurements will be discussed and illustrated.

It will be seen that photodiode arrays have simple and inexpensive control and measurement circuitry and superior blooming and lag performance when compared to most other electronic image sensors. In addition, the signal-integrating capability of the arrays is a flexible and powerful asset for many spectrochemical measurements. These and other characteristics make photodiode arrays effective and versatile sensors for spectrochemical applications.[20]

2. Self-Scanning Silicon Photodiode Arrays

The self-scanning linear silicon photodiode arrays to be discussed in this section were obtained from Reticon Corporation, 910 Benicia Avenue, Sunnyvale, California 94086. We have used arrays of 256, 512, and 1024 elements. The detector elements are on 25.4-μm spacing (0.001 in.), which results in a density of 39.4 diodes/mm. Hence the lengths of the above arrays are 6.50 mm (0.256 in.), 13.00 mm (0.512 in.), and 26.01 mm (1.024 in.). The arrays can be obtained with detector-element heights of 25.4 μm (0.001 in.), 0.432 mm (0.017 in.), and 0.609 mm (0.024 in.). The higher arrays are significantly more sensitive ($\sim$20$\times$) and their slitlike shape provides them with excellent geometry for coupling to monochromators or spectrographs.

Reticon also has available an 1872-element linear array. The diodes are 16-μm high and have a center-to-center spacing of 15 μm (0.00059 in.). However, this device has four output video lines which can considerably complicate the readout and interface circuitry.

Photographs of a 0.001-in.-high 512-element array (Reticon No. RL 512) and a 0.017-in.-high 1024-element array (RL 1024 C/17) are shown in Figures 1 and 2. The thin line across the center of the 512-element sensor is the actual array and the broad band across the center of the 1024-element sensor is the 0.017-in.-high array. The 512 array is contained in an 18-pin integrated circuit package and the 1024 array in a 22-

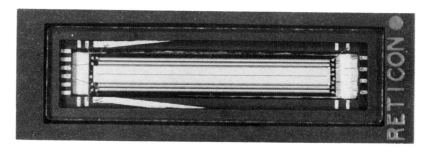

Figure 1. Photograph of a 512-element photodiode array.

Figure 2. Photograph of a 1024-element photodiode array.

pin package. The 1024 array is shown mounted on a driver/amplifier printed-circuit board (Reticon RC 408) which in turn is mounted in the focal plane of a monochromator. The standard window material for the arrays is quartz. At the time of writing, the arrays cost $3.00 per element in unit quantities ($3000 for the 1024-element array). There is no differential in cost for the different heights; a 0.001-in. array costs the same as a 0.017-in. array.

In addition to the photodiode array, the integrated circuit package also contains the necessary scanning circuitry for readout of the array. A *simplified* schematic of the complete integrated circuit is shown in Figure 3. Each photodiode is connected to the output line by a FET switch. The FET switches are controlled by a single bit that is shifted through the shift register. Readout is accomplished using two TTL-level signals, a start pulse, and a clock. The start pulse can be thought of as setting the first bit

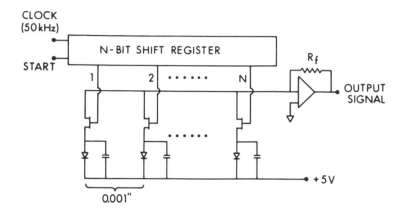

Figure 3. Simplified schematic of the photodiode array integrated circuit.

of the shift register and the clock then cycles the bit through the shift register, reading out the array. The specifications indicate that the array can be clocked anywhere from 10 kHz to 10 MHz, although our experience indicates that the array performs better at higher clock rates (we have used a clock rate of 45 kHz which is limited by our computer data-acquisition system).

Each photodiode in Figure 3 is shown with a parallel capacitor. This is because the photodiodes operate in the charge-storage mode and hence are inherently detectors of the integrating type. When a particular FET switch is closed by the bit in the shift register, the diode is charged up to its full reverse-bias potential, 5 V in this case. The circuitry is such that this is accomplished in about 0.5 μs. This reverse-bias charge stored on the equivalent capacitance of the PN junction can then be discharged between scans by two main mechanisms: by photon-generated charge carriers (light falling on the diode) and by thermally generated charge carriers (dark current). Thus the signal level necessary on the subsequent scan to re-establish reverse bias on the diode is a measure of the total of the light intensity and dark current integrated over the time between scans of the array.

The integration time is controlled by the time between start pulses. Once a scan is initiated by a start pulse, the complete array must be scanned. Another start pulse may be applied immediately after the array has been completely scanned or the start pulse can be delayed for up to several seconds (if the array is cooled) in order to integrate the spectral signal and hence increase sensitivity.

2.1. Control and Measurement Systems

A block diagram of the complete control and measurement circuits for the array is shown in Figure 4. The clock is constructed from a dual monostable integrated circuit (Fairchild 9602) and the $\div N$ counter from presettable 4-bit binary counters (Fairchild 9316). The driver utilizes the start and clock signals to generate four properly phased clocking signals and a start pulse that are necessary for the actual readout of the array. In the discussion of the simplified schematic diagram in Figure 3, the necessity of this driver was not mentioned as its role in operating the array is transparent to the user. The amplifier connected to the video output line of the array is a charge-sensitive amplifier based on an operational amplifier (OA). All these circuits shown inside the dashed line are available from Reticon on two printed-circuit boards (No. RC 400/RC 408 for the RL 1024C array) at a total cost of about \$250. The boards and array require power at +5 V and ±15 V, which was provided with a Heath EU-801-11 Digital Power Module.

The external measurement electronics are quite simple. An OA with unity gain is used for signal inversion and a PAR Model 225 amplifier is used to provide some gain and bandwidth control. Typically, gains of 5 to 10 are used on the PAR amplifier and the signal is dc-coupled with the upper 3-dB point set at 10 kHz.

The $\div N$ counter on the RC 400 circuit card has a maximum modulus of only 4096 (three Fairchild 9316 integrated circuits). In order to maintain a clock rate of 45 kHz and achieve integration times of several

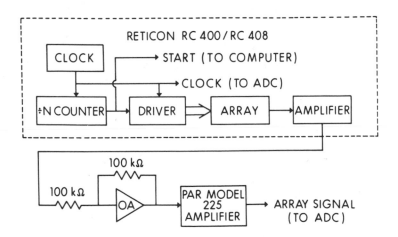

Figure 4. Block diagram of the control and measurement circuits for the array.

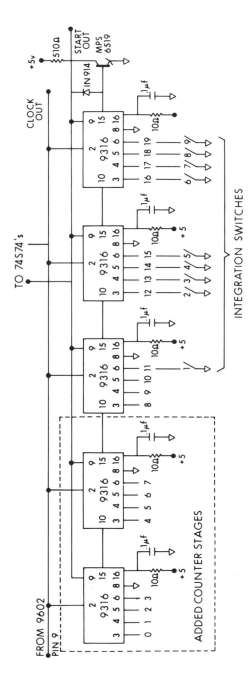

Figure 5. Circuit diagram of the modified ÷N counter.

seconds, two additional 9316s were added to the $\div N$ counter. This circuit modification is shown in Figure 5. The added circuitry is enclosed by the dashed line. As wired for our applications, the minimum integration time is set by closing switch 1, thus providing a modulus of 2^{11} or 2048. With a 45-kHz clock rate, this corresponds to an integration time of 45.5 ms. The minimum modulus that can be used with the 1024-element array is 1028, as the Reticon driver requires 4 extra clock pulses to set it up for a scan. With all switches closed, the modulus is 1,048,575 ($2^{20} - 1$) and the integration time is 23.296 s. With this circuit, a very wide range of accurate integration times can easily be set. At present, the integration time is set by hardware switches, but it could easily be put under software control.

All spectral measurements have been carried out with the array mounted in the exit focal plane of a Heath (GCA-McPherson EU-700) monochromator. This monochromator has a dispersion of about 20 Å/mm. An array density of 39.4 diodes/mm results in an Angstrom-per-diode value of 0.5. Thus the 1024-element array covers about 50 nm (500 Å) when coupled to this monochromator. A spectral peak should be sampled at a rate that provides about five samples across the full width at half height in order to avoid serious sampling errors.[21,22] Thus the spectral lines in the focal plane should be about 2.5 Å wide. With the Heath monochromator, this means that the width of the entrance slit should be set at 100 μm. A schematic representation of the photodiode array in the focal plane is shown in Figure 6.

Two photographs of the photodiode array control and measurement system as coupled to the monochromator are shown in Figures 7a and 7b. The normal exterior view is shown in Figure 7a and for Figure 7b, the

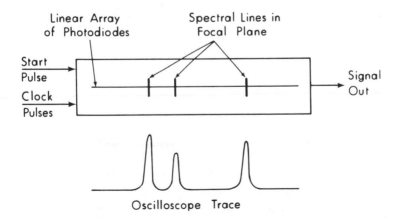

Figure 6. Schematic representation of photodiode array in monochromator focal plane.

Figure 7. Photograph of photodiode control and measurement electronics: (a) exterior view; (b) interior view.

minibox back has been removed and set on top in order that the interior circuitry and mounting system can be seen. The Reticon RC 400 printed-circuit board containing the clock, $\div N$ counter plus extra stages and the driver is mounted inside the minibox (see Figure 7b). The nine integration–time-selection switches and clock adjustment potentiometers are mounted on the outside of the minibox back plate (see Figure 7a). The RC 408 printed-circuit card including the 1024 array and the charge-sensitive output amplifier is mounted in the exit focal plane. The only necessary modification to the monochromator was removal of the exit folding mirror which is held with a large thumbscrew. The water-cooled copper plate which is attached to the hot side of the Peltier cooling module is visible in the lower part of Figure 7b. Note that Figure 2 is a photograph of the 1024 array as seen from the inside of the monochromator. During operation, the monochromator is purged with dry N_2 in order to keep the cooled array from frosting over. The cooling system is described in Section 2.3.

2.2. Readout Systems

A number of systems can be used to measure the output spectral signal from the array. One of the simplest is to photograph an oscilloscope trace of the output signal. Two traces of the complete signal from a 256-element array are shown in Figure 8. The trace in Figure 8A is the chromium triplet at approximately 3600 Å (3605, 3593, 3578 Å) and that

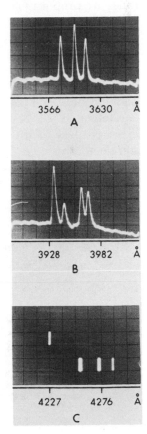

Figure 8. Oscilloscope traces of complete photodiode array signal: (A) chromium triplet, (B) calcium and aluminum doublets, (C) Z-axis modulation of chromium triplet (lower trace) and calcium (upper trace).

in Figure 8B is the calcium lines at 3933 and 3968 Å and the aluminum lines at 3944 and 3961 Å. In each case, the spectral range covered is approximately 110 Å and the signals were obtained using hollow cathode lamps.

The output signal can also be displayed on the oscilloscope in a format analogous to spectra on a photographic plate.[23] In this case, the output signal is applied to the Z-axis input of the oscilloscope, and a triangular waveform is applied to the vertical input so that the trace is intensified at the positions of the spectral lines and blanked out elsewhere. An example of this type of readout is shown in Figure 8C. The lower trace is of the chromium triplet at 4254, 4274, and 4289 Å while the upper trace is of calcium at 4227 Å.

An expanded oscilloscope trace of a single spectral line (Ne 5852 Å) is shown in Figure 9A. This expanded trace represents approximately 9 Å. It is important to remember that the actual form of the output signal from the photodiode array is a series of relatively sharp spikes whose amplitudes are proportional to the integral of the light intensity which has fallen on the individual photodiodes over the integration time, that is, over the time between successive start pulses. This is shown more clearly in Figure 9B where most of the low-pass filtering on the measurement amplifiers has been removed. The spikes from the individual photodiodes can now be clearly seen. The peak is sampled by approximately 11 photodiodes. In addition, the peak at half-height is about 5 photodiodes wide, which is equivalent to 2.5 Å (0.5 Å per diode). This is in agreement with the expected peak width at half-height for this monochromator with a 100-μm slit width.

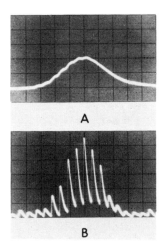

Figure 9. (A) Expanded oscilloscope trace of a single spectral line. (B) Expanded trace with minimal low-pass filtering. (Wavelength coverage is approximately 9 Å.)

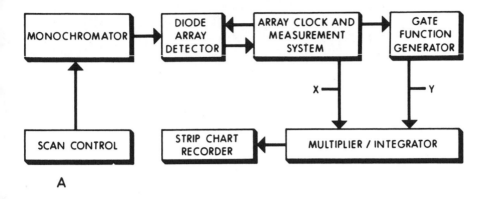

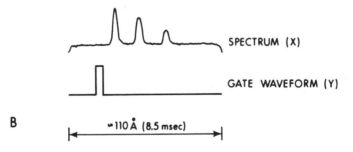

Figure 10. Block diagram of the cross-correlation readout system.

In addition to photographing the output video signal, a boxcar integrator, a single-point sample-and-hold ADC system[24] and a cross-correlation-based system[25] have been used for readout. In particular, cross-correlation-based systems show promise of providing unique and flexible readout.

2.2.1. Cross-Correlation Readout Systems

A block diagram of the cross-correlation readout system is shown in Figure 10. A 256-element array was used for this application and about 110 Å of spectral information could be simultaneously observed. A typical output signal for a triplet spectrum is shown schematically in Figure 10 as it would appear on an oscilloscope at point X. The array was clocked at a rate of 30 kHz so a complete spectrum (110 Å region) was read out in 8.5 ms. The repetition time (time between start pulses) was 60 ms (17 spectra/s).

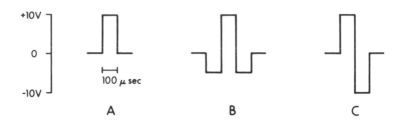

Figure 11. Gate pulse waveforms.

The gate function generator was triggered by the start pulse and various output waveforms could be generated in synchronism with the spectral trace from the photodiode array. These are shown in Figure 11. The simple pulse waveform is shown in Figure 10 as it would appear on an oscilloscope at point Y. Depending on the application, widths of the gate waveforms varied from 100 to 400 μs, which is equivalent to approximately 1.3 to 5 Å on the spectral scale.

To perform the cross-correlation, the spectral signal from the photodiode and a "gate" waveform were multiplied together ($\sim$17 times per s) using an analog multiplier as the monochromator was slowly scanned (0.2 Å/s). The products were integrated using the low-pass filtered output of the multiplier with a time constant of 3 s. This results in the spectrum being slowly scanned past the electronic gate and is exactly analogous to scanning a dispersed spectrum past a mechanical exit slit. Thus the pulse gate amounts to an "electronic exit slit." In contrast to the mechanical slit, the electronic slit can take on unique forms. Useful modifications of the spectral information can be carried out by cross-correlation with certain types of bipolar pulses. A cross-correlation that results in resolution enhancement is shown in Figure 12a and one that approximates a first derivative in Figure 12b.

2.2.2. Digital Computer Readout System

Most spectral measurements with photodiode arrays have been carried out with some form of digital measurement system often combined with a computer. These arrays provide an almost ideal sensor for the digital acquisition of spectra. The array itself, by its mere presence in the focal plane of the spectrometer, accomplishes the sampling operation of digitization. In addition precise repetitive scans of a spectral region can be generated electronically. These scans can easily be time-averaged to improve the signal-to-noise ratio or sequentially stored to provide time resolution, all under computer control.

A computer-coupled photodiode array measurement system is shown in Figure 13.[26,27] The two timing signals used to run the array are also used to sequence the data acquisition: the start pulse and the clock. The start pulse initiates the scan of the array and the start of a data-acquisition cycle by the computer. The clock controls the readout or scan rate of the array and also sequences the digitization and storage of the array output signal. The digitizing system consists of a sample-and-hold module with a 50-ns aperture time coupled to a 10-bit ADC with a conversion time of 10 μs.

The computer is a DEC PDP 8/e with a core of 16,000 and a DEC tape-based OS/8 operating system. The software was written primarily in FORTRAN. Machine language was used to input data from the ADC and output data to the DAC via a DEC DR8-EA 12-channel buffered digital I/O.

Several other workers have used digital and computer-based data-acquisition systems with Reticon photodiode arrays. Dessy[28] has developed a powerful high-speed digital data-acquisition system for rapid scan spectroscopy applications, Walker et al.[29] have interfaced an array-based multichannel spectrometer to a computer, Lovse and Malmstadt[30] have developed a digital shift-register memory (transient recorder)-based acquisition system, and Tull and Nather[31] developed a computerized photodiode array system for an astronomical spectrometer. These systems will be described in more detail in Section 4.

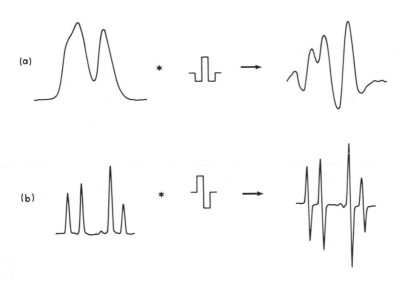

Figure 12. Resolution enhancement (a) and differentiation (b) by cross-correlation. The asterisk implies a correlation operation.

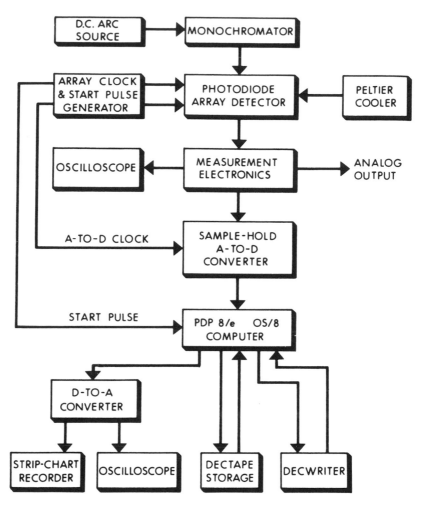

Figure 13. Block diagram of the computer-coupled measurement system for the photodiode array.

2.3. Cooling System

An important block in Figure 13 is the Peltier cooler. As will be seen in the next section, the dark current of the array can be significant at room temperature and, in particular, limit the effective use of the signal-integrating capability of the array. Cooling the array significantly reduces the dark current. A single-stage Peltier cooling module is used to cool a copper buss that contacts the back surface of the array IC package. The

Peltier cooling module is No. CP 1.4-71-06 (1.17 in. square, 0.17 in. high) available from Materials Electronic Products Corporation, 990 Spruce Street, Trenton, New Jersey 08638. The unit cost is about $30. The hot side of the cooling module must be cooled by a water-cooled copper plate in order to achieve effective operation of the cooling module. The cooling module requires ~4 amps at about 6.8 V and cools the array to about −15°C. Additional cooling power could be achieved by stacking cooling modules.

3. Operational Characteristics

3.1. Electronic Background

One of the major operational characteristics of the arrays is their electronic background. This background signal can be seen in the Zn dc arc emission spectrum shown in Figure 14a. This spectrum was measured using the 512-element array. The electronic background consists of two main components, dark-current and fixed-pattern noise. The dark current is the source of the overall height of the pedestal observed in Figure 14a and the noise across the array is due both to diode-to-diode dark-current variations and fixed-pattern noise. The major source of the fixed-pattern noise is feed-through from the clocking signals. This feed-through can result in a very regular fixed-pattern noise. The electronic background signal on the 256-element array is shown in Figure 15. The strong fixed-pattern sinusoid is exactly one-quarter of the clock frequency. Its magnitude can be minimized with an analog notch filter[24] or with a digital filter.[32] An analog notch filter has been used to process the signal from both the 256- and 512-element arrays. The electronic background signal from a 1024-element array is shown in Figure 16. Its basic nature is the same as that from the other arrays. There is a relatively regular oscillation which is attributable to clock feed-through, although it is much less than the other arrays and it has not been necessary to use an analog notch filter with this array. The major fluctuations are due to diode-to-diode dark-current variations.

The electronic background on any of the arrays is exactly repeatable from scan to scan. Thus it can be removed by subtraction utilizing the computer data-acquisition system. The Zn dc arc emission spectrum resulting from the signal shown in Figure 14a when the electronic background is subtracted is shown in Figure 14b. Even if the spectral signal is so small as to be essentially obscured by the electronic background, subtraction can still recover a useful spectrum, as shown in Figures 14c and 14d. In addition, background subtraction will also remove

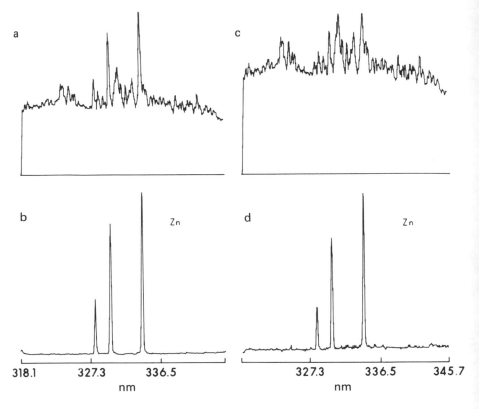

Figure 14. Spectra illustrating the power of electronic background subtraction. Integration time, 2 s; see text for discussion.

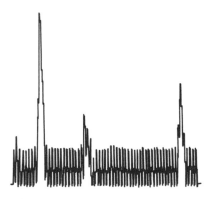

Figure 15. An example of sinusoidal fixed pattern noise on the 256-element array. Integration time, 0.5 s.

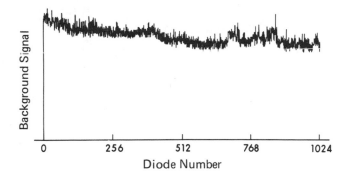

Figure 16. The electronic background signal on the 1024-element array. Integration time, 0.5 s.

the strong fixed-pattern noise signal shown in Figure 15 even when no analog notch filter is used to process the signal. The electronic background subtraction capability of the computer data-acquisition system is clearly a key factor in the measurement of spectra with these arrays. Except where noted, all spectra presented in this chapter have the electronic background subtracted out.

When the spectral signal becomes very small with respect to the electronic background, quantizing noise is often observable in the subtracted spectrum (see Figure 14d). We have found time-averaging to be an effective means of increasing the bit resolution of our data and, hence, of minimizing quantizing noise.[33,34]

The presence of the electronic background can severely limit the effective use of the integrating capability of the arrays. In an integrating detector type such as the arrays, the dark current, which is the major component of the electronic background, increases directly as the integration time and can, at integration times of even less than one second, completely saturate the array (i.e., completely discharge the reverse bias on the diode), leaving no dynamic range left for signal measurement. The background signal of the 1024 array as a function of integration time is shown in Figure 17a. The signal for only about the first 175 diodes is illustrated to more clearly show the nature of the background signal. At an integration time of 0.045 s the array background is close to zero but at an integration time of 0.72 s over one-half of the useful dynamic range is taken up, thus limiting the signal levels that can be measured. With an integration time of 5.76 s the array is strongly saturated by the dark current, making signal measurements impossible. Also note the nature of the background signal shown in Figure 17a. At short integration times

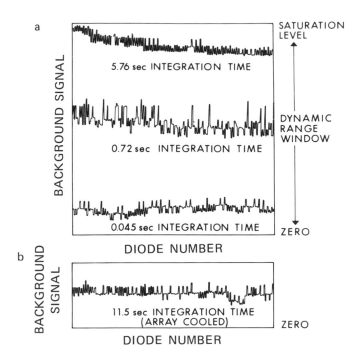

Figure 17. The effect of integration time (a) and cooling (b) on the background signal from the 1024-element array. See text for discussion.

essentially only the fixed-pattern noise is observed. At intermediate integration times the diode-to-diode dark-current variations are superimposed on the fixed-pattern noise, and at long integration times only the fixed-pattern noise is present on top of the saturated dark-current level.

The dark-current level of the array can be significantly reduced by cooling. As mentioned in Section 2.3, the array is cooled to about $-15°C$. At this temperature, the background signal at an integration time of 11.5 s (see Figure 17) is only slightly greater than that for the uncooled array at an integration time of 0.045 s. The effect of temperature on dark current is shown in Figure 18. The dark current decreases by a factor of two for every 11° decrease in temperature, which agrees well with the factor-of-two drop for every 10° decrease expected for silicon.

3.2. Integration Performance and Blooming

One of the key features of the array is the ease with which the integration time can be used to control sensitivity. If, as mentioned above, the array is not cooled, this capability is severely impaired. This is

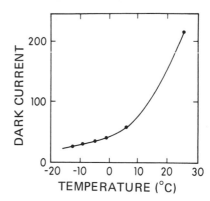

Figure 18. The effect of temperature on dark current.

illustrated in Figure 19. A small portion of the spectrum from a Ne-filled hollow cathode lamp around the Ne 588.2-nm line is shown. The spectra were measured with the 1024 array. The integration time for spectrum a was 0.36 s. As the integration time is increased to 0.73 s (spectrum b) and 1.46 s (spectrum c), the 588.2-nm line saturates, but the weaker lines on the baseline increase in intensity. However, any further attempt to increase the sensitivity by going to longer integration times with the uncooled array (2.9 s for spectrum d and 5.8 s for spectrum e) results in distortion and degradation of the spectral signal. This is simply because the spectral signal at longer integration times is sitting on a large dark-

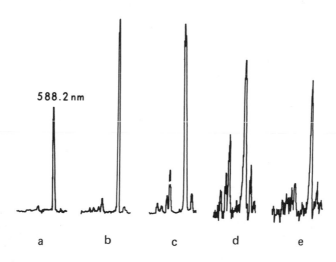

Figure 19. The effect of integration time on sensitivity (uncooled array). (a) 0.36 s, (b) 0.73 s, (c) 1.46 s, (d) 2.9 s, (e) 5.8 s.

current pedestal (which has been subtracted from these spectra) and little or no dynamic range is left to linearly integrate the signal photons.

If the array is cooled, the situation improves drastically. This is shown in Figure 20. The spectrum shown in Figure 20a is an attempt to use an integration time of 5.8 s for the measurement of a Ne spectrum from a hollow cathode lamp source with the uncooled 1024 array. In this case the array background is so large that even after background subtraction it is impossible to recover a useful spectrum. The identical measurement carried out with the cooled array is shown in Figure 20b. In this second case, the dark current is low enough that most of the dynamic range can

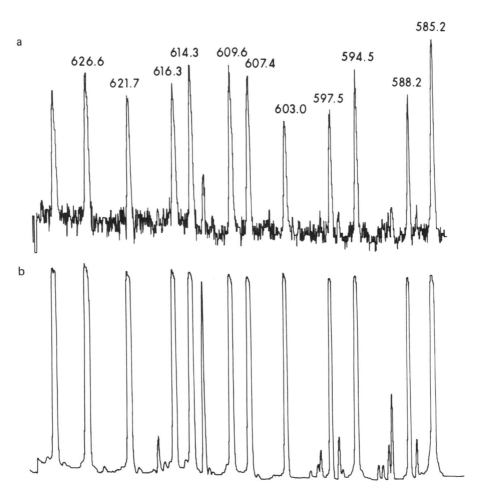

Figure 20. A comparison of the integration performance (5.8 s) for the uncooled (a) and the cooled (b) array.

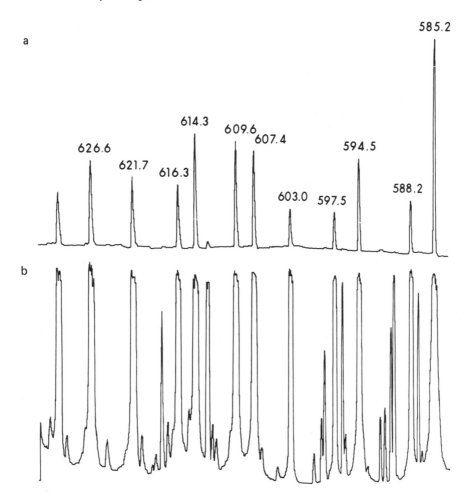

Figure 21. A comparison of the integration performance of the cooled array: (a) 0.18 s, (b) 23.3 s.

be used to integrate the spectral signal while in the first case, essentially no dynamic range is left for signal integration because of the high dark-current level. These two spectra have been computer scaled to the same maximum amplitude to facilitate comparison. The actual magnitude of the measured spectral signal (signal + array background − array background) shown in Figure 20a is much smaller than that of the spectrum shown in Figure 20b because the array background, as mentioned above, utilized most of the array's dynamic range.

An illustration of the range of control that one has over array sensitivity by utilizing the integration time is shown in Figure 21. The spectrum of a Ne-filled hollow cathode lamp measured with a 0.18-s integration time is shown in Figure 21a and the same spectrum measured with a 23.3-s integration time is shown in Figure 21b (which is 128 times the integration time of the first spectrum). These two spectra clearly illustrate the power of the integrating capability of the array.

The spectrum shown in Figure 21b also illustrates a very important characteristic of photodiode arrays. The arrays show essentially no tendency to bloom. Blooming, in electronic image sensors, refers to the situation in which a strong signal spreads to adjacent sensor elements[35,36]; in some cases, a strong signal can bloom out the entire sensor. Thus in spectrochemical applications, even minor blooming of the electronic image sensor can seriously degrade resolution and/or severely limit the use of the integrating capability of the sensor in measuring weak spectral lines in the presence of strong lines. Both silicon vidicons and the CCDs have problems with blooming. Workers using silicon vidicons have had to carefully avoid simultaneously measuring strong and weak signals or develop specific filters to attenuate the strong signals so that blooming would not obscure the desired signals.[37] However, in the spectrum shown in Figure 21b over 12 lines are strongly saturating the array, most being 100 to 200 times more intense than the baseline peaks. In addition, the small, partially resolved peak on the side of the 614.3-nm line would not be seen if blooming were occurring. The 614.3-nm line is about 300 times more intense than this peak. The capability of using the integration time of the array to incrementally build up sensitivity is further illustrated in Figure 22. Thus, with photodiode arrays the integration time can be used

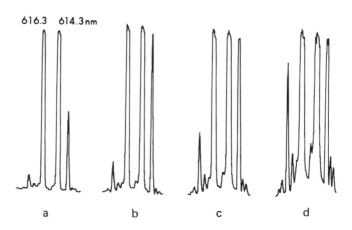

Figure 22. The effect of integration time on sensitivity (cooled array): (a) 2.9 s, (b) 5.8 s, (c) 11.6 s, (d) 23.3 s.

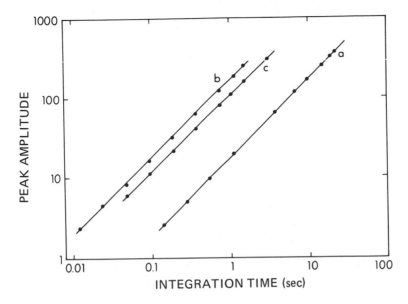

Figure 23. Log–log plot of spectral peak amplitude as a function of integration time: (a, b) 256 array, (c) 512 array.

to increase sensitivity for weak lines; intense lines will not interfere because of blooming even if they strongly saturate the array.

Finally, the increase in signal amplitude as a function of integration time must be linear in order to be fully useful. Log–log plots of spectral peak amplitudes as a function of integration time are shown in Figure 23 for the 256 array (a and b) and the 512 array (c). The two plots for the 256 array were measured in different regions of the array and for spectral lines of different intensity.

3.3. Lag

Another important characteristic of electronic image sensors is lag.[13] Lag refers to image carryover from one frame (integration time) to the next. In most electronic image sensors, the image cannot be completely erased in one readout cycle and thus a certain fraction is carried over as interference into succeeding frames. With the silicon vidicon for example, only 90% of the image may be erased on a readout cycle, leaving 10% to be carried over to the next frame. Thus, it can take up to 10 readout cycles to completely erase the image,[38] and even more are required when the sensor is cooled. In general, lag is undesirable and can be particularly bothersome if the sensor is being used for time-resolution studies.[39]

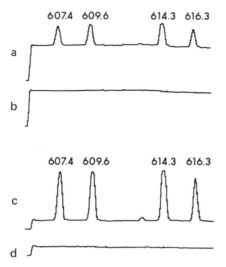

Figure 24. The lag performance of the 256-element array. See text for discussion.

Photodiode arrays, on the other hand, do not appear to exhibit any lag. This is illustrated in Figure 24. The spectrum shown in Figure 24 is again of the Ne-filled hollow cathode lamp, this time measured with the 256-element array. The array was not cooled and the integration time was 3.5 s. The dark-current pedestal is evident, as the electronic background was not subtracted for this measurement. The 614.3-nm line is saturating the array, thus all the peak signals represent relatively strong signals on the array. Just before the end of the integration time, a shutter was closed, thereby blocking any spectral signal from falling on the array during the subsequent integration time. The readout of the array signal from the subsequent integration time is shown in Figure 24b. No carryover of the spectral signal from the previous frame is evident. The array was then cooled to about −15°C and the same measurement repeated at an integration time of 18.1 s, five times longer than in the first case. Again, as shown by Figure 24d, there was no evidence of lag. Thus the lag performance of the photodiode array is an important asset for time-resolution studies.

3.4. Diode-to-Diode Sensitivity Variations

The Reticon RL 1024 C/17 (0.017-in.-high, 1024-array) element array has a specified typical nonuniformity of sensitivity of ±11%. The other arrays are less, the 512 array equal to ±7% and the 256 array equal to

±5%. This characteristic can be measured using a white-light source such as a tungsten bulb. The single-beam spectrum of a tungsten bulb source (GCA/McPherson (Heath) EU-701-56) centered at about 610 nm is shown in Figure 25a. Note that this spectrum has the electronic background subtracted out, thus the variations seen are actual sensitivity variations from diode to diode along the array. The shape of the spectrum is not an accurate measure of the overall sensitivity variation across the array, as this depends on a number of parameters such as the spectrum of the tungsten bulb, the monochromator throughput function, and the spectral response of the Si photodiodes, all of which are difficult to deconvolve.

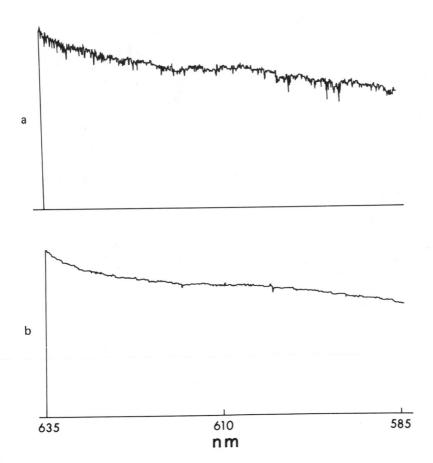

Figure 25. Spectra of a tungsten bulb with the array uncooled (a) and cooled (b) illustrating diode-to-diode sensitivity variations.

We have observed that the diode-to-diode sensitivity variation is significantly minimized when the array is cooled. The spectrum shown in Figure 25b is of exactly the same source but with the cooled array. This further illustrates the benefits of cooling the arrays. It is not known why this occurs but this observation has been verified several times.

In actual practice, this characteristic of the array has not been a problem in any of our analytical measurements. Emission measurements (i.e., flame emission) for the purpose of establishing an analytical curve are not affected by such sensitivity variations and atomic and molecular absorption measurements involve the calculation of a ratio which cancels out sensitivity variations (i.e., double beam).

3.5. Dynamic Range

The dynamic range of the array at a fixed integration time is an important characteristic. The dynamic range is specified by Reticon to be between 2 and 3 orders of magnitude. Log–log plots of peak amplitude vs. %T for the 256- and 512-element arrays are shown in Figure 26. The %T axis was established using neutral density filters and is not highly precise,

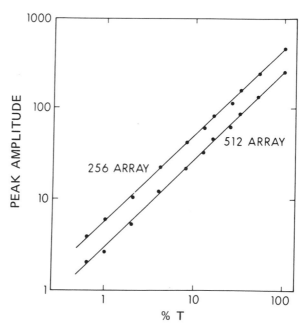

Figure 26. Log–log plots of peak amplitude as a function of %T for the 512 and 256 arrays.

particularly at the low $\%T$ end where two or more filters had to be stacked. However, a dynamic range of somewhat over 2 orders of magnitude is observed. Reticon markets a charge amplifier sample-and-hold circuit (CASH 1B) which they claim provides a dynamic range of 1000:1 when coupled to the array.

4. Measurements and Applications

A large variety of spectrochemical measurements have been carried out with self-scanning photodiode array detectors. The main feature of the arrays for spectrochemical measurements, as with other electronic image sensors, is the simultaneous measurement of spectral information with the transduction of the spectral signal directly to an electronic signal. This capability and other features of the array greatly facilitate simultaneous multielement analysis, the measurement of time-resolved spectra, and the measurement of spectra from transient events. In this section a number of spectrochemical measurements and applications will be described. Fry[19] has recently reviewed several nonspectrochemical applications of arrays, among which are industrial dimension gauging and flaw detection, optical character recognition, facsimile, aerospace guidance systems, and imaging for television and surveillance.

4.1. Simultaneous Multielement Analysis

The development of effective simultaneous multielement analyses based on atomic spectrochemical methods is one of the major goals of a number of academic, government, and industrial laboratories.[40] A key aspect of this development is the design of spectrochemical measurement systems capable of simultaneously measuring spectral information over a wide range of wavelengths. Photodiode arrays are applicable to a number of such measurements.

4.1.1. Flame Emission Spectroscopy

A simple example of the multichannel measurement capability of the arrays is shown in Figure 27 for the flame emission of K and Rb. A mutual ionization interference between the alkali metals is well known. With the array the enhancement effect of Rb on both K lines can be monitored while at the same time measuring the Rb emission intensity.

In general, the array has not proved to be a very sensitive detector for flame emission measurements. The above measurement was carried out with the 0.001-in.-high 512-element array. The new 0.017-in.-high 1024

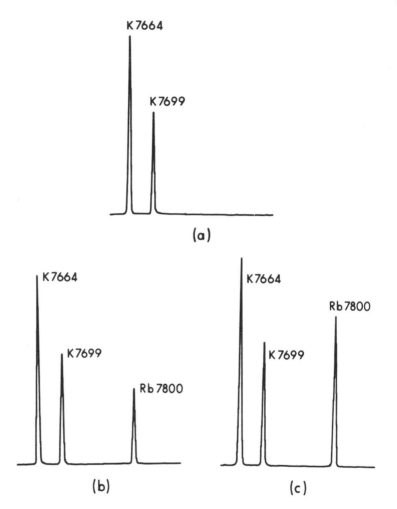

Figure 27. Flame emission spectra of K and Rb. All solutions contain 100 ppm K, (b) contains an added 100 ppm Rb, and (c) contains 200 ppm of added Rb. See text for discussion.

array with its increased sensitivity appears, from our preliminary measurements, to be an excellent detector for the alkali metals (Na, Li, K, Rb, and Cs) and should find some applicability to elements like Ca, Ba, Sr, Ga, In, and Cr. So far, we have only used an air–C_2H_2 flame—use of a N_2O–C_2H_2 flame should extend the flame emission capability.

4.1.2. Atomic Absorption Spectroscopy

To date, atomic absorption analysis has been essentially a single-element technique. The primary reason for this is the single-channel nature of the typical measurement system, which consists of a monochromator coupled to a photomultiplier tube detector. There are a number of ways in which the scope of the atomic absorption technique could be

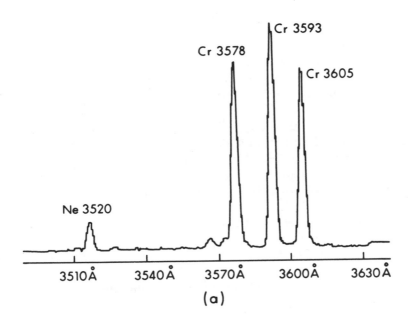

(a)

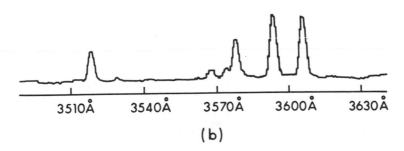

(b)

Figure 28. Reference (a) and sample (b) spectra for the atomic absorption of chromium (100 ppm).

broadened by utilization of a multichannel measurement system. The first and perhaps the most important way is that it would be possible to develop simultaneous multielement analytical procedures for both quantitative and qualitative analysis. Both these capabilities would add new dimensions to the atomic absorption technique. In addition, many types of measurements would be greatly facilitated; for example, the study of complex interferences and interelement effects, the measurement of relative line sensitivities, and the implementation of the two-line method for the correction of background absorbance, to name but a few.

A number of multielement and multiline atomic absorption measurements can be made with the computer-coupled photodiode array spectrometer described earlier in this chapter.[12] The spectrum of a chromium hollow cathode lamp in the region near 3600 Å as measured with the computer-coupled photodiode array spectrometer is shown in Figure 28a. The integration time was 3 s, and four scans were time-averaged, giving a total measurement time of 12 s. The transmission spectrum when 100-ppm Cr was aspirated into the flame is shown in Figure 28b. In terms of an absorbance measurement, the spectrum shown in Figure 28a can be called the reference or source spectrum ($100\%\ T$) and that in Figure 28b the sample transmission spectrum. Both spectra are stored in the computer at the same time, and thus absorbance values can easily be calculated across the complete spectral region. The result is an atomic absorption spectrum for chromium, and it is shown in Figure 29a.

Perhaps the first feature noted in this spectrum is the rather noisy baseline. The reason for this is quite simple. Meaningful absorbance measurements can only be made at wavelengths where the source has a finite intensity, which with a hollow cathode lamp occurs only at specific line locations. Thus the baseline fluctuations are simply a result of calculating the log of the ratio of two small numbers. The calculation of these meaningless absorbance values along the baseline can be eliminated by using a threshold level which the source intensity must exceed before an absorbance value is calculated. Using a threshold value which is 20% of the maximum source intensity, the atomic absorption spectrum shown in Figure 29b is calculated.

Only multiline atomic absorption measurements are illustrated in Figures 28 and 29. In order to do simultaneous multielement analysis, a multielement source is also required in addition to the multichannel measurement system. Possible multielement sources include multielement hollow cathode lamps, optical multiplexing (with mirrors and beamsplitters) of single-element hollow cathode lamps and continuous sources. A multielement hollow cathode lamp (Jarrell-Ash JA45-599) was used for the present work. This lamp contains six elements: Co, Cr, Cu, Fe, Ni, and Mn.

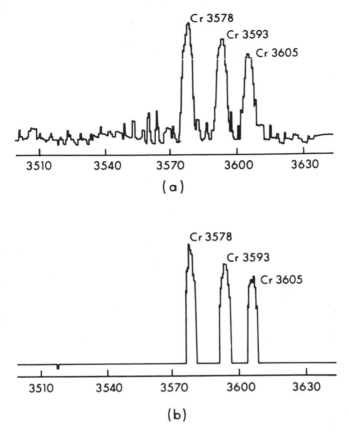

Figure 29. Atomic absorption spectra for chromium. See text for discussion.

Obviously, compromises are necessary in implementing simultaneous multielement atomic absorption analysis with a multichannel detection system of limited wavelength coverage. Perhaps the most serious compromise must be made in the selection of analysis lines. With the 256-element array only about 130 Å of continuous spectral information could be simultaneously observed. Two potential 130-Å regions for atomic absorption measurements with this particular hollow cathode lamp are listed in Table 1 along with possible analysis lines. Reference, sample, and absorbance spectra are shown in Figure 30 for region A. The sample solution was 100 ppm in both Cr and Ni. Reference, sample, and absorbance spectra for region B (Table 1) are shown in Figure 31. Utilizing this region, Fe and Cr can be simultaneously determined.[12]

Table 1. Analysis Regions Used with the JA-45-599 Hollow
Cathode Lamp

Region	Wavelength range (Å)	Possible analysis lines (Å)	
A	3470–3600	Ni 3515	Cr 3578
		Ni 3524	Cr 3598
B	3592–3722	Cr 3593	Fe 3720
		Cr 3605	

The above atomic absorption measurements were carried out using a 256-element array. With the 1024-element array, the measurement capability reported above can be significantly extended. The spectra of two multielement hollow cathode lamps are shown in Figure 32 to illustrate this potential. The spectra shown in Figures 32a and 32b were measured from a Co, Ni, Fe, Cr, Mn, and Cu hollow cathode lamp. In the region illustrated in Figure 32a, Fe, Cr, Ni, and Cu should be able to be determined simultaneously and Mn, Fe, Co, and Ni could be determined in the region illustrated in Figure 32b. Significantly, the UV response of the detector seems sufficient to allow measurements at the Ni 232.0-nm line and thus should also be sufficient for Cd and Zn. An Al–Ca–Mg multielement hollow cathode lamp was the source for the spectrum shown in Figure 32c, which indicates potential simultaneous determination of Mg and Al with this system. The array integration time for the spectrum shown in Figure 32a was 1.46 s; in Figure 32b, 11.6 s; and Figure 32c, 0.73 s. In each case, 10 scans were time-averaged and background-subtracted.

These hollow cathode lamp spectra simply illustrate the potential of the indicated simultaneous multielement determinations. Considerable work remains to be done in studying methods to control mutual interferences as well as determining the best compromise of experimental conditions which is always necessary for simultaneous determinations.

4.1.3. DC Arc Emission Spectroscopy

Simultaneous measurement of a number of analytical curves for elements in a synthetic brass sample has been carried out using a 512-element array.[26] It was necessary to measure two spectral regions in order to obtain useful lines for all elements. The spectrum of a sample containing 2.1% Zn, 1.7% Pb, 2.0% Sn, 0.5% Ni, 0.55% Fe, 1.18% Al, and 92% Cu is shown in Figure 33a for the 2789-to-3070-Å region and in Figure 33b for the 3073-to-3352-Å region. Fifteen scans were time-

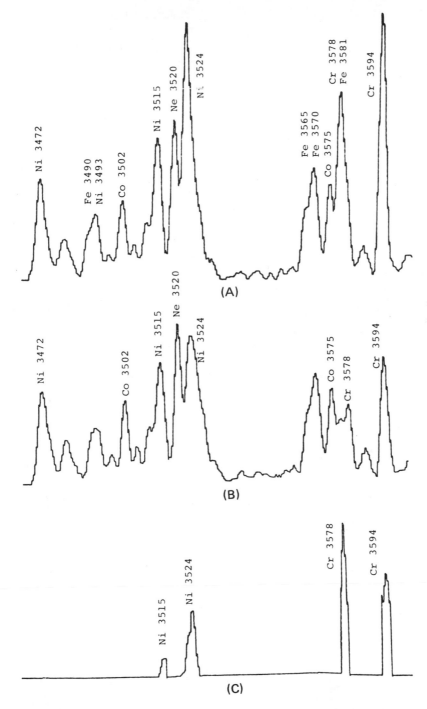

Figure 30. Reference (A), sample (B), and absorbance (C) spectra for Ni (100 ppm) and Cr (100 ppm).

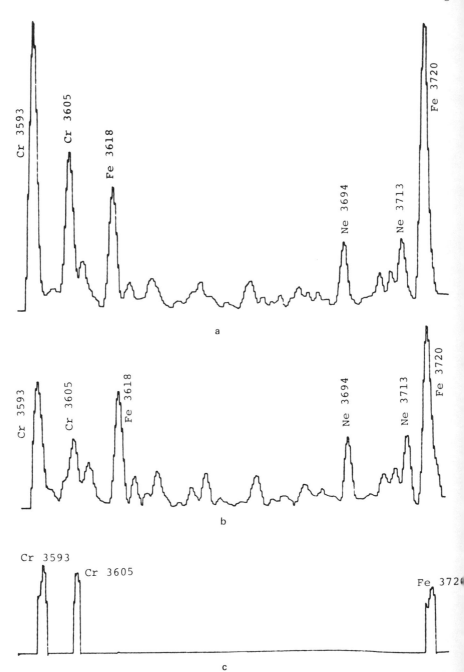

Figure 31. Reference (a), sample (b), and absorbance (c), spectra for Cr (100 ppm) and Fe (100 ppm).

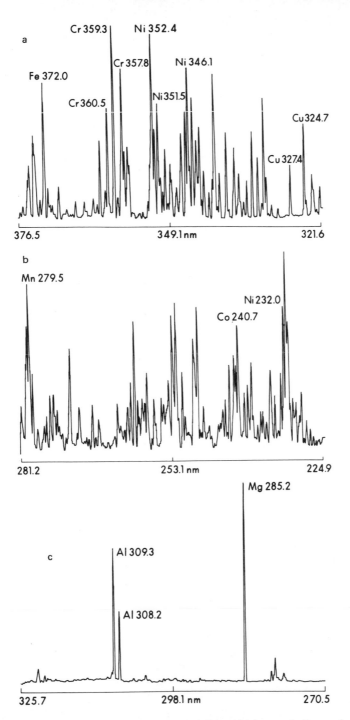

Figure 32. Spectra of a Co, Ni, Fe, Cr, Mn, and Cu multielement hollow cathode lamp (a, b) and a spectrum of a Ca, Al, and Mg multielement hollow cathode lamp (c), taken using 1024-element array.

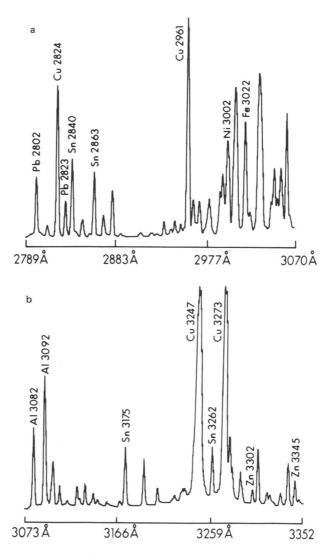

Figure 33. Spectra of brass sample.

averaged, resulting in a total measurement time of 30 s. The lower spectral region (Figure 33a) was used for the simultaneous measurement of Pb, Sn, Ni, and Fe, and the upper spectral region (Figure 33b) for Al, Sn, and Zn. Analytical curves were established without using an internal standard. Two typical analytical curves are shown in Figure 34 for Ni and Pb. The slopes of the log–log plots as determined by linear least-square fits

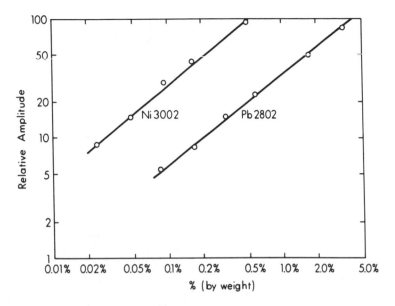

Figure 34. Analytical curves for Ni 3002 and Pb 2802.

for all the analytical lines chosen are listed in Table 2. In most cases the slopes are not unity, which indicates the presence of self-absorption, not unexpected at the concentration ranges of these elements.

DC Arc Time Studies. Time studies are an integral step in the development and implementation of dc arc spectrochemical analyses. In particular, selective volatilization imparts a complex, highly matrix-de-

Table 2. Analytical Curve Data for Brass Samples

Elemental line		Concentration range (%)	Log–log slope
Pb	2802	.09–3.5	0.747 ± 0.015
	2833		0.516 ± 0.010
Sn	2840	.1–2.0	0.700 ± 0.023
	2863		0.802 ± 0.039
Ni	3002	.02–.50	0.793 ± 0.035
Fe	3022	.05–.18	1.029 ± 0.022
Al	3082	.12–2.4	0.652 ± 0.043
	3092		0.580 ± 0.031
Sn	3174	.10–.68	0.869 ± 0.048
	3262	.20–4.0	0.955 ± 0.028
Zn	3303	.1–2.1	0.946 ± 0.042
	3345	.2–4.2	1.081 ± 0.037

pendent time behavior to the emission intensity of spectral lines. Without a knowledge of this time behavior, it is nearly impossible to set up meaningful exposure times or to determine internal standard line compatibility. In addition, fundamental studies of arc characteristics and the development of carrier distillation procedures require the measurement and utilization of time information.

The basic approach to the measurement of time information during a dc arc burn has changed little over the years. The intensity–time data are normally measured photographically by moving the plate either continuously or at regular intervals during the arc burn. This has given rise to a number of terms which are associated with the measurement technique, such as moving plate, racking plate, or jumping plate studies. A rather large amount of important data can be acquired about a sample with this type of measurement. However, the complete workup and utilization of the data, particularly in a quantitative manner, can be very tedious and time-consuming because of the use of the photographic plate as the detector.

The computer-coupled photodiode array spectrometer is almost ideally suited for dc arc time studies. Repetitive measurements of a particular spectral region can easily be generated completely electronically. Sequential spectra, integrated over precisely known time intervals, can be automatically measured and stored in a small computer. Relatively simple software can be used to extract the intensity–time data for several lines from the sequential spectra and automatically plot the results.

A series of time-study spectra measured with the photodiode array spectrometer are shown in Figure 35.[27] The sample was 0.02% Cu and

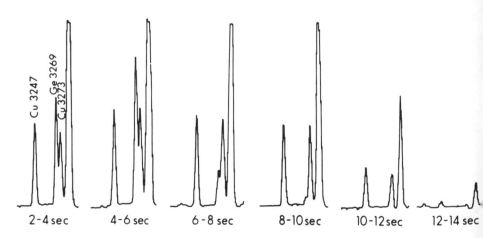

Figure 35. Time-interval spectra for Cu and Ge in a ZnO–graphite matrix.

1% Ge in a 1:1 matrix of ZnO and graphite. The integration time for each spectrum was 2 s and a total of seven spectra were measured, of which six are illustrated. Although a 512-element photodiode covering 260 Å was used, only that part of the spectrum in the vicinity of the copper doublet is shown in Figure 35. The computer-generated intensity-vs.-time plots for Cu 3247 (#) and Ge 3269 (*) are shown in Figure 36. The time for one complete run of the program from arcing to the finished plot was about 2 to 3 min.

4.2. Transient Spectral Events

4.2.1. Dye Laser Spectra

It can be particularly difficult to measure the complete spectrum of a transient spectral event. The photodiode array spectrometer has greatly facilitated the acquisition of the necessary spectral data in a quantitative study of dye laser intracavity enhanced absorption using rare earth absorbers.[41] In this study it was necessary to measure the complete spectrum of a dye laser which extended over about 100 Å. The dye laser was pumped using a frequency-doubled, Q-switched ruby laser at a repetition of about once every 4 s. This low repetition rate and pulse-to-pulse instability made it impossible to obtain reliable data photoelectrically with a PM tube and a scanning spectrometer. The tedium and difficulty of photographic intensity measurements precluded the use of a spectrograph for this application. With the photodiode array spectrometer, the complete spectrum of a single 5-ns dye laser pulse could easily and quantitatively be measured. In addition, using a 16-s integration time on the array up to four spectra of the dye laser output could be summed and averaged directly on the array.

Typical spectra of the dye laser output when a rare earth solution is inserted in the cavity are shown in Figure 37. These spectra are photographs of the photodiode array oscilloscope output. The spectrum of the rhodamine 6G dye laser is shown in Figure 37A. No intracavity absorber was present. Spectra of the same dye laser when 0.096-M and 0.19-M solutions of $Eu(NO_3)_3$ are inserted in the cavity are shown in Figures 37B and 37C. The very weak 5790-Å absorption band is readily observed. The actual absorbance of the 0.096-M solution at 5790 Å is about 0.006. This weak absorbance has been considerably enhanced by insertion of the solution inside the dye laser cavity (Figure 37B). The enhancement in this case is approximately a factor of 35.

Spectra of the 4-methylumbelliferone dye laser when 0.0012-M and 0.0037-M solutions of $Pr(NO_3)_3$ are inserted in the laser cavity are shown

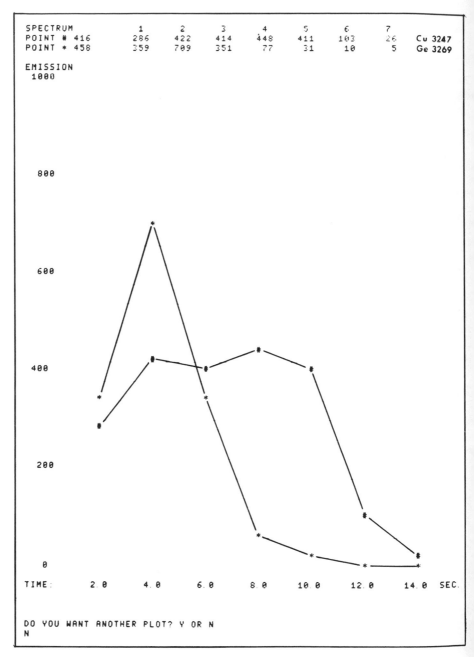

Figure 36. Intensity–time plots for Cu 3247 (#) and Ge 3269 (*) (ZnO–graphite matrix).

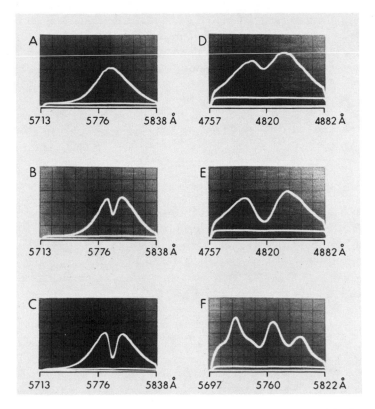

Figure 37. Spectra of a dye laser when a rare earth solution is inserted in the cavity

in Figures 37D and 37E. The 4825-Å absorption band of praseodymium is observed in the center of the dye laser spectrum.

The spectrum of the rhodamine 6G dye laser when a 0.008-M solution of $NdCl_3$ is placed inside the cavity is shown in Figure 37F. The 5755-Å band of neodymium is a multiplet and this can be seen in the complex spectrum that results.

From spectra such as those shown in Figure 37, the apparent absorbance of the rare earth solution inside the cavity could easily be measured. A linear relationship between apparent absorbance and concentration was observed.[41]

4.2.2. Laser Plume Emission Spectra

Q-switched lasers such as the ruby laser have been used as excitation sources to excite atomic emission from a variety of materials. However, the transient nature of the laser plume generated when the Q-switched laser is focused on the sample makes conventional photoelectric measurement of the spectrum of the plume very difficult, and photographic detection is typically used. Again the photodiode array spectrometer is well suited to this type of measurement. The results of a feasibility study using the photodiode array for this measurement are shown in Figure 38. Spectra are shown for the laser plume emission from an Al sample (Al 3961 and 3944 Å), a glass sample (Na 5896 and 5890 Å), and a Cr sample (Cr 4289, 4274, and 4254 Å).

4.3. UV–Vis Molecular Absorption Spectra

The UV–vis molecular absorbance spectrum of benzene as measured with the 1024 array is shown in Figure 39. This measurement was carried out with the standard GCA/McPherson (Heath) UV–vis light source (EU-701-50) and single-beam sample-cell holder (EU-701-11) coupled to the monochromator. The D_2 lamp was the source, 1-cm quartz cells were used, and the spectral bandpass was 1.25 nm. The spectrum shown in

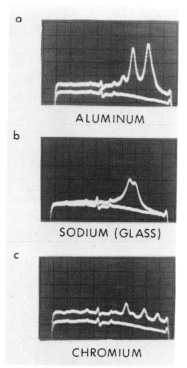

Figure 38. Emission spectra from a laser plume.

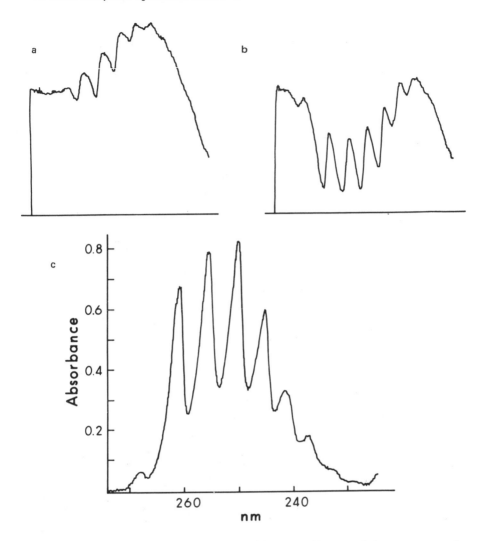

Figure 39. (a) Transmission spectrum of cyclohexane; (b) transmission spectrum of 0.05% benzene in cyclohexane; (c) absorbance spectrum of benzene.

Figure 39a is the transmission spectrum of the solvent, cyclohexane, over a bandwidth of 50 nm centered at about 250 nm. The transmission spectrum of 0.05% benzene in cyclohexane is shown in Figure 39b. These two 1024-point spectra can both be stored in the computer data-acquisition system at the same time. Thus the true double-beam absorbance spectrum of benzene ($\log I_0/I$) over this region can readily be calculated and is shown in Figure 39c. This measurement also indicates that the array has useful sensitivity well into the UV. The array integration time for this measurement was 11.6 s and 10 scans each of the cyclohexane and benzene in cyclohexane spectra were time-averaged.

4.4. Rapid-Scan Spectroscopy

Photodiode arrays are excellent sensors for rapid-scan or time-resolved spectrochemical measurements. Dessy[28] has described a UV-visible absorption spectrometer based on a photodiode array for stopped-flow reaction-rate measurements. Particular emphasis was placed on development of a powerful hardware and software system capable of acquiring spectra at a rate of up to 2000 spectra/s.

A Reticon RL-256/17 photodiode array was used as the detector. Many of the Reticon arrays can be configured so that two adjacent diodes are read out simultaneously. The above array was run in this mode resulting, in effect, in a 128-element array with sensors on 2-mil spacing and twice the sensitivity. The array was clocked at 250 kHz.

The array output signal was clocked directly into a Biomation transient recorder (6 bit × 128 word memory). Most of the above article[28] is concerned with the interface of the Biomation to a PDP 8/L computer and its associated 32,000 fixed-head disk with the goal to maximize data-transfer rates. Two 6-bit words out of the Biomation were first packed into one 12-bit word. The 12-bit words were then transferred to the computer core using direct memory access via a rubber-band memory. The rubber-band memory could hold anywhere from 0 to 16 12-bit words and facilitated timing the data transfers between the Biomation and core (the computer). Data could be transferred from the Biomation to the rubber-band memory at 250,000 words/s (12-bit words and hence 2 data points). The transfer rate from the rubber-band memory to core was 600,000 words/s and from core to the disk was 33,000 words/s. The rate of 2000 spectra/s could be achieved, but because of disk storage limitations only 500 spectra could be acquired at one time (64 words per spectrum, which was equivalent to 128 data points). This somewhat complex and sophisticated system developed by Dessy emphasizes the fact that even though the sensor is capable of rapid spectral measurements it is not a trivial task to design and develop a digital data-acquisition system capable of acquiring and storing these data at high rates.

4.5. Field-Use Multichannel Spectrometer

A compact multichannel spectrometer for field use in the spectral monitoring of minerals, vegetation, water, and air has been developed based on a 256-element photodiode array.[29] The spectrometer resolution, as set by the slit width, was 10 nm and the total spectral range covered was 650 nm (325 to 975 nm). The array was clocked at 2 kHz and integration times (time between start pulses) could be varied from 125 ms to 4 s. In order to achieve the longer integration times, the array was thermoelectrically cooled about 20°C below ambient temperature.

The video output from the array was amplified and digitized using a sample-and-hold amplifier in conjunction with a 12-bit ADC. The digitizing system was interfaced to an Interdata model 4 computer via a DMA input. Data could be displayed on an oscilloscope and recorded permanently on a 9-track magnetic tape. The computer software provided for the subtraction of a dark record before the data were displayed on the oscilloscope.

4.6. Astronomical Spectrometer

Tull and Nather[31] have utilized a self-scanned photodiode array in an astronomical spectrometer. They used a Reticon RL-256A/17 linear array coupled to a spectrograph which had a reciprocal dispersion of 820 Å/mm at the detector (21 Å per diode). The fixed spectral range was from 4300 Å to 9600 Å.

The array video signal was digitized using an 8-bit converter and array scans were time-averaged in a Nova 1200 computer. Complete readout of the 256-element array required slightly over 5 ms. Background subtraction was utilized to remove both sky illumination and fixed-pattern noise. As in experiments by other workers, in order to achieve relatively long integration times cooling of the array was necessary. The array was cooled with the boil-off from liquid nitrogen. At 20°C the leakage and dark-charge accumulation in 0.25 s amounted to about 10% of the saturation charge. At −60°C, the time for a similar accumulation was extended to about 30 min, a factor of 7200.

4.7. Image-Intensified Photodiode Array Systems

In an effort to improve the sensitivity and/or extend the spectral range of photodiode arrays they have been coupled to image-intensifier devices. Riegler and More[42] coupled a 256-element photodiode array to a Chevron channelplate detector in order to develop a position-sensitive detection system for ultraviolet and X-ray photons and particles. The Chevron plate used was a reasonably efficient photon detector from the vacuum UV, approximately 800 Å to the X-ray range at a few keV photon energy. The channelplate provided a gain of 10^7 electrons per detected photon. A P-20 phosphor was used for electron-to-photon conversion and a fiberoptics coupling plate was used to couple directly to the array surface. This device was intended for use in vacuum ultraviolet and X-ray astronomy.

Tull, Choisser, and Snow[43] have developed an image-intensified photodiode array for astronomical spectroscopy. This system is based on the Digicon photon-counting image tube which used nonscanned silicon photodiodes.[44] This new self-scanned Digicon utilizes a 1024-element linear Reticon array which was mated to a standard Digicon tube. In the Digicon tube photoelectrons are emitted from a photocathode upon which the image to be detected is focused. The emitted photoelectrons are accelerated to about 20 keV and then magnetically focused on the array. Thus the array is detecting electrons rather than photons. The accelerated photoelectron should generate one charge carrier for each 3.6 eV of energy deposited in the Si. Hence a single photoelectron of 20 keV should produce a charge amplification in the range of 5000.

Several areas of performance for the self-scanned Digicon were tested by Tull et al.[43] Results are presented for switching transient noise, thermal leakage (dark current), quantum efficiency, spectral response, gain as a function of accelerating potential, output signal as a function of integration time, noise performance, dynamic range, and lag.

With respect to dark current it was found that between $-76°C$ and $+22°C$ thermal leakage varied by a factor of 2 for each $9.7°C$ temperature change. The leakiest diode reached 5% saturation in a 60-s integration time at $-76°C$ while nearly all others have less than half of this leakage. When cooled to $-76°C$ the self-scanned Digicon is capable of detecting sources producing as few as one photoelectron per diode every fourth frame at a 60-s integration time (i.e., every 4 min) or as many as 6000 every 30-ms integration time, resulting in a total effective dynamic range of 4.8×10^7.

In a related study Mende and Shelly have investigated the electron-recording capability of diode arrays.[45] They used a 128-element array and tested its response to electrons using both a Ni^{63} radioactive source and 42-keV electrons generated with an electron accelerator.

4.8. Laser Parameter Measurements

Photodiode arrays have been used to monitor laser parameters such as beam cross section and picosecond pulse duration.[46] These parameters can be more conveniently measured with a photodiode array than with photographic measurements or expensive TV camera systems. Results are reported[46] for the measurement of the laser beam spatial cross section of a single-mode-locked Nd-YAG pulse taken at 1 m from the exit mirror, and TPF and Kerr cell picosecond pulse measurements (35 ps) using a

Reticon RL 256E/17 photodiode array. The array was clocked at 500 kHz, which results in a 500-μs scanning period. Under these conditions the sensitivity of the array in the TPF experiments was equivalent to ASA 200 film. The dynamic range of the array was about 500, as indicated by the measurement of the intensity of the side lobes of the $\sin^2 x/x^2$ diffraction pattern of a slit taken with a He–Ne laser.

4.9. Spectral-Source Profiling

Photodiode arrays offer powerful measurement capability for the spatial profiling of spectral sources. In a recent application[47] a Reticon R 512C/17 photodiode array was used to profile the emission of calcium atoms in a N_2O–C_2H_2 flame and in an argon inductively coupled plasma. The array was mounted vertically just outside the exit slit of a stigmatic monochromator. The spectral source to be profiled (flame or plasma) was imaged onto the entrance slit, which passed a thin vertical slice of the source image. Thus, with the vertical array, one can measure spatial emission information along this vertical slice of the source image at one particular wavelength. The authors also used a Dove prism to rotate the image 90° and hence profile the source horizontally as well as vertically.

The array was interfaced to a Data General Supernova computer and scan rate and integration time were computer-controlled. In general, 10 scans were time-averaged and background-subtracted. The array was not cooled and as a result the maximum useful integration time was limited to about 2 s.

4.10. Solar Magnetographs

Solar magnetographs utilizing Reticon 512-element arrays as dectors have been built.[48,49] The article by Livingston et al.,[48] also contains considerable data on various characteristics of the photodiode array. The spectral response was compared to that of a planar diffused PIN diode and found to be comparable. Uniformity was evaluated under headings of electronic fixed-pattern noise, light-dependent fixed-pattern noise, and interdiode response. In addition, linearity, dark current, modulation transfer function, completeness of readout (lag), and dynamic range were evaluated and several considerations with respect to amplification and signal processing were discussed. Finally, the authors tested the operation of the array at liquid nitrogen temperature (-196°C) where dark current should be virtually zero. The array could be operated successfully at this temperature and they concluded that it has promise as a low-light-level detector with long integration times.

Smithson[49] has also constructed a solar magnetograph using a 512-element photodiode array as the detector. In addition, he measured several characteristics of the array. The saturation exposure at peak intensity was found to be 1.1 μW $s^{-1} \cdot cm^{-2}$, the signal-to-rms-noise ratio at saturation (excluding fixed pattern noise, i.e., background-subtracted) was better than 2000:1, and the signal-to-fixed-pattern-noise ratio was about 50:1. The dark current at room temperature was 1 to 2 pA and decreased by a factor of 2 for each 7.35°C of cooling. At room temperature (26°C), the maximum useful integration time (dark current reaching $\frac{1}{3}$ saturation) was 0.4 s. All measurements with the magnetograph were taken with the array cooled to −40°C.

5. Conclusions

The primary limitations of the array for spectrochemical applications, limitations that it shares with most electronic image sensors, are limited sensitivity (when compared to the photomultiplier tube) and limited wavelength coverage (with reasonable spectral resolution) when coupled to conventional dispersive systems. It does not appear that effective and inexpensive solutions to these problems are close at hand.

As mentioned in the last section, image intensification has been used to improve the performance of arrays. The self-scanned Digicon tube is commercially available (Electronic Vision Co., 11526 Sorrento Valley Road, San Diego, California 92121) but the cost is over $20,000. In addition, wavelength-shifting phosphors are often necessary to provide intensified sensors with UV response, which complicates photometric response and spectral sensitivity.

In order to utilize more effectively the area geometry of many sensors for spectral measurement, coupling to echelle spectrometers has been carried out,[2,11] and interest remains high in the development of such systems. This approach shows good promise but is likely to remain costly.

A related approach is the use of an image-dissector tube coupled to an echelle.[50] The image dissector[2] is not an integrating electronic sensor and the spectral information must be scanned sequentially. However, the image-dissector tube does provide effective electronic slew/scan capability with a detection system analogous to a photomultiplier. The tube alone, with measurement electronics, can cost about $15,000, and a complete system (image dissector, echelle, and computer) is available from Spectroscandia AB Brinkasvagen 1, SF-21660, Naqu, Finland, but at a cost of more than $100,000.

However, despite some of these limitations, the scope and capability of analytical spectrochemical measurements have been significantly extended by the utilization of modern electronic image sensors as detectors.[2,50,51] In this chapter, it has been shown the one type of sensor, the self-scanning linear silicon photodiode array, has many excellent features and characteristics for spectrochemical measurements. Freedom from blooming and lag, flexible signal-integration capability, time-resolution capability, and simple low-cost control and measurement electronics can all be considered assets. It was seen that the key operational features necessary in an effective array measurement system included cooling of the array, facility for electronic background subtraction, and repetitive time-averaging of the array signal. From the wide variety of measurements to which photodiode arrays have been applied, it is clear that sensors of this type will be increasingly incorporated into many spectrochemical, scientific, and industrial measurement systems.

6. References

1. R. M. Barnes, Emission spectrometry, *Anal. Chem., 46,* 150R (1974).
2. Yair Talmi, Applicability of TV-type multichannel detectors to spectroscopy, *Anal. Chem., 47,* 658A (1975).
3. Yair Talmi, TV-type multichannel detectors, *Anal. Chem., 47,* 697A (1975).
4. R. E. Benn, W. S. Foote, and C. T. Chase, The image orthicon in spectroscopy, *J. Opt. Soc. Am., 39,* 529 (1949).
5. J. T. Agnew, R. G. Franklin, R. E. Benn, and A. Bazarian, Combustion studies with the orthicon spectrograph, *J. Opt. Soc. Am., 39,* 409 (1949).
6. J. E. Anderson, Image orthicon slit spectrograph, *Rev. Sci. Instrum., 37,* 1214 (1966).
7. K. M. Aldous, D. G. Mitchell, and K. W. Jackson, Simultaneous determination of seven trace metals in potable water using a vidicon atomic absorption spectrometer, *Anal. Chem., 47,* 1034 (1975)
8. K. W. Busch, N. G. Howell and G. H. Morrison, Elimination of interferences in flame spectrometry using spectral stripping, *Anal. Chem., 46,* 2074 (1974).
9. M. J. Milano, H. L. Pardue, T. E. Cook, R. E. Santini, D. W. Margerum, and J. M. T. Raycheba, Design and evaluation of a vidicon scanning spectrometer for molecular absorption and atomic emission spectrometry, *Anal. Chem., 46,* 374 (1974).
10. D. O. Knapp, M. Omenetto, L. P. Hart, F. W. Plankey and J. D. Winefordner, Simultaneous multielement atomic emission flame spectrometry with an image vidicon detector, *Anal. Chem. Acta., 69,* 455 (1974).
11. D. L. Wood, A. B. Dargis, and D. L. Nash, A computerized television spectrometer for emission analysis, *Appl. Spectrosc., 29,* 310 (1975).
12. G. Horlick and E. G. Codding, Simultaneous multielement and multiline atomic absorption analysis using a computer-coupled photodiode array spectrometer, *Appl. Spectrosc., 29,* 167 (1975).
13. L. M. Biberman and Sol Nudelman (eds.), *Photoelectronic Imaging Devices*; Vol. 1, *Physical Processes and Methods of Analysis*; Vol. 2, *Devices and Their Evaluation,* Plenum Press, New York (1971).
14. G. P. Weckler, Charge storage lights the way for solid-state image sensors, *Electronics, 40*(9), 75 (1967).

15. P. H. Wendland, A charge-storage diode vidicon camera tube, *IEEE Trans. Electron Devices, ED-14*, 285 (1967).
16. M. H. Crowell, T. M. Buck, E. F. Labuda, J. V. Dalton, and E. J. Walsh, A camera tube with a silicon diode array target, *Bell Syst. Tech. J., 46*(2), 491 (1967).
17. R. H. Dyck and G. P. Weckler, Integrated arrays of silicon photodetectors for image sensing, *IEEE Trans. Electron Devices, ED-15*, 196 (1968).
18. P. J. W. Noble, Self-scanned silicon image detector arrays, *IEEE Trans. Electron Devices, ED-15*, 202 (1968).
19. P. W. Fry, Silicon photodiode arrays, *J. Phys. E., 8*, 337 (1975).
20. G. Horlick, Characteristics of photodiode arrays for spectrochemical measurements, *Appl. Spectrosc., 30*, 113 (1976).
21. G. Horlick and W. K. Yuen, Fourier domain interpolation of sampled spectral signals, *Anal. Chem., 48*, 1643 (1976).
22. H. V. Malmstadt, C. G. Enke, S. R. Crouch, and G. Horlick, *Optimization of Electronic Measurements*, p. 79, W. A. Benjamin, Menlo Park, Ca. (1974).
23. R. K. Brelm and V. A. Fassel, Direct reading spectrochemical analysis with a rapid-scanning spectrometer, *Spectrochim. Acta., 6*, 341 (1954).
24. G. Horlick and E. G. Codding, Some characteristics and applications of self-scanning linear silicon photodiode arrays as detectors of spectral information, *Anal. Chem., 45*, 1490 (1973).
25. G. Horlick and E. G. Codding, Analog cross correlation readout system for a spectrometer using a silicon photodiode array detector, *Anal. Chem., 45*, 1749 (1973).
26. E. G. Codding and G. Horlick, Simultaneous multielement quantitative spectrochemical analysis using a DC arc source and a computer-coupled photodiode array spectrometer, *Spectrosc. Lett., 7*(1), 33 (1974).
27. G. Horlick, E. G. Codding, and S. T. Leung, Automated direct current arc time studies using a computer-coupled photodiode array spectrometer, *Appl. Spectrosc., 29*, 48 (1975).
28. R. E. Dessy, Transient recorders in automated T-jump/stopped-flow equipment and rapid scanning spectrometers, in *Computers in Chemical and Biochemical Research*, Vol. 2 (C. E. Klopfenstein and C. L. Wilkins, eds.), p. 177, Academic Press, New York (1974).
29. G. A. H. Walker, V. L. Buckholy, D. Camp, B. Isherwood, J. Glaspey, R. Coutts, A. Condal, and J. Gower, A compact multichannel spectrometer for field use, *Rev. Sci. Instrum., 45*, 1349 (1974).
30. D. Lovse and H. V. Malmstadt, Electronic Circuit Considerations for a Solid State Diode Array Spectrometer, FACSS Meeting, Indianapolis, Ind., Oct. 1975, Paper No. 35.
31. R. G. Tull and R. E. Nather, Experimental use of self-scanned photodiode arrays in astronomy, in: *Astronomical Observations with Television-Type Sensors*, p. 171, (G. A. H. Walker and J. Glaspey, eds.), The Institute of Astronomy and Space Science, University of British Columbia, Vancouver, B.C. Canada (1973).
32. K. R. Betty and G. Horlick, A simple and versatile Fourier domain digital filter, *Appl. Spectrosc. 30*, 23 (1976).
33. H. V. Malmstadt, C. G. Enke, S. R. Crouch, and G. Horlick, *Optimization of Electronic Measurements*, p. 124, W. A. Benjamin, Menlo Park, Ca. (1974).
34. G. Horlick, Reduction of quantization effects by time averaging with added random noise, *Anal. Chem., 47*, 352 (1975).
35. B. M. Singer and J. K. Kostelec, Theory, design and performance of low blooming silicon diode array imaging targets, *IEEE Trans. Electron Devices, ED-21*, 84 (1974).

36. E. C. Douglas, High light-level blooming in the silicon vidicon, *IEEE Trans. Electron Devices, ED-22,* 224 (1975).

37. K. W. Busch, N. G. Howell, and G. H. Morrison, Simultaneous determination of electrolytes in serum using a vidicon flame spectrometer, *Anal. Chem., 46,* 1231 (1974).

38. S. A. Colgate, E. P. Moore, and J. Colburn, SIT vidicon with magnetic intensifier for astronomical use, *Appl. Opt., 14,* 1429 (1975).

39. M. J. Milano and H. L. Pardue, Evaluation of a vidicon scanning spectrometer for ultraviolet molecular absorption spectrometry, *Anal. Chem., 47,* 25 (1975).

40. J. D. Winefordner, J. J. Fitzgerald, and N. Omenetto, Review of multielement atomic spectroscopic methods, *Appl. Spectrosc., 29,* 369 (1975).

41. G. Horlick and E. G. Codding, Dye laser intra-cavity enhanced absorption measured using a photodiode array direct reading spectrometer, *Anal. Chem., 46,* 133 (1974).

42. G. R. Riegler and K. A. More, A high resolution position sensitive detector for ultraviolet and X-ray photons, *IEEE Trans. Nuclear Sci., NS-20,* 102 (1972).

43. R. G. Tull, J. P. Choisser, and E. H. Snow, Self-scanned digicon: a digital image tube for astronomical spectroscopy, *Appl. Opt., 14,* 1182 (1975).

44. E. A. Beaver and C. E. McIlwain, A digital multichannel photometer, *Rev. Sci. Instrum., 42,* 1321 (1971).

45. S. B. Mende and E. G. Shelly, Single electron recording by self-scanned diode arrays, *Appl. Opt., 14,* 691 (1975).

46. W. Seka and J. Zimmermann, Photodiode arrays: A convenient tool for laser diagnostics, *Rev. Sci. Instrum., 45,* 1175 (1974).

47. M. L. Franklin, C. Baber, and S. R. Koirtyohann, Spectral source profiling with a photodiode array, *Spectrochim. Acta* (in press).

48. W. C. Livingston, J. Harvey, C. Slaughter, and D. Trumbo, Solar magnetograph employing integrated diode arrays, *Appl. Opt., 15,* 40 (1976).

49. R. C. Smithson, The Lockheed diode array magnetograph, *Solar Phys., 40,* 241 (1975).

50. A. Danielson, P. Lindblom, and E. Soderman, Image dissector echelle spectrometer system for spectrochemical analysis, *Chem. Scripta, 6,* 5 (1974).

51. K. W. Busch and G. H. Morrison, Multielement flame spectroscopy, *Anal. Chem., 45,* 712A (1973).

52. R. E. Santini, M. J. Milano, and H. L. Pardue, Rapid scanning spectroscopy: Prelude to a new era in analytical spectroscopy, *Anal. Chem., 45,* 915A (1973).

5

Application of ESCA to the Analysis of Atmospheric Particulates

T. Novakov, S. G. Chang, and R. L. Dod

1. Introduction

Aerosol particles (also referred to as particulate matter or particulates) play a major role in the overall air pollution problem. They are responsible for reduction of visibility and acidification of waters, and in certain size ranges are deposited in the lungs where they can cause a variety of health effects. The air pollution aerosol particles can consist of solid and/or liquid substances. Some of these, such as windblown dust, soot, and fly ash, originate in sources outside the atmosphere and are known as "primary" particles. Others are formed directly in the atmosphere by reactions among the primary particulate and gaseous species and are known as "secondary" pollutants. The nature of both these primary and secondary particulate species, as well as the extent to which atmospheric chemistry is governed by (photochemical) gas-phase or heterogeneous gas-particle reaction mechanisms, is presently a very active area of research.

In order for an effective control strategy to be devised, the chemical species responsible for the adverse environmental and health effects must be selectively identified and the process by which these species are formed in the source effluents or in the atmosphere must be ascertained. The key to the successful accomplishment of this important task is the determination of the exact chemical composition of the compounds and species associated with these particles. It is important to determine the chemical composition of particle constituents as they actually exist in aerosol form

T. Novakov, S. G. Chang, and R. L. Dod • Energy and Environment Division, Lawrence Berkeley Laboratory, University of California, Berkeley, California 94720

and not, for example, as they may appear in solutions. What is required is a truly *in situ* method for chemical analysis, where the analysis could be performed without the need for collecting the particles on filters or other collection media. Reliable methods of this kind are unavailable now, however.

One must further distinguish the bulk from the surface composition of aerosol particles. Some particles or some of their major bulk constituents may eventually be soluble in body fluids, so that practically their entire content can be toxicologically harmful. On the other hand, particle surfaces may be chemically dissimilar to the interiors of the particles and may carry microscopic layers of contaminants that make contact with lung membranes, leading to exposures that have little to do with the bulk composition of the particles. An intermediate situation may exist when harmful contaminants cover the surface of extremely small particles (of the order of 100 Å or less) which coagulate into larger clusters in the atmosphere. In this case the surface composition may be undistinguishable from the bulk composition of such composite particles.

Obviously, no single method can provide satisfactory answers to all of the above problems. Wet chemical and other microanalytical procedures have to be complemented by nondestructive physical methods. X-ray photoelectron spectroscopy, also known as ESCA (Electron Spectroscopy for Chemical Analysis),* whose application to chemical characterization of pollution particles is described in this chapter, is one such method.[1] For example, application of this method has helped to uncover the presence of significant concentrations of reduced nitrogen species other than ammonium in ambient aerosol particles.[2] This group of species contains certain amines and amides[3] which are not soluble in water or such solvents as benzene and therefore could not be detected by wet chemical methods. Most analyses of pollution aerosol particles have employed wet chemical methods; and on the basis of this kind of measurement, different workers have concluded that the principal particulate nitrogen species are ammonium and nitrate ions[4] and have suggested that the most likely combination of these is ammonium nitrate and ammonium sulfate.[5] The species uncovered with the aid of electron spectroscopy have thus escaped observation by means of wet chemistry.

In this chapter we will describe the use of X-ray photoelectron spectroscopy for chemical characterization of ambient and source-enriched aerosol particles. These analyses involve measurement of the chemical shift, core electron level splitting, relative concentrations, and

---

* For the principles of the ESCA method see Ref. 23. For the application of the method to chemical characterization of particulates, see Ref. 1. For a review of analytical applications, consult Ref. 35.

volatility (in vacuum) of different particulate species. Because the method of photoelectron spectroscopy has been described in great detail in a number of papers and monographs, only the fundamentals of the technique relevant to this topic will be reviewed here. Because most of the mass of airborne pollution particles consists of compounds of carbon, nitrogen, and sulfur, special emphasis will be placed on characterization of C, N, and S species. Attempts to chemically characterize some trace metals such as lead and manganese, both originating in fuel additives, will also be described.

2. Method

X-ray photoelectron spectroscopy is the study of the kinetic energy distribution of photoelectrons expelled from a sample irradiated with monoenergetic X-rays. The kinetic energy of a photoelectron E_{kin}, expelled from a subshell i, is given by $E_{kin} = h\nu - E_i$, where $h\nu$ is the X-ray photon energy and E_i is the binding energy of an electron in that subshell. If the photon energy is known, experimental determination of the photoelectron kinetic energy provides a direct measurement of the electron binding energy.

The electron binding energies are characteristic for each element. The intensity of photoelectrons originating from a subshell of an element is related to the concentration of atoms of that element in the active sample volume. In principle, this feature enables the method to be used for quantitative elemental analysis. The binding energies, however, are not absolutely constant but are modified by the valence electron distribution, so that the binding energy of an electron subshell in a given atom varies when this atom is in different chemical environments. These differences in electron binding energies are known as the chemical shift. The origin of the chemical shift can be understood in terms of the shielding of the core electrons by the electrons in the valence shell. A change in the charge of the valence shell results in a change of the shielding which affects the core electron binding energies. For example, if an atom is oxidized, it donates its valence electrons and thus becomes more positively charged than the neutral configuration. Some of the shielding contribution is removed, and in general the binding energies of the core electron subshells are increased. Conversely the binding energies will show an opposite shift for the reduced species. Therein lies the usefulness of chemical shift determinations in the analysis of samples of unknown chemical composition. In practice, the measurements of the chemical shifts are complemented by the determination of relative photoelectron intensities, from which the stoichiometric information can be inferred.

The relation $E_{\text{kin}} = h\nu - E_i$ is unambiguous for gaseous samples. In solid samples, however, the photoelectron has to overcome the potential energy barrier at the surface of the sample. This potential energy barrier is known as the work function of the sample ϕ_{sample}. However, if the solid sample is in electrical contact with the electrically "grounded" spectrometer, the Fermi levels of the sample and of the spectrometer are equalized. On entering the spectrometer a photoelectron is accelerated by $e[\phi_{\text{sample}} - \phi_{\text{spect}}]$, and as it reaches the detector it acquires the kinetic energy $E_{\text{kin}} = h\nu - E_{i,f} - \phi_{\text{spect}}$. In the experiment, therefore, the kinetic energy is determined by the spectrometer work function ϕ_{spect} and by the binding energy referenced to the Fermi level of the spectrometer $E_{i,f}$.

In order to use photoelectron spectroscopy as an analytical tool it is important to understand how different factors influence the chemical shift, and how the observed chemical shifts relate to the chemist's intuitive conception of bonding and molecular structure. We shall briefly describe some of the theoretical results on chemical shifts, insofar as they pertain to the subject of analytical applications of photoelectron spectroscopy.

The electron binding energy is defined as the difference between the total energies of the final and the initial states. An exact calculation of the total energy of even the simplest multielectron systems is impractical. Self-consistent field methods which yield rather accurate total energies are used instead. In this type of calculation the total energy of the initial state—when each core orbital is doubly occupied—is calculated first. The total energy of the final state is calculated in accordance with Koopmans' theorem,[6] i.e., assuming the same wave functions as those used for the initial state calculation except one of the orbitals is now singly occupied. This approach assumes that the photoelectron is suddenly removed, leaving the passive orbitals "frozen." Physical justification for this approach relates to the fact that the time required for the photoelectron to leave the atom is much shorter than the lifetime of the core-level hole.

The electrons in the outer shells, however, can respond quickly to the formation of the positive core hole by "shrinking" their orbitals adiabatically. This relaxation process is fast because it involves no change in the quantum state. The additional Coulomb repulsion of the shrunken orbitals causes the calculated photoelectron kinetic energy to be greater than the one obtained under the frozen-orbital approximation. In other words, the effect of the relaxation is to reduce the calculated binding energy with respect to the one calculated under the assumption of strict validity of Koopman's theorem. To take the relaxation effects into consideration, the total energy of the final state is calculated for the core hole state instead of the ground state where all orbitals are doubly occupied. Several approaches to this problem have been investigated by

Bagus,[7] Rosen and Lindgren,[8] Schwartz,[9] Brundle *et al.*,[10] and Siegbahn[11] among others.

Alternative theoretical approaches have been developed to calculate the binding energies from the frozen-orbital model corrected for relaxation effects. Hedin and Johansson[12] formulated a correction to the binding energy obtained from Koopmans' theorem by including the polarization potential created by the presence of a hole in the core orbital. Snyder[13] considered the problem of orbital relaxation in an atom from which an inner-shell electron has been ejected and discussed the hole effect in terms of atomic shielding constants. Liberman,[14] and Manne and Åberg[15] have considered the relationship between hole-state and frozen-orbital calculations based on physical insight.

Theoretical approaches such as those listed above are valuable in elucidating the contributions of various factors to the chemical shift, but such calculations are of limited value to calculate theoretically the binding energies or binding-energy shifts for molecules of practical interest.

Fortunately, however, the experimentally observed chemical shifts are well correlated with certain parameters directly related to conventional ideas about chemical structure. For example, even in the early stages of photoelectron spectroscopy it was realized that the chemical shifts can be related to the oxidation state of the atom in a molecule. Subsequently, attempts to correlate the binding energy shifts with the estimated effective atomic charges were made. Electronegativity difference methods,[16,17] the extended Hückel molecular orbital method,[18] and the CNDO method[19-21] were used to calculate the effective charge. Only rough correlations were obtained, however. It was subsequently found that the poor correlations result from neglect of the potential generated by all charges in the molecule.

Chemical shifts can be adequately described by the electrostatic-potential model in which the charges are idealized as point charges on atoms in a molecule. The electron binding-energy shift, relative to the neutral atom, is equal to the change in the electrostatic potential resulting from all charges in the molecule as experienced by the atomic core under consideration. Different approaches to the potential model calculations were used by Gelius *et al.*,[22,23] Siegbahn *et al.*,[17,24] Ellison and Larcom,[20] and Davis *et al.*[21] Detailed theoretical analyses of the potential model were given by Basch[25] and Schwartz.[26]

In practice, line broadening and the small magnitude of the chemical shifts may make the determination of even the oxidation state difficult in some cases. Because of this difficulty in cases of transition metal compounds, the "multiplet splitting" effect can be employed to infer the oxidation state. Multiplet splitting of core electron binding energies[27] is observed in the photoelectron spectra of paramagnetic transition metal compounds. In any atomic or molecular system with unpaired valence electrons, the 3s–3d exchange interaction affects the core electrons differently, according to the orientation of their spin. This causes the 3s core level to be split into two components. For example, in an Mn^{2+} ion whose ground state configuration is $3d^5 \; ^6S$, the two multiplet states will be ^7S and ^5S. In an Mn^{4+} ion having the ground state configuration $3d^3 \; ^4F$, the two spectroscopic states will be ^5F and ^3F. In the first approximation, the magnitude of the multiplet splittings should be proportional to the number of unpaired 3d electrons. Hence the Mn(3s) splitting will be greatest for Mn^{2+} ions and least for Mn^{4+}.

Because of the low energy of photoelectrons induced by the most commonly used Mg or Al X-rays, the effective escape depth for electron emission without suffering inelastic scattering is small. Recent studies have given electron escape depths of 15 to 40 Å for electron kinetic energies between 1000 and 2000 eV.[28] This renders the ESCA method especially surface sensitive and thus useful in surface-chemical studies. Because of the high energy resolution, the electrons which have escaped the solid sample without energy loss are well separated from the low-energy electrons whose energy has been degraded by the inelastic collisions. The chemical shift measurements are performed only on electrons with no energy loss.

3. Analytical Aspects of ESCA

During recent years, a number of papers and reviews have discussed the qualitative and quantitative analytical aspects of ESCA. Generally, it appears that the principal sources of uncertainty and error in ESCA analytical determinations are related to: (1) electron energy-loss processes, (2) electron escape-depth variations, (3) binding-energy calibration procedures, and (4) sample exposure to vacuum and X-rays during analysis. The first two of these problems may have major repercussions for the determination of concentrations of elements and species in chemically heterogeneous samples. The third problem influences the validity of the assignment of chemical states through the determination of chemical shifts. Finally, the spectrometer vacuum and heating of the sample by the X-ray source may be the cause of losses of volatile species. These problems are of a general nature and obviously have to be taken into account in the analysis of atmospheric particulate matter by ESCA.

In this section experimental results and conclusions on some of the problems that relate to the analytical application of ESCA will be reviewed.

These results deal with different specific objectives which are felt to be equally applicable to the topics of this chapter.

Wagner[29] has determined a table of relative atomic sensitivities, which enables the conversion of photoelectron peak intensities to relative atomic concentrations of elements in a sample. Wagner's study, as well as the one by Swingle,[30] indicates that ESCA can be used as a semiquantitative ($\leq 50\%$ relative error) or even as a quantitative ($\leq 10\%$ relative error) method when comparing chemically similar samples. Unfortunately, both of the above authors found that the relative photoelectron intensities obtained with chemically dissimilar samples show wide variations. For example, Wagner's[29] data on the Na(1s)/F(1s) intensity ratios (corrected for stoichiometry) for a number of sodium- and fluorine-containing compounds were found to vary by as much as a factor of 2.

Swingle[30] has attributed the observed variations in the apparent sodium and fluorine atomic sensitivities to the differences in the structure of the photoelectron energy-loss spectrum. In cases of chemically similar samples, however, the same mechanism should be responsible for inelastic electron scattering.

The effect of the chemical form of an atom on the relative intensity of the photoelectron peak arising from that atom has been studied in detail by Wyatt, Carver, and Hercules.[31] They have demonstrated that different lead salts show different atomic sensitivities and have suggested that the escape depth for lead salts depends on the crystalline frame surrounding the lead cation and its coordination number, because the observed differences in sensitivities could not be accounted for by inelastic electron scattering alone. These findings imply that elemental analysis must be done with great caution, unless the chemical form of the elements in the sample is well defined. Since in many "real world" systems the exact chemical form of elements is not necessarily known, the above authors suggest that all of the elements to be measured should be converted into the same crystalline form. However, this approach, even if feasible, would of course eliminate the nondestructive nature of ESCA analysis.

It was already mentioned that quantitative analysis by ESCA is reliable when chemically similar materials are used. Furthermore, in cases of homogeneous mixtures ESCA can be used to determine bulk composition, in spite of the surface sensitivity of the method. Thus Siegbahn et al.[24] determined the percentages of copper and zinc atoms in various alloys. Larson[32] has performed a similar study on various gold–silver alloys. Swartz and Hercules[33] were able to do quantitative analysis of MoO_2–MoO_3 mixtures by measuring the chemically shifted molybdenum peaks. Obviously, in order to accurately determine the concentration ratios of the bulk constituents, the inhomogeneities throughout the effective sample volume must be of the order of the electron escape depth or less.

The bulk sensitivity of ESCA can be estimated at approximately 0.15% based on bulk percentage. Therefore ESCA is not a "trace element" technique in the usual sense. However, because of the possibility of detecting as little as 0.1% of a monolayer[34] (about 10^{12} atoms), the absolute sensitivity of ESCA is in the picogram range. Thus ESCA is a unique trace method if the analysis is confined to preferentially surface-located species. A potentially very rewarding approach to the analysis of solid substances is to perform ESCA and bulk-type measurements (for example, X-ray fluorescence) on the same sample. A proper analysis of such data would enable differentiation between preferentially surface and bulk species.

The topics discussed thus far in this section pertain essentially to the determination of relative concentrations of elements and species. The importance and uniqueness of ESCA, however, are in its capability to measure the chemical shifts which contain implicit information about the molecular forms. The principal difficulty encountered in chemical-shift measurements pertains to the calibration procedure used to account for the electrostatic charging of the sample. Hercules and Carver[35] list four basic calibration approaches which have been used. These make use of (1) contamination carbon, (2) admixed species, (3) vapor deposition of noble metals, and (4) sample constituents as internal calibration standards.

Because of low concentrations and interfering lines, it may be difficult in some cases to measure the binding energy accurately. However, as pointed out by Brinen[36] in referring to the application of ESCA to catalysts, differences in binding energies between treated or untreated catalysts and differences in relative intensities are often more informative. Subtle changes in chemical bonding, resulting either in a broadened line or in the appearance of "shoulders" on the peaks of spectra from starting material, imply formation of new species which may be crucial to the understanding of catalytic activity. Problems associated with ESCA analysis of atmospheric matter and the surface-chemical reactions leading to its formation are similar to the problems mentioned above.

4. Application of ESCA to Particulate Analysis

4.1. Effect of Sample Composition at Relative Intensities

Effects of the kind described above may seriously limit the applicability of ESCA for the analysis of atmospheric particulates. Therefore, it is important to assess the usefulness of (1) photoelectron peak intensities for inferring the likely stoichiometry of certain compounds, for example, to distinguish between ammonium sulfate and ammonium bisulfate; and (2) the relative photoelectron peak intensities for determination of relative concentrations of elements and chemical species in particulate matter. We have investigated the possible effects of the chemical composition and the surrounding matrix on photoelectron peak intensities, using heterogeneous samples that reasonably simulate the situation found in ambient particulates.

The first set of these experiments investigated the constancy of the intensity ratios of S(2p) and N(1s) photoelectron peaks, corrected for stoichiometry, for a number of pure sulfur- and nitrogen-containing compounds. (The nitrogen and sulfur peak intensities were normalized to the number of corresponding atoms in the molecular unit.) The resulting peak intensity ratios are displayed in Figure 1, in the order of increasing molecular weight per nitrogen atom. The mean value for peak intensity ratios of these samples is 1.65 ± 0.21 (1σ).

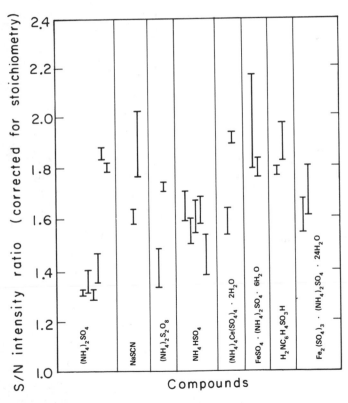

Figure 1. Intensity ratios of sulfur (2p) and nitrogen (1s) photoelectron peaks, corrected for stoichiometry, for a number of pure sulfur- and nitrogen-containing compounds.

The results indicate that within an uncertainty of ±15% the photo-electron peak intensities of nitrogen and sulfur reflect the relative abundances of these atoms in the compounds. These results are consistent with results reported by others, considering that the compounds examined are primarily ionic compounds containing ammonium and sulfate ions. Nitrogen and sulfur atoms in these compounds are surrounded with a constant sphere of nearest neighbors and would thus be expected to exhibit similar atomic sensitivities from one component to the next. Similarly, the organic nitrogen and sulfur compounds are also expected to yield constant atomic sensitivities for these elements.

The second set of experiments investigated the effects of matrix dilution on photoelectron peak intensities of nitrogen and sulfur from ammonium bisulfate. Activated carbon, lead chloride, and graphite mixtures were used as the diluent matrix. As before, such mixtures were assumed to simulate the conditions normally found in atmospheric particulates.

Ammonium bisulfate was used because, along with ammonium sulfate, it is a common form of atmospheric sulfate. Since ammonium bisulfate is moderately deliquescent, it was found advantageous to first dry the substance in air at 80–90°C. This treatment resulted in an increase of $S(2p)/N(1s)$ photoelectron peak ratio from ~1.6 to ~2.2, indicating the loss of ammonium. The dried ammonium bisulfate, however, was found to be stable in vacuum, both in terms of signal strength and the photoelectron peak intensity ratio.

In one series of experiments, a mixture of equal weights of Norit A activated carbon and lead chloride was used as the dilution matrix material. Samples containing different amounts of ammonium bisulfate were prepared by weighing pure substances which had been ground in an agate mortar and thoroughly mixed with a high-speed dental amalgam blender. The mixtures, containing 2, 10, 30, and 50% ammonium bisulfate, were examined by ESCA both immediately after preparation and after being exposed to laboratory air for approximately one day. The latter samples were used for the study because exposure to air allowed the ammonium bisulfate to deliquesce and perhaps more closely approximated an ambient particulate sample collected in a humid atmosphere.

The results of this dilution experiment demonstrated a constancy in the sulfate-to-ammonium peak ratio to within ±10%, i.e., well within the error limit for peak intensity determination.

In another dilution matrix effect study different diluents were used with a constant (dried) ammonium bisulfate concentration of 10% by weight. In addition to the activated-carbon–lead-chloride mixture, pure Norit A activated carbon and pure graphite powder were used as diluents. The mean value of the sulfur-to-nitrogen photoelectron peak intensity ratio of 2.18 ± 0.30 was determined in this series of experiments. This value is within the error limits, in agreement with pure dried ammonium bisulfate.

Thus, based on these results, it seems justifiable to use relative photoelectron peak intensities to infer the apparent stoichiometry of sulfur- and nitrogen-containing compounds of the kind commonly associated with air pollution particulates. It will be shown later in the text that photoelectron peak intensities can be used to determine the concentration ratios of certain elements and species.

4.2. Binding-Energy Calibration

Before outlining the results for chemical states of sulfur, nitrogen, and carbon species in atmospheric particulates, we shall first justify the use of apparent carbon (1s) binding energy to correct for sample charging. As discussed later in more detail, carbon peaks from atmospheric particulates appear essentially as a single peak with a binding energy corresponding to a neutral charge compatible with condensed hydrocarbons and/or sootlike material. Since carbon is by far the most abundant element in particulates, practically the entire ESCA C(1s) signal is due to the sample itself, rather than to hydrocarbon contamination of the sample in the spectrometer.

In order to test the validity of using the carbonaceous content of a sample as the internal binding energy reference, the apparent binding energies of C(1s) and Pb(4f$_{7/2}$) from a number of ambient samples collected near a freeway were determined. The results are shown in Figure 2, where these binding energies are plotted against time of day of sample collections. (The samples were collected for 2 h on silver membrane filters.) The figure shows that the variations in the carbon and lead peak positions are similar. Assuming that the chemical composition of lead and carbon species are similar throughout the episode, we can conclude that the binding-energy error caused by the use of C(1s) as the reference should not exceed ±0.25 eV. It appears, therefore, that use of the C(1s) peak binding energy as an internal reference is adequate to determine the chemical states of major species associated with atmospheric particulates.

4.3. Chemical States of Sulfur and Nitrogen from Chemical Shift Measurements

The feasibility of using X-ray photoelectron spectroscopy for the chemical characterization of particulates was first explored by Novakov, Wagner, and Otvos.[37] The samples used were collected on glass fiber filters, without particle size separation, at several locations in the San

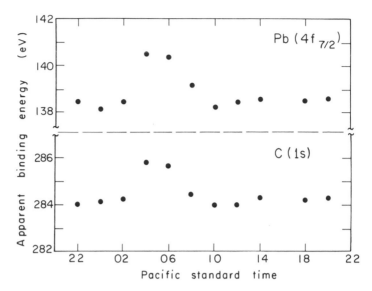

Figure 2. Apparent binding energies (not corrected for sample charging) of carbon (1s) and lead ($4f_{7/2}$) photoelectrons from a number of 2-h samples collected near a freeway. The apparent binding energies are plotted against time of day corresponding to sample collection.

Francisco Bay Area. These authors have determined the elemental composition of the particulate samples and have found that most of the particulate nitrogen is in the reduced chemical state. The observed nitrogen (1s) photoelectron lines were of complex structure, indicating the presence of several different reduced nitrogen species. The sulfur (2p) peak appeared to have a single component, consistent with sulfate.

A more systematic ESCA study of the chemical states of sulfur and nitrogen as a function of particle size and time of day was performed by Novakov *et al.*[2] as part of the 1969 Pasadena Smog Experiment. The particle separation was accomplished by use of a Lundgren cascade impactor. The samples for analysis were collected on Teflon films covering the rotating impactor drums. Two "size cuts," 2.0–0.6 μm and 5.0–2.0 μm, corresponding to the fourth and third impactor stages, were used in this experiment.

The sulfur (2p) spectra indicated the presence of two components which were assigned to sulfate and to sulfite. The 6+ and the 4+ state were present in both size ranges. Their relative abundance changed with particle size and time of day, however.

A more complex situation was encountered in the study of chemical states of particulate nitrogen species. The ESCA spectra from this study have revealed the presence of four different chemical states of particulate nitrogen: nitrate, ammonium, and two reduced species tentatively assigned to an amino-type and a heterocyclic nitrogen compound. Surprisingly, the nitrate was observed only in the larger particle size range but not in the smaller one.

The results just described were presented at the 161st National ACS Meeting in Los Angeles at which Hulett et al.[38] also reported on the use of ESCA for chemical characterization of coal fly ash and smoke particles. These authors have analyzed specimens of smoke particles collected on filter paper from coal burned in a home fireplace and found three distinct components in the sulfur (2p) peak: a single reduced state assigned to a sulfide and two species of higher oxidation, corresponding to sulfite and sulfate. The sulfur (2p) peak of fly ash particles was also found to consist of two components. These were assigned to sulfates and/or adsorbed SO_3, since their binding energies were much higher than those for sulfite ions.

Araktingi et al.[39] used ESCA to analyze particulates collected on glass fiber filters in Baton Rouge. A number of elements were identified in this work, the most abundant of which were sulfur, nitrogen, and lead. The N(1s) peak appeared to consist of at least two components with binding energies at about 399.4 and 401.5 eV. The authors did not propose an assignment to these peaks. It would appear, however, that these are similar to the nitrogen peaks identified in Pasadena samples by Novakov et al. The peak at 399.4 eV would correspond to the amino-type nitrogen, while the peak at 401.5 eV would correspond to ammonium. No appreciable nitrate was found in the study of Araktingi et al.[39] The S(2p), according to these authors, was entirely in the form of sulfate.

In summary, these early experiments have demonstrated that there is a considerable variety in the chemical states of sulfur and nitrogen associated with air pollution particulates. For example, particulate sulfur may exist in both oxidized (sulfate, adsorbed SO_3, and sulfite) and reduced (sulfide) states, while nitrogen species include nitrate, ammonium, and two other, previously unrecognized, reduced forms. This obviously suggests a more complex situation than the one inferred from wet chemical results, i.e., that the only significant sulfur and nitrogen species are sulfate, nitrate, and ammonium ions.

Craig et al.[40] have attempted to formulate an "inventory" of chemical states of particulate sulfur based on their examination of ESCA spectra of more than a hundred samples collected at various sites in California. The samples used in that study were collected both with and without particle size segregation. All samples were collected for 2 h as a function of time of day. The sulfur spectra of these samples were of varying degrees of complexity, sometimes covering a wide range of binding energies. The binding energies deduced from the analysis of particulate samples were assigned to certain characteristic chemical states with the help of ESCA results obtained with a number of pure compounds and with certain surface species produced by the adsorption of SO_2 and H_2S on several solids. These authors[40] caution that the chemical species were assigned in terms of reference compounds, presumably similar to those present in or on ambient particulates. This ambiguity, however, should not influence the correct assignment of major species such as sulfate, whose binding energy is not very sensitive to the choice of cations. On the other hand, it was also pointed out that species designated as adsorbed SO_3 may actually be a different compound, such as a sulfate that produces a chemical shift similar to adsorbed SO_3.

The chemical states of particulate sulfur identified in the work of Craig *et al.* are, in terms of the chemical shift, equivalent to adsorbed SO_3, SO_4^{2-}, adsorbed SO_2, SO_3^{2-}, S^0, and possibly two kinds of S^{2-}. The assignment of S^0 as neutral (elemental) sulfur was made because of the similarity of its binding energy to that of elemental sulfur. It is possible, however, that sulfur in species designated by S^0 is actually in the -2 oxidation state. Differences in binding energies between different sulfides are expected because of the differences in the corresponding bond ionicity. Naturally not all of these species did occur at all times and locations. In all instances, sulfates were found to be the dominant species, although concentrations of other forms of sulfur were at times comparable to the sulfate concentrations.

As we mentioned earlier, ESCA analyses of ambient particulates uncovered the existence of previously unsuspected reduced nitrogen species[2] whose binding energies are similar to certain amines and/or heterocyclic nitrogen compounds. For simplicity we shall denote these species by N_x. Further studies on the chemical structure of N_x species were performed by Chang and Novakov[3] by means of temperature-dependent ESCA measurements. The experimental procedure consisted in measuring ESCA spectra of ambient samples and gradually increasing sample temperatures. The samples were collected on silver membrane filters which could withstand the temperatures used in the experiment.

The results of one such measurement for an ambient particulate sample, collected in Pomona, California, during a moderate smog episode (24 October 1972), are shown in Figure 3. The spectrum taken at a sample temperature of 25°C shows the presence of NO_3^-, NH_4^+, and N_x. At 80°C the entire nitrate peak is lost, accompanied by a corresponding loss in the ammonium peak intensity. The shaded portion of the ammonium peak in the 25°C spectrum represents the ammonium fraction volatilized between 25 and 80°C. The peak areas of the nitrate and the volatilized ammonium are approximately the same, indicating that the nitrate in this sample is mainly in the form of ammonium nitrate. The

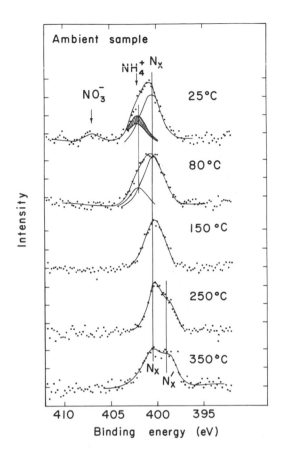

Figure 3. Nitrogen (1s) photo-electron spectrum of an ambient sample as measured at 25, 80, 150, 250, and 350°C (from Ref. 3).

ammonium fraction still present at 80°C but absent at 150°C is associated with an ammonium compound more stable than ammonium nitrate, such as ammonium sulfate. At 150°C the only nitrogen species remaining in the sample is N_x. At 250°C the appearance of another peak, labeled N_x', is seen. The intensity of this peak continues to increase at 350°C. The total $N_x + N_x'$ peak area at 150, 250, and 350°C remains constant, however, indicating that a part of the N_x is transformed into N_x' as a consequence of heating.

N_x' species will remain in the sample even if its temperature is lowered to 25°C, provided that the sample has remained in vacuum. However, if the sample is taken out of vacuum and exposed to the humidity of the air, N_x' will be transformed into N_x. It was concluded that N_x' species are produced by dehydration of N_x:

$$N_x \underset{+H_2O}{\overset{-H_2O}{\rightleftharpoons}} N_x'$$

Based on the described temperature behavior and on laboratory studies[3] of reactions (see below) that produce species identical to those observed in the ambient air particulates, N_x was assigned to a mixture of amines and amides. (N_x photoelectron peaks are broad indications of the presence of more than one single species.) Dehydration of the amide results in the formation of a nitrile, N_x'.

Application of temperature-dependent ESCA measurement[3,41] also revealed the presence of a hitherto unrecognized form of ammonium characterized by its relatively high volatility in vacuum. Temperature-dependent studies have also indicated that nitrate in ambient samples may occur in a volatile form different from common nitrate salts. Tentatively, such a nitrate is assigned to nitric acid adsorbed on the filter material and/or on the particles.

These conclusions are illustrated with the aid of the spectra shown in Figure 4. The spectrum shown in Figure 4a (collected in 1973 in West Covina, California) has been obtained with the sample at $-150°C$. The sample was kept at a low temperature in order to prevent volatile losses in

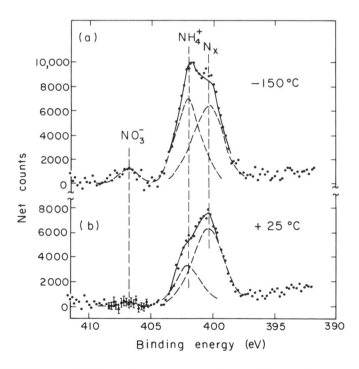

Figure 4. (a) Nitrogen (1s) photoelectron spectrum of an ambient sample as measured at $-150°C$; (b) the spectrum of the same sample as measured at 25°C (from Ref. 3).

the ESCA spectrometer vacuum. It will be shown later that volatilization in vacuum was suspected as one reason for the apparent inconsistency between the ammonium and nitrate determination by ESCA and by wet chemical techniques. Individual peaks corresponding to nitrate, ammonium, and N_x (amines and amides) are clearly seen in the spectrum.

Figure 4b shows the same spectral region of the same sample after its temperature was raised to 25°C. This spectrum shows only a trace of the original nitrate and about a 60% decrease in the ammonium peak intensity. Considering the nitrate and ammonium peak intensity in the lower temperature case, it is estimated that at most about 15% of the total ammonium could be associated with nitrate as NH_4NO_3. The volatile ammonium component is therefore not ammonium nitrate. Ammonium sulfate and ammonium bisulfate were found to be stable in vacuum at 25°C during time intervals normally used to complete the analysis. Furthermore, since no detectable decrease in the sulfate peak was observed over the same temperature range, it was also concluded that ammonium sulfate (and/or bisulfate) is not being volatilized in the spectrometer vacuum.

The limited volatility of ammonium salts and the behavior of the ambient samples suggests that a major fraction of ammonium in these samples is present in a previously unrecognized form. The volatility properties of nitrate in this and other samples suggest the possibility of the existence of adsorbed nitric acid in accordance with the wet chemical results of Miller and Spicer.[42]

4.4. Chemical Characterization of Particulate Carbon

ESCA has also been used in attempts to chemically characterize particulate carbon.[3] In most instances the carbon (1s) peak of ambient particulates appears essentially as a single peak with a binding energy compatible with either "elemental" carbon or condensed hydrocarbons or both. As seen in Figure 5, where the carbon (1s) spectrum of an ambient air particulate sample is shown, chemically shifted carbon peaks, due to suspected oxygen bonding, are of low intensity compared with the intense neutral chemical state peak.

From the standpoint of air pollution it is important to distinguish the hydrocarbon-type (mostly secondary species) carbon from sootlike (primary species) carbon. Unfortunately, because of the nature of chemical bonding in hydrocarbons, these cannot be distinguished from the sootlike carbon by ESCA shift measurements under the realistic conditions of sampling and analysis.

Chang and Novakov[3] have therefore employed other supplementary measurements to estimate the relative abundance of primary particulate

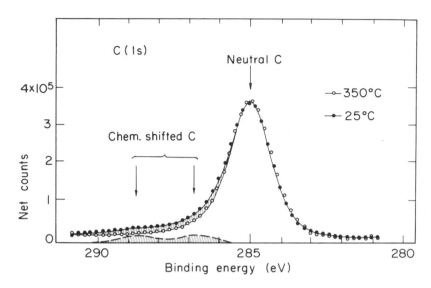

Figure 5. Carbon (1s) photoelectron spectrum of an ambient sample as measured at 25 and 350°C. The shaded area represents the difference between low- and high-temperature spectra. The apparent volatile losses are mainly confined to the chemically shifted component of the carbon peak (from Ref. 3).

carbon. This was attempted by comparing the carbon (1s) peak obtained from the sample at 25°C with the carbon (1s) peak from the sample at 350°C. The difference between the low-temperature and the high-temperature runs should give the fraction of volatile carbon. Figure 5 shows the result of one such experiment for a sample collected in 1975 from West Covina. This experiment suggests that most of the ambient particulate carbon is nonvolatile in vacuum at 350°C. Assuming that the secondary hydrocarbons will have a substantial vapor pressure at 350°C, the authors have suggested that a substantial fraction of the total particulate carbon is of a primary (sootlike) nature.

The conclusion about a high nonvolatile carbon content could be erroneous if a large fraction of particulate carbon volatilizes even at 25°C in vacuum. This seems unlikely, however, because reasonably good agreement has been found between the total carbon concentration as measured by ESCA and a combustion technique[41] (see Section 4.6).

4.5. Chemical States of Trace Metals in Particulates

It is important to know the chemical states of trace metals in atmospheric particulates. The application of chemical shift measurements to the chemical characterization of metals has encountered difficulties because of the small differences in binding energies between different metal compounds. Chemical characterization of particulate lead species was attempted by Araktingi *et al.*[39] These authors have attempted to determine the relative abundance of lead oxide and lead halide in samples collected in Baton Rouge, Louisiana. Because of the small chemical shift between oxide and halide, the two suspected components could not be resolved, however. The authors reached the tentative conclusion that the lead in these samples is probably present as a mixture of a lead halide and an oxide.

Harker *et al.*[43] used ESCA to determine the chemical state of manganese in particulate emissions from a jet turbine combustion burning 2-methyl cyclopentadiene, manganese tricarbonyl (MMT) as jet fuel additive.

Manganese (3p) peak binding energies for an exhaust sample and for the compounds MnO, Mn_2O_3, Mn_3O_4, and MnO_2 were determined. These values, corrected for sample charging using the hydrocarbon contaminant,

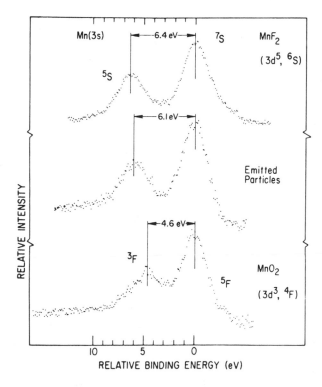

Figure 6. Manganese (3s) spectra of MnF_2, exhaust particles, and MnO_2 showing the multiplet splitting of the 3s core level (from Ref. 43).

Table 1. Measured Mn(3p) Electron Binding Energies[a]

	Exhaust	MnO[b]	Mn$_3$O$_4$	Mn$_2$O$_3$	MnO$_2$
Binding energy (eV)	48.9 ± 0.2	48.8 ± 0.2	48.9 ± 0.2	49.0 ± 0.2	49.9 ± 0.2

[a] From Ref. 43.
[b] Powdered reagent-grade MnO which undergoes rapid surface oxidation at ambient conditions was used.

are listed in Table 1. Clearly, the oxidation state of manganese in the exhaust sample cannot be determined on the basis of chemical shift alone.

The oxidation-state assignment was made by examining the multiplet splitting of the (3s) core level in the exhaust sample and some manganese compounds. In addition to oxides, MnF$_2$ was also studied, since it is the most ionic compound of divalent manganese, and therefore its Mn^{2+} ion should exhibit the largest possible (3s) splitting. Table 2 gives a summary of the measured splittings along with values from the literature. In Figure 6 comparative (3s) spectra for MnF$_2$, exhaust particulates, and MnO$_2$ are shown. Based on the magnitude of the splitting, it is concluded that the oxidation state of the manganese in the combustor exhaust is +2 as MnO. Other +2 manganese compounds are eliminated by the fact that oxygen is the only negatively charged species present in sufficient concentrations to balance the manganese.

Table 2. Multiplet Splitting of Mn(3s)[a]

Compound	Splitting (eV)			
	This work	Wertheim *et al.*[b]	Fadley *et al.*[c]	Carver *et al.*[d]
MnF$_2$ (2+)	6.4 ± 0.2	6.50 ± 0.02	6.5	6.3
Exhaust particulate	6.1 ± 0.2	—	—	—
MnO (2+)	—	6.05 ± 0.04	5.7[e]	5.5[e]
Mn$_3$O$_4$ (2+ and 3+)	5.5 ± 0.2	—	—	—
Mn$_2$O$_3$ (3+)	5.5 ± 0.2	5.50 ± 0.10	—	5.4
MnO$_2$ (4+)	4.6 ± 0.2	4.58 ± 0.06	4.6	—

[a] From Ref. 43.
[b] G. K. Wertheim, S. Hutner, and H. J. Guggenheim, *Phys. Rev. B*, 7(1), 556 (1973).
[c] C. S. Fadley, D. A. Shirley, A. J. Freeman, P. S. Bagus, and J. V. Mallow, *Phys. Rev. Lett.*, 23, 1397 (1969).
[d] J. C. Carver, T. A. Carlson, L. C. Cain, and G. K. Schweitzer, in: *Electron Spectroscopy* (D. A. Shirley, ed.), p. 803, North-Holland, Amsterdam (1972).
[e] Earlier lower values for MnO were due to surface oxidation of available analytical grade MnO. The MnO reference compound used in the study by Wertheim *et al.* was produced by oxidation of clean Mn *in vacuo*.

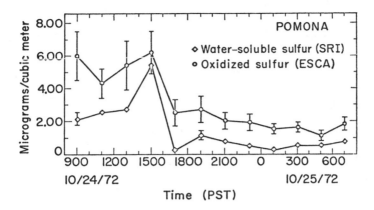

Figure 7. Diurnal variations of sulfate as measured by ESCA and by a wet chemical method (SRI). The data are for a 24-h sampling period during a smog episode in southern California (from Ref. 41).

4.6. Quantitative Aspects of ESCA Analyses

An intermethod comparison of ESCA with analytical methods of proven accuracy and precision was undertaken by Appel et al. [41] in order to validate the quantitative aspects of ESCA analysis (Figure 7). This work focused on validation of sulfate, nitrate, ammonium, and carbon data as obtained by ESCA in a program to characterize ambient California particulates.

From an analytical standpoint the principal reasons for concern about the validity of ESCA analysis are:

1. Peaks in ESCA spectra corresponding to species in the same oxidation state for a given element are often poorly resolved.
2. To obtain quantitative results by ESCA, an element in the sample must be analyzed by another technique in order to provide an internal standard. For example, the ESCA analysis ratio determines the sulfate-to-lead peak ratio which is normalized to the lead concentration as determined by X-ray fluorescence (XRF).
3. ESCA analyzes only the surface of the exposed particles and thus may not yield analyses representative of the average composition of particles.
4. ESCA requires maintaining the sample in high vacuum, which may cause the loss of volatile species.

Samples for this study[41] were collected on 47-mm Gelman GA-1 cellulose acetate filters for sulfate analysis and on 47-mm, 1.2-μm-pore-size silver membrane filters for carbon and nitrogen species analysis. Two-hour samples were collected in addition to 24-h high-volume (Whatman 41) filters.

ESCA analyses were conducted only on the membrane filters while, depending on sensitivity, wet chemical methods employed either the 2-h or 24-h high-volume filters. Direct comparisons involved analyses of sections from the same filters, while indirect comparison required comparison of calculated 24-h average values from ESCA analysis of twelve 2-h filters with the wet chemical analysis of the high-volume filter.

Three wet chemical methods[44] were used:

1. Stanford Research Institute (SRI) used a microchemical method, which measures total water-soluble sulfur, to analyze 2-h filters.
2. Barium chloride turbidimetric analysis was used to analyze water-soluble sulfate in high-volume samples.
3. Air and Industrial Hygiene Laboratory (AIHL), California State Department of Health, uses a method for 2-h samples which employs an excess of a barium–dye complex in acetonitrile–water solution. A decrease in absorbance due to the formation of barium sulfate is measured.

ESCA analysis included the determination of relative concentrations of sulfate and lead from the measured peak areas corrected for elemental sensitivity. Relative concentrations were converted to $\mu g/m^3$ of sulfate by normalization to lead concentration (in $\mu g/m^3$) which were determined by XRF analysis.[45]

Table 3 lists the results of the comparisons of ESCA sulfate determinations with those by the SRI and AIHL methods and by the $BaCl_2$ turbidimetric method. The ratio of means between wet chemical methods and ESCA varies from 0.5 to 1.0. It was concluded that ESCA provides

Table 3. Comparison of Sulfate Data (in $\mu g/m^3$ Sulfate), ESCA vs. Wet Chemistry[a]

Site	Date	High-vol wet chemistry	$\bar{\Sigma}$ 2-h low-vol ESCA	High-vol/$\bar{\Sigma}$ 2-h low-vol
San Jose	8-17-72	1.6 ± 0.4	1.1 ± 0.2	1.5 ± 0.5
San Jose	8-21-72	1.0 ± 0.2	1.4 ± 0.3	0.7 ± 0.2
Fresno	8-31-72	4.2 ± 1.0	4.0 ± 1.1	1.1 ± 0.4
Riverside	9-19-72	5.9 ± 1.5	6.4 ± 4.6	0.9 ± 0.7

Ratio of means = 1.0
Spearman's ρ = 0.80
Linear regression slope = 1.0
(intercept ≡ 0)

[a] From Ref. 1. ESCA, cellulose ester filters, $\bar{\Sigma}$ 2-h low volume; wet chemistry, Whatman 41, 24-h high volume.

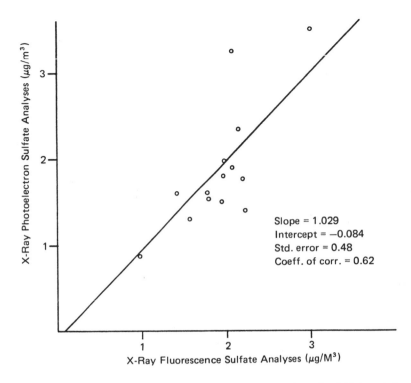

Figure 8. Comparison of quantitative particulate sulfate analyses by ESCA and X-ray fluorescence spectroscopy (from Ref. 46).

sulfate analyses which are correct within a factor of 2. A qualitative indication of the precision of ESCA results is obtained by comparing the diurnal patterns for sulfate determined both by ESCA and by alternate procedures. Figure 7 shows sulfate diurnal patterns as measured by ESCA and by the SRI method. The data are for a 24-h sampling period during a moderate smog episode in Pomona, California. ESCA and SRI procedures yield strikingly similar diurnal patterns suggesting a sufficient precision for the ESCA method.

More recently, Harker[46] compared the results of ESCA sulfate analysis to the results obtained by XRF and a wet chemical method. A number of 4-h samples of fluoropore filters were collected in the Los Angeles area. The results of Harker's study are shown in Figures 8 and 9. These results obviously prove that the ESCA technique can give a good representation of the bulk composition of atmospheric particulates in spite of its surface sensitivity.

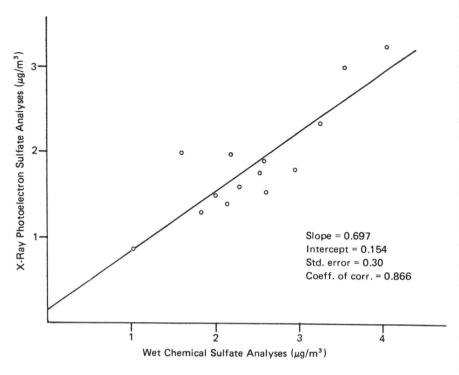

Figure 9. Comparison of quantitative particulate sulfate analyses by ESCA and a colorimetric wet chemical procedure (from Ref. 46).

Carbon analysis by ESCA has also been validated, and the results are described in the above-mentioned paper by Appel.[41] Twenty-nine samples collected on silver membrane filters were analyzed for total carbon, both by ESCA and by a combustion technique. A mean ratio of 0.9 ± 0.1 was found between the combustion method and ESCA, suggesting that average carbon analyses are reasonably accurate.

The results of ESCA analyses for nitrate were also compared with the results of wet chemical procedures conducted on the same filters and on 24-h filters.[41] With nitrate, using both comparative wet chemical techniques, the ESCA results were lower by a factor of about 5. This result is consistent with the volatilization of adsorbed nitric acid, as discussed above.

Similarly, ESCA analyses systematically underestimate the ammonium concentrations. The reasons for this discrepancy are related to the volatility of certain ammonium species in vacuum.

4.7. Determination of Molecular Forms by ESCA

ESCA analysis of particulates enables not only the possibility of detection of specific ions and functional groups, but also their mutual relationship. This is achieved by the measurement of the ESCA chemical shift augmented by the determination of relative concentrations and by study of the volatility properties of certain particulate species.

The capability of ESCA for a straightforward differentiation of different forms of atmospheric sulfates was recently demonstrated by Novakov *et al.*[47,48] Figure 10 shows the nitrogen (1s) and sulfur (2p) regions in ESCA spectra of two ambient samples. One was collected in West Covina, California, in the summer of 1973, and the other was collected in St. Louis, Missouri, in the summer of 1975. The peak

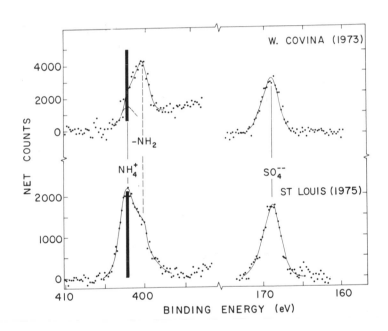

Figure 10. Nitrogen (1s) and sulfur (2p) regions in X-ray photoelectron spectra of two ambient samples. The peak positions corresponding to NH_4^+, $-NH_2(N_x)$, and SO_4^{2-} are indicated. The solid vertical bar represents the ammonium intensity expected under the assumption that the entire sulfate is in the form of ammonium sulfate. The differences in the relative ammonium content of the two samples is obvious. The sulfate and ammonium intensities in the St. Louis sample are compatible with ammonium sulfate. The ammonium content in the West Covina sample is insufficient to be compatible with ammonium sulfate. Both samples were exposed to the spectrometer vacuum for about one hour (from Ref. 48).

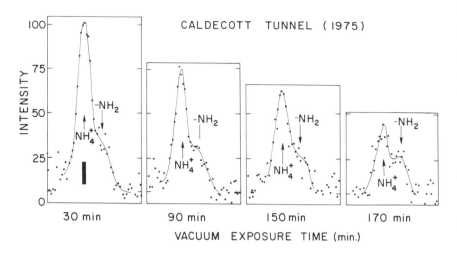

Figure 11. The variation in the observed ammonium peak intensity with vacuum exposure for a sample collected in a highway tunnel. The decrease in the peak intensity is caused by the volatilization of the ammonium species present in the sample. The solid vertical bar represents the ammonium intensity expected under the assumption that the sulfate in this sample is in the form of ammonium sulfate. The nitrate in this sample is also small compared with ammonium. The ammonium in this sample is considerably in excess of that expected for ammonium sulfate or ammonium nitrate (from Ref. 48).

positions corresponding to NH_4^+, $—NH_2$, and SO_4^{2-} are indicated. The solid vertical bar indicates the ammonium peak intensity expected under the assumption that the entire sulfate is in the form of ammonium sulfate. Obviously, the observed ammonium content in the West Covina sample is insufficient to account for the sulfate by itself. This is in sharp contrast with the St. Louis sample where the observed ammonium intensity closely agrees with that expected for ammonium sulfate.

These results demonstrate that ammonium sulfate in the aerosols can easily be distinguished from other forms of sulfate such as the one found in the West Covina case. However, wet chemical analyses[49] performed on West Covina samples collected simultaneously with the ESCA samples resulted in ammonium concentrations substantially higher than those suggested by the ESCA measurements. As mentioned earlier, this apparent discrepancy between the two methods was subsequently explained by the volatility of some ammonium species in the ESCA spectrometer vacuum. That these volatile losses are not caused by the volatilization of ammonium sulfate is evidenced by the St. Louis case, where no volatile losses were observed. Similarly, ammonium nitrate (negligible in these samples) and ammonium bisulfate were found to be stable in vacuum during the time periods usually required to complete the analysis.

Therefore, species other than these have to be responsible for the apparent loss of ammonium in vacuum.

That the volatile ammonium is not necessarily associated with sulfate or nitrate ions is illustrated by means of results[47,48] represented in Figure 11. Here the changes in the nitrogen (1s) spectrum of a sample collected in a highway tunnel are shown as a function of the sample vacuum exposure time. Obviously, the ammonium peak intensity decreases with the vacuum exposure time of the sample. The amino-type nitrogen species intensity remains constant, however. The amount of nitrate in this sample was negligible compared with ammonium. The maximum ammonium peak expected under the assumption that the entire sulfate is ammonium sulfate is indicated by the solid vertical bar in Figure 11. It is obvious, therefore, that the counterions for this ammonium are neither nitrate nor sulfate.

Figure 12 summarizes the findings about ammonium volatility in the

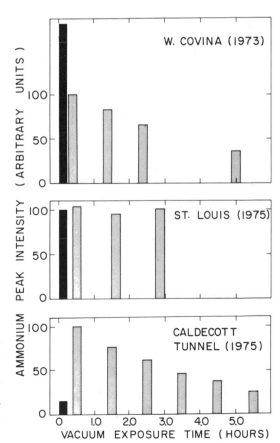

Figure 12. Volatility properties of West Covina, St. Louis, and automotive ammonium aerosol. The shaded bars on the far left of the figure indicate the expected ammonium intensity if the entire sulfate were ammonium sulfate (from Ref. 48).

three samples discussed above. The shaded bars at the far left of the figure indicate the expected ammonium intensity based on the assumption that all of the sulfate in the sample is in the form of ammonium sulfate. It is evident from the figure that only the St. Louis sample contains ammonium sulfate, while the West Covina and the tunnel samples contain a different kind of ammonium which volatilizes in the spectrometer vacuum.

Novakov *et al.*[47,48] have applied this procedure routinely to analyze a number of ambient samples. The results of such measurements for six St. Louis samples are shown in Figure 13 where the ratio of the observed ammonium peak intensity to the peak intensity expected of the ammonium in the form of ammonium sulfate is plotted as a function of the sample vacuum-exposure time. From inspection of this figure, it is evident that in addition to the cases of practically stoichiometric ammonium sulfate (samples 913 and 914), there are cases where the observed

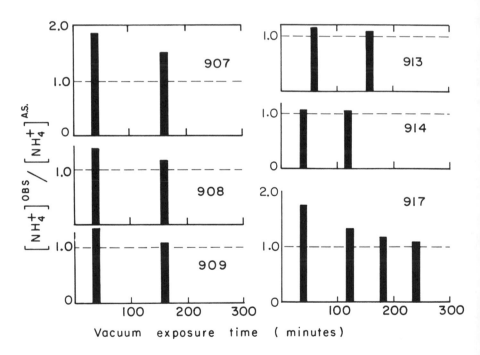

Figure 13. Volatility property of ammonium in six ambient St. Louis samples. The ratio of the observed ammonium peak to the one expected under the assumption that the entire sulfate in these samples is ammonium sulfate vs. vacuum exposure time is shown. Note the cases of apparently stoichiometric ammonium sulfate (samples 913 and 914) and the cases where the volatile ammonium component is found in excess of that required for ammonium sulfate (from Ref. 48).

ammonium is found in excess of ammonium sulfate. The excess ammonium consists of volatile ammonium species which decay away until ammonium sulfate is the only ammonium species left (sample 917).

The anions corresponding to the volatile ammonium species cannot be identified with certainty at this time. One possibility is that these species are produced by the adsorption of ammonia on fine soot particles to form carboxyl- and hydroxyl-ammonium complexes which have been shown to have similar volatility properties to ambient particulates.[3] Another possibility is that these species could be due to ammonium halides which are also volatile in vacuum.

In conclusion, ESCA analysis of ambient samples allows for a straightforward differentiation of different forms of atmospheric sulfate- and ammonium-containing species. The following distinctly different cases have been identified:

1. Ammonium sulfate accounts for the entire ammonium and sulfate content of the sample.
2. Ammonium appears in concentrations above those expected for ammonium sulfate (and nitrate). The "excess" ammonium is volatile in vacuum.
3. Ammonium appears mostly in a volatile form independent of sulfate and nitrate.

4.8. Use of ESCA in Reaction Mechanism Studies

ESCA has also been applied to characterize the particulate products of both heterogeneous (gas-particle) and homogeneous (gas-phase) reactions performed under laboratory conditions. The analysis of the reaction products is essential in order to assess the relevance of such reactions to the processes that actually occur in the "real" atmosphere.

The ESCA method has been most extensively used in conjunction with heterogeneous aerosol reactions, especially those involving fine combustion-generated carbonaceous (soot) particles. These particles are the inevitable products of even seemingly complete combustion. Furthermore, the chemical properties of soot particles are similar to those of activated carbon, which is a well-known catalytically and surface-chemically active material. Soot particles are formed as the result of pyrolysis, polymerization, and condensation of gases and vapors produced in the incomplete combustion of fossil fuels.[50] Fine soot particle diameters are of the order of 100 Å. The crystallites consist of several layers which exhibit the hexagonal graphitic structure.

In addition to carbon, soot particles contain about 1 to 3% hydrogen and 5 to 15% oxygen by weight.[51] Oxygen associated with soot particles is

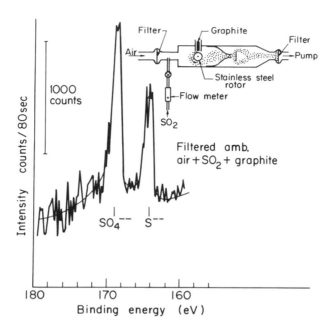

Figure 14. Experimental arrangement used to study the interaction of SO_2 with graphite particles. The rotation of the stainless steel rotor produces small graphite particles, which can be collected on a filter after they have interacted with SO_2. The ESCA spectrum of graphite particles reveals two sulfur (2p) peaks corresponding to sulfate and to sulfides (from Ref. 53).

located in surface carbon–oxygen complexes, typically of the carboxyl, phenolic hydroxyl, and quinone carbonyl type.[52] The nature and the abundance of these surface complexes depend upon the combustion regime. Therefore, soot particles from different sources may differ not only in their physical properties such as size, but also in their surface-chemical reactivity. An extreme case would be lamp black or carbon black. This material consists of relatively large particles and has a very low oxygen-to-carbon ratio. Fine soot particles which are invisible to the naked eye, on the other hand, have a much higher oxygen-to-carbon ratio. In this sense, fine soot particles can be regarded as an oxidized form of carbon black.

The laboratory experiments where ESCA was used to detect the reaction products involved surface-chemical reactions of various carbon particles and gases such as sulfur dioxide, nitric oxide, and ammonia.

The following experiment[53] may serve as a demonstration of the oxidizing and reducing properties of carbon particles. The experiment was performed with the setup shown in Figure 14, by means of which small graphite particles (diameters $\simeq$ 20 μm) can be collected on a filter after they have interacted with sulfur dioxide and air. The ESCA spectrum of such graphite particles clearly demonstrates the presence of two sulfur (2p) peaks corresponding to sulfate (binding energy of 168.4 eV) and sulfide. Blank filters without graphite particles, under identical sulfur dioxide exposure conditions, do not produce measurable amounts of sulfate (or sulfide).

An insight into what governs the oxidative and reductive properties of carbon particles can be gained from the results of the following experiments.[54] We have observed that the ESCA spectrum of "atomically" clean graphite (i.e., showing no trace of oxygen on its surface) exposed to about 10^{-6} Torr-s sulfur dioxide reveals only the presence of reduced sulfur species (sulfide). However, if the graphite surface were initially oxygenated and subsequently exposed to sulfur dioxide under similar conditions, sulfate formation occurred. The oxidation (or oxygenation) can be achieved by *in-situ* exposure of a hot graphite surface to water vapor or by cutting the graphite in air to expose the fresh surface to oxygen and moisture. These results indicate that surface carbon–oxygen complexes are responsible for the oxidative catalytic activity of carbonaceous materials. The sulfide species, on the other hand, could be produced by chemisorption of sulfur dioxide on carbon surface.

Sulfur dioxide oxidation by soot particles is demonstrated by the following experiment.[53] Soot specimens from a premixed propane–oxygen flame were prepared on silver membrane filters. In order to assure the equivalency of soot substrates to be used for experiments with different sulfur dioxide exposure conditions, small sections of each filter were cut and used in the apparatus shown in Figure 15 (inset). Dry and prehumidified particle-free air or nitrogen was used with a sulfur dioxide concentration of about 300 ppm and an exposure time of 5 min. The entire chamber was kept at about 150°C to prevent water condensation. Typical ESCA spectra of sulfur-dioxide-exposed soot samples are shown in Figure 15. Sulfate peaks were always found to be more intense in the case of prehumidified air than in the case of dry air. Blank filters without soot particles exposed to sulfur dioxide and prehumidified air under identical conditions showed negligible sulfate levels. When dry and prehumidified nitrogen were used instead of air, very low levels of sulfate were produced. This indicates the importance of the oxygen in air for sulfur dioxide oxidation. Water molecules enhance the observed sulfate concentration in the air + sulfur dioxide + soot system. The only possible competing mechanism of relevance to the described experiment is sulfur dioxide oxidation via dissolved molecular oxygen in water droplets. This alternative is ruled out by the experiment with blank filter and prehumidified air, which does not result in significant sulfate formation. Further-

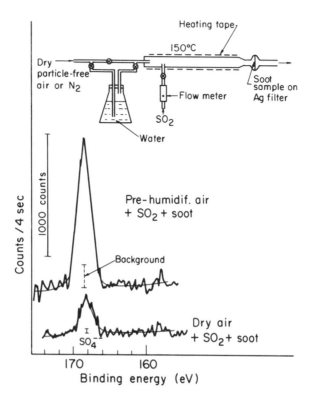

Figure 15. Apparatus used to study the interaction of SO_2 with soot particles collected on filters from a propane–oxygen flame. Dry or prehumidified particle-free air was used. Typical ESCA spectra of SO_2-exposed soot are also shown. Sulfate peaks were always found to be more intense in the case of prehumidified air. Blank filters, that is, filters without soot particles, exposed to SO_2 and prehumidified air under identical conditions showed only low, background-level sulfate peaks (from Ref. 53).

more, the elevated temperature prevents the formation of liquid water droplets.

The above-described experiment is a "static" experiment, i.e., with soot particles precollected on a filter and subsequently exposed to SO_2 and air. Demonstration of SO_2 oxidation on soot particles can also be done with a flow system, where SO_2 can be introduced downstream from the flame.

Photoelectron spectra representing the sulfur (2p) and carbon (1s) regions of propane soot particles produced by a Bunsen burner[48] are shown in Figure 16a. The S(2p) photoelectron peak at a binding energy of 169 eV corresponds to sulfate. The C(1s) peak appears essentially as a single component line and corresponds to a substantially neutral charge state consistent with the soot structure. It is of interest to note that even the combustion of very low sulfur content fuels (0.005% by weight) results in the formation of an easily detectable sulfate emission.

The specific role of soot particles as a catalyst for the oxidation of SO_2 is demonstrated with the aid of Figure 16b. Here we show the S(2p) and C(1s) photoelectron peaks of soot particles, generated in an analogous manner, but exposed to additional SO_2 in a flow system. A marked increase in the sulfate peak intensity, relative to carbon, is evident. The atomic ratios of sulfur to carbon in Figures 16a and 16b are about 0.15 and 0.50 respectively.

The sulfate associated with soot particles is water-soluble and contributes to the acidification of the solution. These suggest that this sulfate is actually surface-bonded sulfuric acid. The hydrolysis product of this sulfate behaves chemically as sulfuric acid because it easily forms ammonium sulfate in reaction with gaseous ammonia.[48] This is illustrated in Figure 17. Figure 17a shows the N(1s) and S(2p) ESCA regions of a soot sample exposed to SO_2. Here again, the solid bar represents the ammonium intensity expected if the entire sulfate were ammonium sulfate. The nitrogen content of this sample is low compared with sulfate. In Figure 17b the ESCA spectrum of a soot sample exposed first to SO_2 and

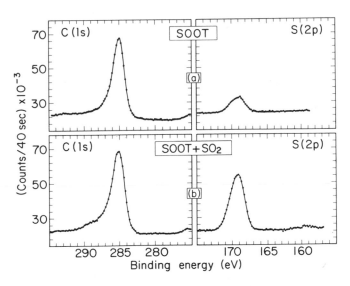

Figure 16. Carbon (1s) and sulfur (2p) photoelectron spectrum of (a) soot particles produced by combustion of propane saturated with benzene vapor (sulfur content of this fuel, 0.005% by weight), and (b) soot particles generated in a manner analogous to (a), but exposed to additional SO_2 in humid air (from Ref. 48).

subsequently to NH_3 in humid air is shown. It is obvious that ammonium sulfate is the principal form of sulfate in the sample.

ESCA was also used to study the formation of reduced particulate nitrogen by gas-particle reactions. It was shown that reduced nitrogen species identical to those observed in ambient particulates, i.e., N_x and a volatile ammonium species, can be produced by surface reactions of common pollutant gases such as ammonia and nitric oxide with soot particles.[3] Nitrogen species produced in these reactions at elevated temperature occur in the form of surface complexes such as amines, amides, and nitriles. Ammonium ions may be associated with soot particles in the form of carboxyl-ammonium, hydroxyl-ammonium salts, and/or physically adsorbed ammonia species produced by soot–NH_3 and soot–NO reactions at ambient temperature.

The temperature-dependent ESCA measurements of "synthetic" samples demonstrated that the nitrogen species produced by the above reactions are equivalent to the ambient air particulate species. The procedure employed is the same as that used for ambient samples.

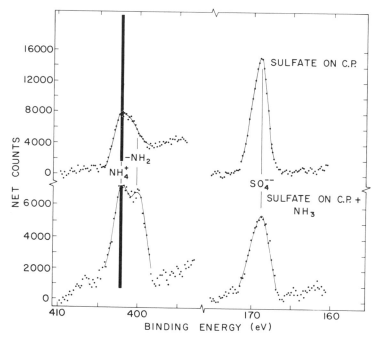

Figure 17. Top: ESCA spectrum of sulfate produced by catalytic oxidation of SO_2 on soot particles. Note that the nitrogen content in this sample is low compared with sulfate. Bottom: ESCA spectrum of a soot sample exposed first to SO_2 (i.e., as in Fig. 16a), and subsequently to NH_3 in humid air. Formation of ammonium sulfate is evident.

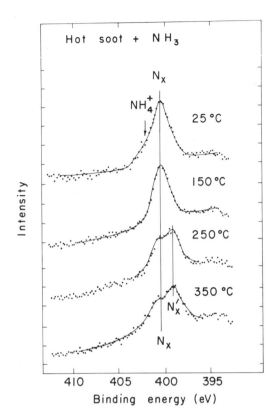

Figure 18. Nitrogen (1s) spectrum of a soot sample exposed to NH_3 at elevated temperature as measured at 25, 150, 250, and 350°C (from Ref. 3).

Results in Figure 18 show that N_x species produced by surface reactions of hot soot with NH_3 have the same kind of temperature dependence as the ambient samples. The spectrum taken at room temperature shows that most nitrogen species in this sample are of the N_x type. Heating the sample in vacuum to 150°C does not influence the line shape or intensity. At 250°C, however, the formation of N_x' is evident. Further transformation of N_x to N_x' occurred at 350°C.

Synthetic N_x' species will remain unaltered even when the temperature is lowered back to room temperature if the sample remains in vacuum. However, if the sample is taken out of vacuum and exposed to moisture, N_x' will be transformed back to the original N_x compound. The behavior of the synthetic N_x species is thus identical with the behavior of ambient species.

It was proposed that at ambient temperatures surface carboxyl and phenolic groups will react with ammonia to form carboxyl ammonium or phenolic ammonium salts. Ammonia may also physically adsorb by hydrogen bonding to these same functional groups.

At elevated temperatures ammonia will react by a nucleophilic substitution reaction with carboxyl groups to produce an amide, which may become a nitrile by further dehydration. Alternatively, carboxyl and phenolic hydroxyl ammonium salts may dehydrate at elevated temperatures to produce amides and/or nitriles, and amines, respectively.

Acknowledgment

This work was done with support from the U.S. Energy Research and Development Administration and the National Science Foundation—RANN.

5. References

1. T. Novakov, in: *Proceedings of the Second Joint Conference on Sensing Environmental Pollutants*, pp. 197–204, Instrument Society of America, Pittsburgh (1973).
2. T. Novakov, J. W. Otvos, A. E. Alcocer, and P. K. Mueller, Chemical composition of photochemical smog aerosol by particle size and time of day; chemical states of sulfur and nitrogen by photoelectron spectroscopy, *J. Colloid, Interface Sci., 39,* 225 (1972).
3. S. G. Chang and T. Novakov, Formation of pollution particulate nitrogen compounds by NO-soot and NH_3-soot gas-particle surface reactions, *Atmos. Environ., 9,* 495 (1975).
4. C. E. Junge, *Atmospheric Chemistry and Radioactivity*, Academic Press, New York (1973).
5. G. M. Hidy and C. S. Burton, Atmospheric aerosol formation by chemical reactions, *Int. J. of Chem. Kinetics, Symposium, 1,* 509 (1975).
6. T. Koopmans, Über die Zuordnung von Wellenfunktionen und Eigenwerten zu den einzelnen Elektronen eines Atoms, *Physica, 1,* 104 (1933).
7. P. S. Bagus, Self-consistent field wave functions for hole states of some Ne-like and Ar-like ions, *Phys. Rev., 139,* A619 (1965).
8. A. Rosen and I. Lindgren, Relativistic calculations of electron binding energies by a modified Hartree-Fock-Slater Method, *Phys. Rev., 176,* 114 (1968).
9. M. E. Schwartz, Direct calculation of binding energies of inner-shell electrons in molecules, *Chem. Phys. Lett., 5,* 50 (1970).
10. C. R. Brundle, M. B. Robin, and H. Basch, Electronic energies and electronic structures of the fluoromethanes, *J. Chem. Phys., 53,* 2196 (1970).
11. P. Siegbahn, *Ab initio* calculations on furan with a new computer program, *Chem. Phys. Lett., 8,* 245 (1971).
12. L. Hedin and G. Johansson, Polarization corrections to core levels, *J. Phys. B, 2,* 1336 (1969).
13. L. C. Snyder, Core-electron binding energies and Slater atomic shielding constants, *J. Chem. Phys., 55,* 95 (1971).
14. D. Liberman, Improvement on Koopmans' theorem, *Bull. Am. Phys. Soc., Ser. 9,* 731 (1964).

15. R. Manne and T. Åberg, Koopmans' theorem for inner-shell ionization, *Chem. Phys. Lett.*, 7, 282 (1970).
16. T. D. Thomas, X-ray photoelectron spectroscopy of halomethanes, *J. Am. Chem. Soc.*, 92, 4184 (1970).
17. K. Siegbahn, C. Nordling, G. Johansson, J. Hedman, P. F. Heden, K. Hamrin, U. Gelius, T. Bergmark, L. O. Werme, R. Manne, and Y. Baer, *ESCA Applied to Free Molecules*, North-Holland, Amsterdam (1969).
18. M. E. Schwartz, Correlation of core electron binding energies with the average potential at a nucleus: carbon 1s and extended Hückel theory valence molecular orbital potentials, *Chem. Phys. Lett.*, 7, 78 (1971).
19. J. A. Pople, D. P. Santry, and G. P. Segal, Approximate self-consistent molecular orbital theory. I. Invariant procedures, *J. Chem. Phys.*, 43, S129 (1965).
20. F. O. Ellison and L. L. Larcom, ESCA: A new semi-empirical correlation between core-electron binding energy and valence-electron density, *Chem. Phys. Lett.*, 10, 580 (1971).
21. D. W. Davis, D. A. Shirley, and T. D. Thomas, K-electron binding energy shifts in fluorinated methanes and benzenes: Comparison of a CNDO potential method with experiment, *J. Chem. Phys.*, 56, 671 (1972).
22. U. Gelius, P. F. Heden, J. Hedman, B. J. Lindberg, R. Manne, R. Nordberg, C. Nordling, and K. Siegbahn, Molecular spectroscopy by means of ESCA, *Phys. Ser.*, 2, 70 (1970).
23. U. Gelius, B. Roos, and P. Siegbahn, *Ab initio* MO SCF calculations of ESCA shifts in sulphur-containing molecules, *Chem. Phys. Lett.*, 4, 471 (1970).
24. K. Siegbahn, C. Nordling, A. Fahlman, R. Nordberg, K. Hamrin, J. Hedman, G. Johansson, T. Bergmark, S.-E. Karlsson, I. Lindgren, and B. J. Lindberg, ESCA-atomic, molecular and solid state structure by means of electron spectroscopy, *Nova Acta Regiae Soc. Sci. Upsaliensis Ser. IV*, Vol. 20 (1967).
25. H. Basch, On the interpretation of K-shell electron binding energy chemical shifts in molecules, *Chem. Phys. Lett.*, 5, 337 (1970).
26. M. E. Schwartz, Correlation of 1s binding energy with the average quantum mechanical potential at a nucleus, *Chem. Phys. Lett.*, 6, 631 (1970).
27. C. S. Fadley, Multiplet splittings in photoelectron spectra, in: *Electron Spectroscopy* (D. A. Shirley, ed.), p. 781, North-Holland, Amsterdam (1972).
28. See C. S. Fadley, R. Baird, W. Siekhaus, T. Novakov, and S. A. L. Bergstrom, Surface analysis and angular distribution measurements in X-ray photoelectron spectroscopy, *J. Electron Spectrosc. Relat. Phenom.*, 4, 93 (1974), and references therein.
29. C. D. Wagner, Sensitivity of detection of the elements by photoelectron spectrometry, *Anal. Chem.*, 44, 1050 (1972).
30. R. S. Swingle, II, Quantitative surface analysis by X-ray photoelectron spectroscopy, *Anal. Chem.*, 47, 21 (1975).
31. D. M. Wyatt, J. C. Carver, and D. M. Hercules, Some factors affecting the application of electron spectroscopy (ESCA) to quantitative analysis of solids, *Anal. Chem.*, 47, 1297 (1975).
32. P. E. Larson, Quantitative measurements on gold–silver alloys by X-ray photoelectron spectroscopy, *Anal. Chem.*, 44, 1678 (1972).
33. W. E. Swartz, Jr., and D. M. Hercules, X-ray photoelectron spectroscopy of molybdenum compounds. Use of electron spectroscopy for chemical analysis (ESCA) in quantitative analysis, *Anal. Chem.*, 43, 1774 (1971).
34. C. R. Brundle, and M. W. Roberts, Surface sensitivity of ESCA [electron spectroscopy for chemical analysis] for submonolayer quantities of mercury adsorbed on a gold substrate, *Chem. Phys. Lett.*, 18, 380 (1973).

35. D. M. Hercules and J. C. Carver, Electron spectroscopy: X-ray and electron excitation, *Anal. Chem., 46,* 133R (1974).
36. J. S. Brinen, Applying electron spectroscopy for chemical analysis to industrial catalysis, *Accounts Chem. Res., 9,* 86 (1976).
37. T. Novakov, C. D. Wagner, and J. W. Otvos, Analysis of atmospheric particulates by means of a photoelectron spectrometer, paper presented at the Pacific Conference on Chemistry and Spectroscopy, *Abstracts,* p. 46, San Francisco, October 1970.
38. L. D. Hulett, T. A. Carlson, B. R. Fish, and J. L. Durham, Studies of sulfur compounds adsorbed on smoke particles and other solids by photoelectron spectroscopy, *Proceedings of the Symposium on Air Quality,* p. 179, Plenum, New York (1972).
39. Y. E. Araktingi, N. S. Bhacca, W. G. Proctor, and J. W. Robinson, Analysis of airborne particulates by electron chemistry for chemical analysis (ESCA), *Spectrosc. Lett., 4,* 365 (1971).
40. N. L. Craig, A. B. Harker, and T. Novakov, Determination of the chemical states of sulfur in ambient pollution aerosols by X-ray photoelectron spectroscopy, *Atmos. Environ., 8,* 15 (1974).
41. B. R. Appel, J. J. Wesolowski, E. Hoffer, S. Twiss, S. Wall, S. G. Chang, and T. Novakov, An intermethod comparison of X-ray photoelectron spectroscopic analysis of atmospheric particulate matter, *Intern. J. Environ. Anal. Chem., 4,* 169 (1976).
42. D. F. Miller and C. W. Spicer, Measurement of nitric acid in smog, *J. Air Pollut. Control Assoc., 25,* 940 (1975).
43. A. B. Harker, P. J. Pagni, T. Novakov, and L. Hughes, Manganese emissions from combustors, *Chemosphere, 6,* 339 (1975).
44. For the description of wet chemical methods see Reference 41 and references therein.
45. R. D. Giauque, data presented in: Characterization of Aerosols in California (ACHEX), Vol. III, Science Center, Rockwell International, final reports to Air Resources Board, State of California (September 1974).
46. A. B. Harker, Quantitative Comparison of the XPS Technique with XRF and Wet Chemical Sulfur Analyses, Science Center, Rockwell International, Internal Technical Report (1976); and private communication.
47. T. Novakov, R. L. Dod, and S. G. Chang, Study of air pollution particulates by X-ray photoelectron spectroscopy, *Fresenius' Z. Anal. Chem., 282,* 287 (1976).
48. T. Novakov, S. G. Chang, R. L. Dod, and H. Rosen, Chemical characterization of aerosol species produced in heterogeneous gas-particle reactions, APCA Paper 76-20.4, presented at the Air Pollution Control Association Annual Meeting, Portland, Oregon, June 1976. Also Lawrence Berkeley Laboratory Report LBL-5215 (June 1976).
49. C. W. Spicer, private communication.
50. J. B. Edwards, *Combustion-Formation and Emission of Trace Species,* Ann Arbor Science, Ann Arbor (1974), and references therein.
51. A. Thomas, Carbon formation in flames, *Combustion and Flame, 6,* 46 (1972).
52. H. P. Boehm, Chemical identification of surface groups, *Advan. Catal. Relat. Subj., 16,* 179 (1966).
53. T. Novakov, S. G. Chang, and A. B. Harker, Sulfates as pollution particulates: Catalytic formation on carbon (soot) particles, *Science, 186,* 259 (1974).
54. T. Novakov and W. Siekhaus, unpublished data.

6

Using the Subject as His Own Referent in Assessing Day-to-Day Changes of Laboratory Test Results

Per Winkel and Bernard E. Statland

1. Introduction

Clinical chemistry can be defined as that part of the health services which has as its goals the measuring of analytes in body fluids and tissues of subjects and the interpretation of these measurements to assess the health status of human subjects. To accomplish these objectives, the clinical chemist must relate the laboratory results obtained on a subject to an appropriate reference.

Classically a reference interval has been used for each quantity based on the values obtained on a group of healthy subjects, i.e., one specimen is collected from each subject, and then all the specimens are assayed for the analyte in question. The results are then analyzed and a *group-specific reference interval* can be established. This reference interval is used on a more or less intuitive basis in assessing the clinical chemistry quantity values of any subject. As will become apparent later in this review, we have found that the classical group-specific approach is often very insensitive in assessing the health status of a given subject.

An alternative approach is to use the subject's previous values as a referent for any future value. In this latter approach a number of specimens are obtained from a subject over a stated period of time during which the subject is in the same well-defined state of health. All specimens

Per Winkel and Bernard E. Statland • Department of Hospital Laboratories, North Carolina Memorial Hospital, Chapel Hill, North Carolina 27514

are assayed for the analyte in question, and the results are used to estimate an interval which will contain the quantity value as measured in a future specimen with a specified probability p subject to the condition that the subject is still in the same state of health. We shall refer to such an interval as a *subject-specific prediction interval*. An estimate of such an interval we shall denote as a *subject-specific reference interval*.

The present review will focus on the subject-specific reference interval. It is divided into five major sections. The first part of the review will assess the need for using the subject as his own referent by comparing the intraindividual and the interindividual variations as measured in healthy subjects. In the second part we will develop a general theoretical framework within which it should be feasible to analyze the problems involved in the computation of subject-specific reference intervals. In the third part of the review, we will present the various mathematical models which have been suggested as a basis for the computation of subject-specific reference intervals for healthy subjects and we will discuss some of the problems involved in their application. In the fourth part, we will present the results of many investigators who have estimated the day-to-day variation in healthy subjects, and in the fifth part we will assess the influence of the analytical variation on the total intraindividual variation.

2. Assessing the Need for Using the Subject as His Own Referent

One of the major problems in computing a subject-specific reference interval is choosing the appropriate biological model which explains the variation of laboratory test results as a function of time elapsed. This review will present various models which have been proposed. One of the models, the homeostatic model, assumes that the mean of the values over time is constant. Using this model one is able to determine the variation occurring within an individual over time over a series of values. This variation is referred to as the *intraindividual variation*. In addition when one obtains specimens from a number of individuals in a group of subjects over a number of occasions, one can determine the *interindividual variation* of mean values of the subjects in the group. A third term, which has been suggested by Harris,[1] the *ratio value*, is computed from the other two, and is defined as the ratio of the intraindividual variation over the interindividual variation when both variations are in terms of standard deviation. Over the past four years, we have determined the ratio values of a number of analytes found in the serum or in the whole blood of healthy subjects. The majority of the experiments lasted between 10 and 20 days with venipunctures taking place every second or third day on the average and always at the same time of day. The groups studied consisted

Table 1. Ratio Values (Intraindividual over Interindividual Variation) for Selected Analytes in the Blood of Healthy Subjects

Analyte	Ratio value	Analyte	Ratio value
Group I		*Group III*	
IgM	0.06	Lactate dehydrogenase	0.44
IgA	0.09	Alanine aminotransferase	0.45
Haptoglobin	0.12	Creatinine	0.47
Alkaline phosphatase	0.14		
IgG	0.15	Total WBC	0.48
γ-Glutamyl transferase	0.16	Cortisol	0.50
Complement C3	0.17	Creatine kinase	0.56
Complement C4	0.19	Neutrophils	0.62
α_1-Antitrypsin	0.19	Thyroxine	0.63
α_2-Macroglobulin	0.19	Urea	0.73
		Aspartate aminotransferase	0.90
Group II		Uric acid	1.00
Basophils	0.24		
Cholesterol	0.25	*Group IV*	
Transferrin	0.26	Total protein	1.04
Orosomucoid	0.26	Chloride	1.05
Eosinophils	0.28	Iron	1.15
Hemoglobin	0.29	Potassium	1.19
Hematocrit	0.29	Sodium	1.75
Platelets	0.30	Calcium	2.83
Lymphocytes	0.31	Albumin	5.60
Monocytes	0.34		
Triglycerides	0.36		

of 11–20 healthy subjects in the age range of 20–40 years. Table 1 presents the ratio values for 39 of the analytes found in blood. As noted we have arranged the ratios in ascending order and have divided the analytes into four groups on the basis of ratio values: Group I, 0.01–0.20; Group II, 0.21–0.40; Group III, 0.41–1.00; Group IV, 1.01–10.00.

Figure 1 depicts the subject-specific ranges* of the concentration values of haptoglobin for each of 14 healthy subjects who were bled on six "occasions" over a six-day period. For this analyte the ratio value was only 0.12. Figure 2 depicts the subject-specific ranges for the activity values of aspartate aminotransferase for each of the same 14 healthy subjects.[3] In this case (Figure 2) the ratio value was 0.90. The smaller the ratio value, the greater is the need for using subject-specific reference intervals rather than relying upon group-specific reference intervals. For the analytes in Groups I and II, the group-specific reference interval would be much too insensitive in identifying a subject in whom a "present value" is outside of

---

* The smallest interval including the highest and the lowest values observed in a subject.

his personal, subject-specific interval. As noted in Table 1, the analytes found in Groups I and II consist of the specific proteins, many of the hematological quantities, the enzymes alkaline phosphatase and γ-glutamyl transferase, and the lipids cholesterol and triglycerides. The analytes found in Group IV are those analytes which are known to affect physiological functions when they are only slightly abnormal, e.g., calcium, potassium, sodium, and albumin.

On the basis of the above information, it now becomes important to investigate the magnitudes and nature of the intraindividual variation of various analytes in healthy subjects. The total intraindividual variation is a variation resulting from the effect of three factors: (1) preparation of the subject, (2) physiological variation as a function of time elapsed, and (3) analytical variation. The preparation of the subject includes prior diet,[4] previous physical activity,[5] prior ethanol ingestion,[6,7] posture of the subject prior to venipuncture,[8] duration of tourniquet application before venipuncture,[9] etc. Recently Statland and Winkel[10] have reviewed the various factors involved in the preparation of the subject, and thus this present review will not deal with them specifically. However, when subject-specific reference intervals are computed and used for predictive pur-

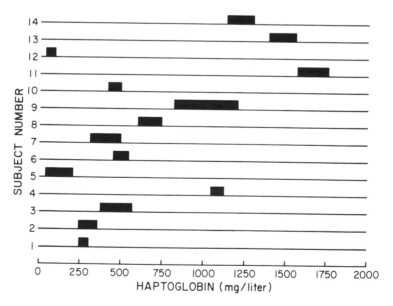

Figure 1. Range of concentration values for haptoglobin in sera of 14 healthy subjects followed over 10 days. (From B. E. Statland, P. Winkel, and L. M. Killingsworth, *Clin. Chem.*, 22, 1635, 1976, with permission.)

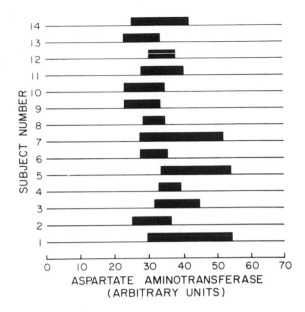

Figure 2. Range of activity values of aspartate aminotransferase in sera of 14 healthy subjects followed over 10 days.

poses, it is necessary that the measurements used for the computation of subject-specific reference intervals as well as the value to be predicted are all compatible with regard to the effect of the preparation of the subject and analytical procedure. In the following section we will develop a theoretical framework which defines these conditions more precisely.

3. A Theoretical Framework Defining the Conditions for the Application of Models of Biological Time Series

In this section we will attempt a more rigorous definition of the term subject-specific prediction interval. Furthermore we will introduce some concepts which will allow us to state explicitly the conditions to be fulfilled by previous observations if they are to be used for the computation of a subject-specific prediction interval. Assume that we are to obtain n measurements of the quantity Q in subject S and use these observations to compute a prediction interval for subject S at time t_{n+1}: Then $T = (t_1, t_2, \ldots, t_i, \ldots, t_n)$ is the vector of time points corresponding to the n measurements (the value of $t_i <$ the value of t_{i+1}, $i = 1, \ldots, n$).

3.1. Subject-Specific Prediction Interval

Let H be a statement pertaining to the health status of subject S. Our goal is to make inferences regarding the truth value of this statement at time t_{n+1} based on the present observation and previous observations made when H was true. We therefore first define a function $h(t)$ which is equal to 1 if H is true at time t and equal to zero if H is false at time t. If $q_s(t)$ denotes a possible value of the quantity Q as measured in a specimen obtained from subject S at time t, the set $M = \{q_s(t_{n+1}H)\}$ is the set of all possible values of the quantity Q as measured in a specimen obtained from subject S at time t_{n+1}. $\mathcal{P}$ is a probability defined on the set M and Z denotes the identity function the domain of which is M. We now define a function $F(p)$. The domain of this function is the closed interval $[0,1]$ and the range is a set of sets of real-valued bounded closed intervals or $]-\infty, +\infty[$ and it is defined as follows: $F(p) = \{I \mid \mathcal{P}(Z \in I \mid h(t_{n+1}) = 1) = p\}$ the set of subject-specific p-prediction intervals. Each interval within this set has the property that the quantity Q as measured in a specimen obtained from subject S at time t_{n+1} will fall within this interval with the probability p subject to the condition that the statement H is true at time t_{n+1}.

3.2. The Reference-Value Vector

A quantity value is influenced by the preparation of the subject prior to the venipuncture as well as by the procedure of obtaining, handling, and assaying the specimen. We shall refer to the totality of these influences as the experimental procedure. $\mathcal{E}$ is the set of all possible experimental procedures. By E we denote a set function the domain of which is $\mathcal{E}$ and the range of which is a subset of R_k (the set of all possible k-dimensional vectors of real values). E thus is a prescription according to which we may characterize an experimental procedure. For example, an experimental procedure may be characterized by three values: (a_1, a_2, a_3) where a_1 denotes the length of the time period during which the subject was fasting prior to the venipuncture, a_2 the length of the time period prior to venipuncture during which the subject assumed the sitting position, and a_3 the duration of tourniquet application after venipuncture. In this example the experimental procedure is not characterized in any great detail. Thus E may assign the same set of values to many different experimental procedures, i.e., each element in $\mathcal{E}$ is mapped into one and only one element in R_k but several elements from $\mathcal{E}$ may be mapped into the same element in R_k. Let P_i be a probability defined on $\mathcal{E}$ at time t_i. For each subset ϵ of $\mathcal{E}$. $P_i(\epsilon)$ defines the probability that the actual experimental procedure ϵ_i to be employed at time t_i will belong to this subset. Implicitly P_i defines the probability that $E(\epsilon_i) \in r$ at time t_i where the set $r = E(\epsilon)$. It should be noted that it is only the latter probability we can estimate. Let $Q_s(T) = (q_s(t_1), q_s(t_2), \ldots, q_s(t_i), \ldots, q_s(t_n))$ be a vector of possible values of the quantity Q to be measured in specimens obtained from subject S where $q_s(t_i)$ is a possible value of the quantity Q to be measured in a specimen to be obtained at time t_i from subject S. $K = \{Q_s(T)\}$ then is the set of all such vectors of possible values. X is the identity function the domain of which is K.

Let the vector $x = (x_1, x_2, \ldots, x_i, \ldots, x_n)$ be the observed value vector of X; x then is a reference value vector for subject S at time t_{n+1} if (1) $h(t) = 1$ for all $t \in [t_1, t_n]$, (2) $P_i = P_{n+1}$ for $i = 1, 2, \ldots, n$.

3.3. Subject-Specific Reference Interval

A subject-specific reference interval for subject S at time t_{n+1} is a subject-specific prediction interval for subject S at time t_{n+1} computed from the values of a reference value vector for subject S at time t_{n+1}.

4. Selection of an Appropriate Model of Biological Time-Series

4.1. Definitions of Various Models

Before we can define and compute a subject-specific reference interval, we must adopt some notion as to the appropriate model of biological time-series. Must we assume a constant mean value (set point) over a long time interval (homeostatic model)? Should we allow a changing set point over time in a healthy subject (nonstationary model)? If we assume that the former model (fixed set point) is the more correct biological model, then we should place equal weight on all values observed in a subject over past time. If we assume a nonstationary set point, then we would rely most heavily on the more recent values.

Harris[11,12] reviewed three major statistical time-series models which belong to the class of so-called "autoregressive models." Two extreme models, the "homeostatic model" and the "random-walk model," and an "intermediate," more general model, were introduced. In the homeostatic model it is assumed that the quantity values fluctuate at random around a fixed set point. It is also assumed that these fluctuations are independent of each other. Such models are called *deterministic* since the random components in the models do not create any dependency among the measurements. Random biological fluctuations relative to a fixed set point μ take place. However, the organism responds to these fluctuations by changing the concentration value back to the set point. The salient point is that this counterregulation is so fast relative to the time interval between consecutive measurements that the fluctuation influencing any given measurement is without influence on the value of the subsequent measurement. However, if we shorten the length of the time interval (i.e., increase the frequency of timed measurements), we will reach a point where measurements are no longer independent.

In the second model, the random-walk model, the biological state or the true concentration value at time t (m_t) is equal to the biological state at time ($t-1$) (m_{t-1}) ($t-1$ is the time at which the previous measurement was made) plus the net effect of the random biological fluctuations taking place between the two times of measurements (Δ_t). Thus we have

$$m_t = m_{t-1} + \Delta_t \tag{1}$$

The measured value (x_t) is subject to analytical error (a_t) so we have

$$x_t = m_t + a_t \tag{2}$$

Combining (1) and (2) we get $x_t = m_{t-1} + \Delta_t + a_t$. The expected values of Δ_t and a_t are both zero, and they are independent of each other and of previous fluctuations. However, a complete *lack* of regulation does not seem to be a reasonable assumption, at least from a physiological point of view.

The third model, the intermediate model, bridges the gap between the homeostatic model and the random-walk model. Here, it is assumed that the quantity value is regulated toward a set point. However, consecutive observations are no longer independent in that the state of the system at time t depends on the previous state of the system at time $t-1$. In explicit mathematical terms the model states that the deviation of the biological state from the set point μ at time t is equal to the effect of random disturbances of the system (e_t) taking place between the two observation times plus the previous deviation from the set point ($m_{t-1} - \mu$) multiplied by a factor ρ, the absolute value of which lies between zero and one, i.e.,

$$m_t - \mu = \rho(m_{t-1} - \mu) + e_t \tag{3}$$

The factor ρ reflects the effectiveness of the organism in counteracting the disturbances present at time $t-1$, the two extremes being $\rho = 0$ (the homeostatic model) implying that previous disturbance of the system has been nullified at time t, and $\rho = 1$ (the random-walk model) implying that the previous disturbance has not been modified at all by the organism. Since the observed value x_t is equal to the biological state m_t plus the effect of analytical error a_t, we obtain the following equation when combining (2) and (3): $x_t - \mu = \rho(m_{t-1} - \mu) + e_t + a_t$.

4.2. The Choice of a Model and the Problem of Specifying an Alternative Hypothesis to That Implied by the Model

Once a model has been chosen, the next step is to predict future quantity values on the basis of values already observed in a given subject. The clinician may have introduced a factor, e.g., a new therapeutic maneuver, that may cause a change of the quantity value greater than that induced by random biological and analytical factors (i.e., the statement H would be: the quantity value is not influenced by the therapeutic maneuver), or he may suspect that a significant shift has occurred in a patient secondary to a change in the course of the disease (i.e., the statement H would state that the disease is in some well-defined state). In either of the two situations the observed value should be compared with that predicted by the model chosen and the decision be made whether or not one should act on the assumption that the observed change is due to the influence of random biological and analytical factors alone. For the model-specific computations involved in this process, the reader is referred to References 11 and 12 for a detailed derivation and application. However, in principle the computations are the same for all the models in that the difference between the "expected value" and the "observed value" divided by the standard deviation of this difference under the hypothesis is computed to obtain a quantity which at least for large n is assumed to follow the standardized Gaussian distribution with mean zero and variance 1. Granted the absolute change is due to random biological and analytical factors alone, the chance that this quantity will exceed 1.96, which is the 5% significance limit for the standardized Gaussian distribution, is approximately 5%.

For illustrative purposes let us apply the above approach to a specific example. The following example is taken from a recent study of the day-to-day changes of serum iron values in healthy subjects.[13] Figure 3 depicts the serum iron values as measured in one of the subjects on four consecutive days. The standard deviation (SD) for this subject is 3.17 μmol/liter. According to the homeostatic model the serum iron value predicted for Day 5 is the mean value of the first four observations, i.e., 17.6 μmol/liter. The standard deviation of the difference between the observed and the predicted value (i.e., the mean) is SD $\times \sqrt{1 + 1/n} =$ 3.17 $\times \sqrt{5/4} = 3.54$ (n is the number of previous observations, in this case, four). Thus a change in serum iron of 6.94 μmol/liter relative to the previously estimated mean value would be declared significant at the 5% level since 6.94/3.54 = 1.96. When we use the random-walk model (Figure 4) the predicted value is 14.64 μmol/liter and the standard deviation of the difference between predicted and observed value is 3.28 (the analytical standard deviation used is the total within-batch variation, which is 0.707

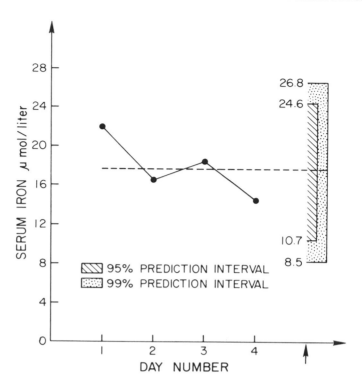

Figure 3. Estimated 95% and 99% prediction intervals for serum iron value in healthy subject using homeostatic model (see text). Since the Gaussian distribution was used for the computation of the intervals and not the Student's t distribution, which would be more correct, the true probability levels are approximately 0.1 and between 0.1 and 0.05 rather than 0.05 and 0.01.

μmol/liter). Figures 3 and 4 depict the differences between the two models. Figure 3 presents the estimated 95% and 99% prediction intervals assuming a homeostatic model, i.e., the most likely value on Day 5 would be the mean of the previous four values, or 17.6 μmol/liter. Figure 4 presents the estimated 95% and 99% prediction intervals using the random-walk model, i.e., the most likely value for Day 5 is approximately the same as the value of Day 4, or 14.6 μmol/liter. The serum iron value actually observed on the fifth day was 25 μmol/liter, which is of borderline significance according to the homeostatic model but significant according to the random-walk model. It should be noted though that it is only correct to use the Gaussian distribution when n is large. In the present example where n is quite small it would be more correct to use Student's t distribution. Therefore the true probability level is approximately 0.1

rather than 0.05. However, the purpose of the example is only to illustrate the differences between the two models. In applying these statistical techniques the traditional approach then would be to distinguish significant changes from insignificant ones. Such an approach, however, is less than adequate in the clinical setting as may be realized from the following:

Using the homeostatic model we found that only in 5% of the cases will random factors alone induce a difference between the predicted and the observed value which exceeds 6.94 μmol/liter or approximately 7.0 μmol/liter. Now, let us assume that a change of 7.0 μmol/liter relative to the set point has in fact been induced by a pathological factor. The resultant overall observed change will reflect the net effect of the pathological factor and the random factors. In 50% of the cases this additional change induced by random factors will be positive (see Figure 5) and consequently the observed change will exceed the 5% significance

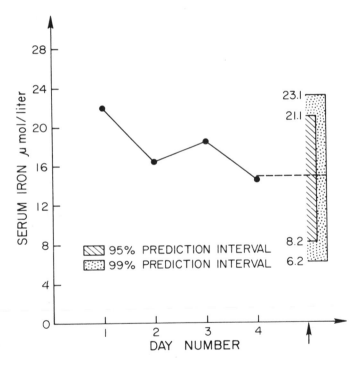

Figure 4. Estimated 95% and 99% prediction intervals for serum iron value in healthy subject using random-walk model (see text). Since the Gaussian distribution was used for the computation of the intervals and not the Student's t distribution, which would be more correct, the true probability levels are approximately 0.1 and between 0.1 and 0.05 rather than 0.05 and 0.01.

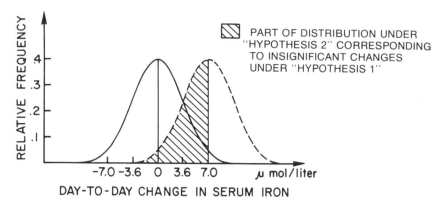

DAY-TO-DAY CHANGE IN SERUM IRON

—— DISTRIBUTION OF DAY-TO-DAY CHANGES IN
SERUM IRON VALUES UNDER THE "HYPOTHESIS
1" THAT CHANGES ARE INDUCED BY RANDOM
BIOLOGICAL AND ANALYTICAL FACTORS ALONE.

--- DISTRIBUTION OF DAY-TO-DAY CHANGES IN
SERUM IRON VALUES UNDER THE "HYPOTHESIS
2" THAT A CHANGE OF 7.0 μmol/liter HAS TAKEN
PLACE.

Figure 5. The distributions of day-to-day change in serum iron value under two
alternative hypotheses.

limit (7.0 μmol/liter) and be declared significant. In the remaining 50% of
the cases, however, the additional change due to random factors will be
negative, implying that the net change will be declared insignificant since it
is less than 7.0 μmol/liter. In other words there is a 50% chance that a
change of 7.0 μmol/liter caused by a pathological nonrandom factor will
be ignored if one pays attention to significant overall changes. If the
change induced by a pathological factor is less than 7.0 μmol/liter, the
chances are even greater than it will pass unnoticed. Thus in addition to
the predicted value and the significance of the observed deviation from
this value the clinician should also be given information pertaining to the
hypothesis alternative to the one stating that only random factors are
responsible for the observed change. However, in order to do this
intelligently it is necessary that the clinician specify his problem in
quantitative terms.*

_____

* In fact the *a priori* probability that hypothesis 1 is true should also be known and the
risks involved in making wrong decisions should be assigned relative weights in order to
truly quantitate the whole decision procedure. It is difficult to assess, though, how
rigorously one should really adhere to this scheme.

Thus we recommend that such statistical methods not be implemented without some reservations. The result may very well be that too many absurd, albeit significant, changes are encountered. Before the methods are used, two conditions should at least be fulfilled, in our opinion: a well-defined alternative hypothesis should be formulated by the clinician, and the preparation of the patient and the specimen-collection technique should both be under control. A difficulty, at least in case of the random-walk model and the intermediate model, is that it is necessary that the time intervals between consecutive measurements be constant. Furthermore, in case of the intermediate model a considerable number of observations is necessary because ρ has to be estimated in addition to the usual parameters. The models apply to the clinical situation in the sense that the disturbances of the physiological mechanisms which influence the quantity values in blood are not controlled but occur more or less at random. However, in many cases we do have some knowledge about the response to various types of disturbances and this knowledge should be built into the models in some way. Ideally a physiologic model should be formulated. An example is the model stated by Winkel and co-workers.[14] They related the changes in plasma progesterone values during pregnancy to changes in the growth rate of placenta and suggested that this model be used for the prediction of spontaneous abortions.

5. Estimates of the Day-to-Day Variance in Healthy Subjects

For all studies of biologic day-to-day variation in healthy subjects the homeostatic model has been assumed to apply and the average of the variances, each computed for each subject relative to his estimated set point, has been computed. For many constituents the homeostatic model may be valid as a crude approximation and in these cases such data may be valuable when one wants to predict the value of future observations based on only a few or perhaps only one previous observation. Even when the homeostatic model is not entirely valid, such data may be useful since it is often impossible to obtain measurements equally spaced in time, which is necessary in order to apply the random-walk or the intermediate model. For the following we will review the literature on the day-to-day biologic variation in healthy subjects for the electrolytes, metabolites, iron, lipids, enzymes, proteins, and hematological quantities.

5.1. Electrolytes

Table 2 compares various studies (15–19) of the day-to-day variation of several serum constituents including a number of electrolytes. It is

Table 2. Comparison of Five Studies, Each Estimating the Mean Physiological Day-to-Day Intraindividual Variation (Expressed as 100 × Coefficient of Variation)

Constituent	Winkel et al[15] (100 × CV)[a]	Winkel et al[16] (100 × CV)[b]	Williams et al[17] (100 × CV)[c]	Young et al[18] (100 × CV)[d]	Glenn et al[19] (100 × CV)[e]
Sodium	0.7	0.5	<0	1.4	2.8
Potassium	4.3	3.2	5.0	4.6	4.1
Calcium	1.7	1.1	1.7	1.6	—
Magnesium	—	—	1.3	2.3	—
Chloride	2.1	1.3	1.4	2.1	2.3
Phosphate	5.8	6.2	7.5	9.6	—
Urea	12.3	14.2	11.9	13.6	—
Creatinine	4.3	4.2	—	4.4	—
Uric acid	7.3	6.0	10.1	8.5	—
Glucose	—	—	5.6	6.5	—
Bilirubin	22.0	—	—	—	—

[a] Eleven healthy subjects, aged 21–27 years, male; five samples every third day for two weeks, all samples stored at −20°C and run on one occasion.
[b] Same subjects as in Study A; five samples every fifth week for a total of four months, all samples run the day after venipuncture (the within-day variations eliminated by use of three-way ANOVA).
[c] Sixty-eight subjects, aged 20–59 years, male; weekly samples over 10–12 weeks, analyzed on several different runs.
[d] Nine healthy subjects, aged 27–34 years, male; weekly samples over 10 weeks, all samples stored at −20°C and run on one occasion.
[e] Twenty healthy subjects, aged 20–30 years, male; 12 samples over 4 weeks, all samples stored and assayed within 4 days.

noted that for the electrolytes as well as for the majority of other constituents the day-to-day variation appears to be independent of the length of the time interval between venipunctures and of differences among demographic characteristics of the subjects studied.

Ionized serum calcium may be more tightly controlled than total serum calcium. Pedersen[20] compared the day-to-day variation of total serum calcium with that of ionized serum calcium in healthy volunteers subjected to physical exercise prior to the blood drawing. While the day-to-day standard deviation for total serum calcium was 0.041 mmol/liter, it was only 0.034 mmol/liter in the case of ionized calcium. The coefficient of variation of the day-to-day variation for total calcium in this study would be 1.6%; this compares favorably with the studies of Winkel et al.,[15,16] Williams et al.,[17] and Young et al.[18] (see Table 2). Frank and Carr[21] studied the month-to-month variation over one year of serum sodium, potassium, chloride, calcium, magnesium, bicarbonate, and phosphate in healthy subjects. They found a systematic month-to-month variation for all constituents except potassium and chloride in males and potassium in females. The results are difficult to interpret in that (1) the biological month-to-month variation was confounded with the long-term analytical variation and (2) the inter- and intraindividual variation were not separated in the analysis of variance, making the statistical test unduly crude. In a four-month experiment Winkel et al.[16] were not able to demonstrate a systematic month-to-month variation for the electrolytes sodium, chloride, calcium, and phosphate when the significant long-term analytical variation was separated from the biological month-to-month variation. However, in case of potassium, the month-to-month variation was statistically significant. The latter month-to-month variation may be related to season in that a positive correlation between serum potassium and temperature has been demonstrated.[22] McLauglin et al.[23] found a significant within-year variation of 25-hydroxycholecalciferol which was related to bright sunshine hours. However, this variation was not accompanied by a general within-year variation of plasma calcium or plasma phosphate, confirming the findings of Winkel et al.[16]

5.2. Metabolites

The day-to-day physiological variations for serum urea, creatinine, uric acid, glucose, and bilirubin as studied by various investigators are reported in Table 2. While the coefficient of variation (CV) for creatinine is approximately 4% in all cases, the CVs for urea (11.9% to 14.2%) and uric acid (6.0% to 10.1%) are much higher than for creatinine, probably reflecting dietary influences. Although the subjects were in a fasting state at the time of venipuncture, they were allowed freedom in choosing the dietary composition of lunches, dinners, and snacks during the study period.

In a study lasting from February to May, Winkel et al.[16] could not demonstrate any effect of time of year on the concentration values of the metabolites creatinine, urea, and uric acid. A study by Milby et al.[24] indicates that at least in newborns season may effect the serum bilirubin. They observed that in their own institution an excess of cases of neonatal jaundice above 100 mg/liter indirect-reacting serum bilirubin occurred during the fourth quarter of each of 4 years reviewed, and in another hospital 2 of the 3 years reviewed also indicated the same phenomenon.[24]

5.3. Iron

A substantial day-to-day variation of serum iron in healthy subjects has been reported by several workers.[13,25-27] Hoyer[26] found that the variations from day to day were of the same order of magnitude as the variations from week to week. His reported coefficients of variation are quite consistent and range from 22.5% to 26.6% with no statistically significant difference between the males and the females studied.

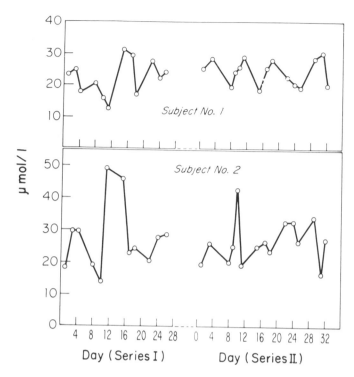

Figure 6. Daily serum iron concentrations for two healthy young women (1 and 2) in two series. (From B. E. Statland, and P. Winkel, *Amer. J. Clin. Pathol.*, *67*, 84, 1977, with permission.)

Figure 6, which is taken from the study of Statland et al.,[27] presents some typical results of serum iron concentration values over the day-to-day study for two of the female volunteers. Statland et al. noted that the magnitude of the physiological day-to-day CV for serum iron binding capacity was 8.8%, i.e., much less than that found for the CV of serum iron values (26.6%).

The mechanisms for the day-to-day as well as for the within-day variations of serum iron concentrations have been studied using computer-simulated ferrokinetic models.[28,28] However, the regulation of the shifts of iron from one compartment to another is not completely understood. It should be noted that the total amount of iron in the plasma compartment of the blood represents less than 0.1% of the total body iron.[27]

5.4. Lipids

Munkner[30] reported a range of 10–30% CV for day-to-day variation in the concentration values of free fatty acids in plasma of healthy fasting subjects. Hammond et al.[31] found a mean day-to-day coefficient of variation of 4.8% for cholesterol and 25.0% for triglycerides. Warnick and Albers[32] studied the week-to-week variation of triglycerides in fasting healthy subjects. They found a coefficient of variation of 18%. Hollister et al.[33] determined cholesterol and triglyceride concentration values every third day for a total of 12 venipunctures in each of 28 healthy males aged 36–75 years. They noted that the mean CV for cholesterol was 6.6% (range 2.8–14.9%) and for triglycerides was 15.9% (range 7.1–34.8%). They found that in the case of triglycerides, the CV was directly related to the mean level. They suggest that the *maximum ranges* of spontaneous short-term variation be used as criteria for a significant therapeutic (hypolipidemic drug) effect; that is, 15% reduction in the case of serum cholesterol levels and 35% in the case of serum triglycerides.

Leonhardt and co-workers[34] compared the short-term (day-to-day) and long-term (month-to-month and year-to-year) variation of triglyceride and cholesterol concentrations in serum in healthy subjects and found that the physiological variation increases as the time interval between venipuncture increases. Winkel et al.[16] studied the month-to-month variation of serum cholesterol and total lipids in the sera of healthy subjects. The study lasted from February to May. They observed a month-to-month variation for total lipids common for the whole group with a minimum in the late winter months. Fuller et al.[35] found a significant fall from summer to winter in the case of serum triglycerides. In neither study could significant month-to-month changes in serum cholesterol be demonstrated. However, Thomas et al.[36] did find higher serum cholesterol values in the winter than in the summer in young male prisoners. Their experimental design did not allow a distinction between long-term analytical variation and biological month-to-month variation; however, their findings were confirmed by Winkelman et al.[37]

5.5. Enzymes

Statland et al.[3] studied the day-to-day variation of the activity values of alkaline phosphatase (AP), gamma-glutamyl transferase (GGT), lactate dehydrogenase (LDH), aspartate aminotransferase (AST), alanine aminotransferase (ALT), and creatine kinase (CPK) in 14 healthy subjects. To overcome the long-term analytical variation, they collected all the sera obtained from duplicate specimens over the study (14 volunteers × 6 days

Table 3. Physiologic Day-to-Day Variation, Biologic Variation in Mean Values, and
Ratios of Intraindividual over Interindividual Variations for Enzyme Activity Values
in Serum of Healthy Subjects

Constituent	Physiologic day-to-day variation (100 × CV)	Biologic intersubject variation (100 × CV)	Ratio values (intra/inter)
Alkaline phosphatase	3.5	24.8	0.14
γ-Glutamyl transpeptidase	3.9	23.8	0.16
Lactate dehydrogenase	5.5	12.6	0.44
Aspartate aminotransferase	10.9	12.1	0.90
Alanine aminotransferase	13.2	29.6	0.45
Creatine kinase	25.7	46.0	0.56

× 2 replicates), and stored the sera at −80°C until the actual assay, and on
one occasion all the specimens from each of the volunteers (12 specimens)
were assayed in one disk run of a fast centrifugal analyzer. Table 3
presents the results of the study, i.e., the mean day-to-day CV × 100 for
each of the six enzymes. As noted in Table 3, the day-to-day variations of
the enzymes found in muscle (AST, LDH, and CPK) are much greater
than the variations of the enzymes AP and GGT. Figure 7 presents the
activity values of AP for each of four subjects for each of the 6 days of
venipuncture. The remarkable constancy of the values over the time
interval studied is very obvious. Wroblewski and LaDue[38] studied the
day-to-day variation of LDH activity values in healthy subjects. They noted
a mean daily variation of 30%. Winkelman et al.[37] reported a 20% higher
mean LDH value in the summer months as compared to the winter.

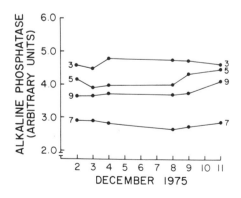

Figure 7. Alkaline phosphatase activity
in serum of each of four healthy subjects on each of six days.

Table 4. Physiologic Day-to-Day Variation, Biologic Variation in Mean Values, and Ratios of Intraindividual over Interindividual Variations for Concentration Values of Specific Proteins in Serum of Healthy Subjects

Constituent	Physiologic day-to-day variation (100 × CV)	Biologic intersubject variation (100 × CV)	Ratio values (intra/inter)
Total protein	2.9	2.8	1.04
Albumin	2.8	0.5	5.60
Transferrin	2.5	9.5	0.26
α_1-Antitrypsin	2.9	15.7	0.19
α_2-Macroglobulin	3.1	16.6	0.19
IgG	2.7	18.1	0.15
IgA	3.5	41.1	0.09
IgM	3.1	54.0	0.06
Complement C3	3.8	19.7	0.19
Complement C4	5.9	35.6	0.17
Haptoglobin	8.8	70.5	0.12
Orosomucoid	11.1	43.1	0.26

5.6. Proteins

Statland et al.[2] studied the intraindividual variations of concentration values of 10 specific proteins in sera of 14 healthy subjects who were subjected to venipuncture at 8:00 A.M. on 6 separate mornings over a 10-day interval. Table 4 presents the mean physiological day-to-day intraindividual CV × 100 for each of the proteins. Figure 8, which presents the actual values of concentration of IgM for 9 of the subjects, illustrates how small the intraindividual variation is relative to the interindividual variation for this constituent.

5.7. Leukocytes and Other Hematological Quantities

Statland et al.[39,40] used a very precise automated leukocyte analyzer, Hemalog D™ (Technicon Instruments Corp., Tarrytown, N.Y. 10591) to assay for leukocyte types. They determined the mean day-to-day (daily over 5 days), and the mean week-to-week (weekly over 5 weeks) physiological intraindividual variation of concentration values of leukocyte cell types, platelet count, hematocrit, and hemoglobin in a group of 20 healthy adult volunteers. The mean physiological day-to-day and week-to-week intraindividual variations in terms of 100 × coefficient of variation include: hemoglobin, 2.6, 2.2; hematocrit, 2.2, 2.7; platelets, 3.6, 6.6; total

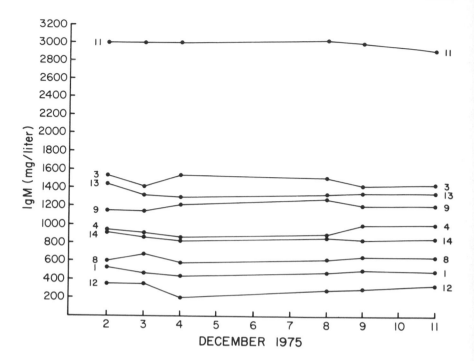

Figure 8. IgM concentration in serum of each of nine healthy subjects on each of six days. (From B. E. Statland, P. Winkel, and L. M. Killingsworth, *Clin. Chem., 22,* 1635, 1976, with permission.)

WBC, 15.5, 15.7; neutrophils, 23.2, 26.0; lymphocytes, 9.7, 15.0; large unstained cells, 10.9, 16.0; and high-peroxidase cells, 49.4, 25.7.

Figures 9–13 presents the subject-specific ranges for each of 20 healthy volunteers for concentration values of neutrophils, lymphocytes, monocytes, eosinophils, and basophils, respectively as determined on ten occasions over a 29-day time interval.

Various investigators[41–43] have reported a relationship between leukocyte concentration values and the phases of the menstrual cycle. Cyclic variations in the daily excretion of the adrenocorticoids during the phases of the menstrual cycle have been implicated as causing the fluctuations seen day-to-day in the eosinophil counts.[42,43] Pepper and Lindsay[43] followed the eosinophil counts and the 17-hydroxycorticosteroid excretion over twelve complete menstrual cycles. Although they found that menstrual eosinopenia occurred in nine cycles (75% of the cases), the lowest eosinophil values of the entire cycle were usually found in the middle of the cycle. In eleven of the twelve subjects, the steroid peaks occurred during or shortly before the day of eosinopenia.

Rorsman[44] studied the day-to-day fluctuations of the absolute basophil count in a group of healthy subjects over a short time period (three
days) and monthly fluctuations over an eight-month period. A total of
2000 leukocytes were counted per specimen to obtain the fraction number
of basophils which was then multiplied by the total leukocyte count. The
mean physiological coefficient of variation for the group of subjects
examined was 14% over the three-day period and 24% during the eight-
month period. For comparison it may be noted that Statland *et al.*[40]
found a 9.8% CV for the day-to-day and 15.0% CV for the week-to-week
physiological variation of basophils.

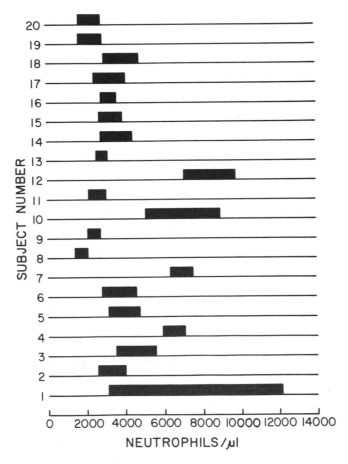

Figure 9. Range of concentration values of neutrophils in blood of 20 healthy subjects
followed over 29 days. (From B. E. Statland, P. Winkel, S. C. Harris, M. S. Burdsall, and
A. M. Saunders, *Advances in Automated Analysis*, Technicon International Congress,
Mediad Incorporated, Tarrytown, N.Y., December, 1976, with permission.)

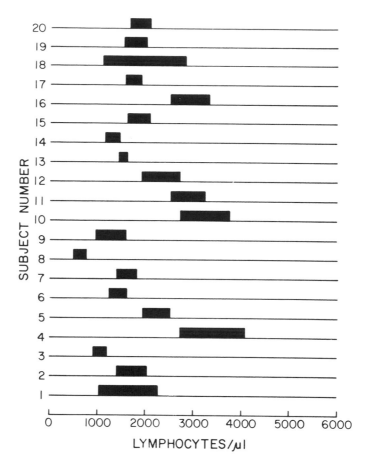

Figure 10. Range of concentration values of lymphocytes in blood of 20 healthy subjects followed over 29 days. (From B. E. Statland, P. Winkel, S. C. Harris, M. S. Burdsall, and A. M. Saunders, *Advances in Automated Analysis,* Technicon International Congress, Mediad Incorporated, Tarrytown, N.Y., December, 1976, with permission.)

6. Contribution of Analytical Variation to the Intraindividual Day-to-Day Variation

In the clinical setting the clinical chemist must contend with the variation due to the analytical procedure, which will tend to blur the variation due to physiological causes. This analytical variation can be separated into two components, one due to preinstrumental sources and one due to instrumental sources. The former sources include in case of an assay performed on serum, the specimen collection, the transportation of the whole blood, the centrifugation procedure, the decanting of the serum and perhaps the freezing and thawing of the serum, and the storage of the specimen before the actual assay begins. The instrumental variation is of two types: within-batch and batch-to-batch. The within-batch variation includes the variation occurring within a batch run and thus includes the variation due to uncertainty in the temperature control, spectrophotometric reading, and fluid dispensing. The batch-to-batch variation of the

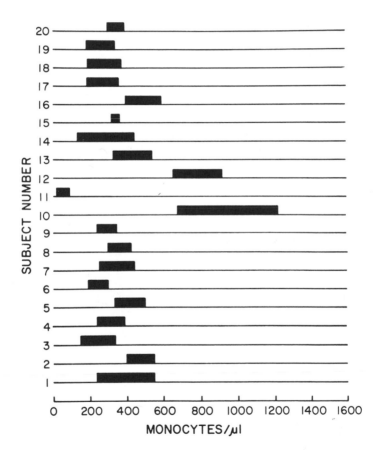

Figure 11. Range of concentration values of monocytes in blood of 20 healthy subjects followed over 29 days. (From B. E. Statland, P. Winkel, S. C. Harris, M. S. Burdsall, and A. M. Saunders, *Advances in Automated Analysis,* Technicon International Congress, Mediad Incorporated, Tarrytown, N.Y., December, 1976, with permission.)

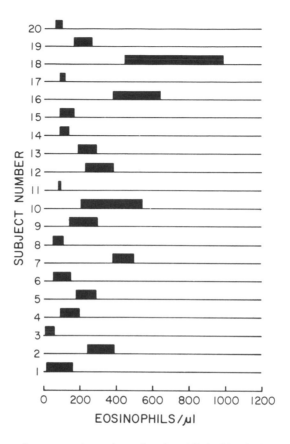

Figure 12. Range of concentration values of eosinophils in blood of 20 healthy subjects followed over 29 days. (From B. E. Statland, P. Winkel, S. C. Harris, M. S. Burdsall, and A. M. Saunders, *Advances in Automated Analysis*, Technicon International Congress, Mediad Incorporated, Tarrytown, N.Y., December, 1976, with permission.)

instrumental assay involves the variation in making up the reagents from day-to-day, the variation due to lack of stability of the reagents and that induced by the calibration procedure. Today most routine laboratories rely on control materials to estimate and monitor the batch-to-batch variation. Bokelund et al.[45] have recommended the use of randomized paired blood duplicates to estimate the within-batch analytical variation and also to identify the aberrant result which otherwise may not be picked up by an occasional control specimen. Thus, before the clinical chemist can deal with the issue of the contribution of analytical variation, he must be able to estimate it with some certainty. Assuming that one can make a reasonable estimate of the analytical variation, what is the effect of various magnitudes of analytical variation on known biological variation? Figure 14 compares the increase (× 100) in intraindividual variation (standard deviation) due to analytical variation as a function of the ratio of the analytical variance over the biological variance where the biological variance is equal to the physiological day-to-day intraindividual variance. For example, when the ratio of the two variances is 0.25 (a in Figure 14), the increase is 11.8%. As noted in the figure, when the ratio reaches 8.0 (e in the figure), the increase is 200%. The computations are explained in the legend to Figure 14.

What should be the maximum allowable analytical variation for measuring any analyte? Cotlove et al.[46] suggest that the maximum

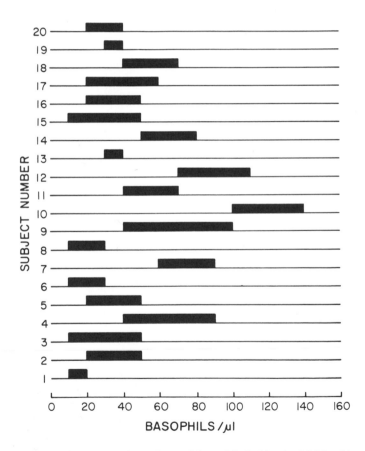

Figure 13. Range of concentration values of basophils in blood of 20 healthy subjects followed over 29 days. (From B. E. Statland, P. Winkel, S. C. Harris, M. S. Burdsall, and A. M. Saunders, *Advances in Automated Analysis,* Technicon International Congress, Mediad Incorporated, Tarrytown, N.Y., December, 1976, with permission.)

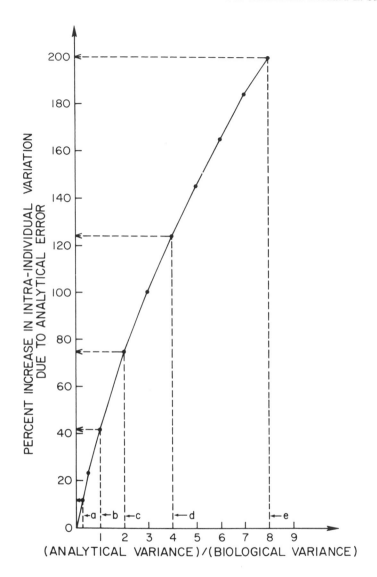

Figure 14. 100 × Increase in intraindividual variation due to analytical error as a function of the ratio between analytical variance and biological intraindividual variance. The letters indicate 100 × increase for various ratios of analytical and biological variance. (a) the ratio is 0.25, which corresponds to an increase of 11.8%; (b) the ratio is 1.0, which corresponds to an increase of 42%; (c) the ratio is 2.0, which corresponds to an increase of 76%; (d) the ratio is 4.0, which corresponds to an increase of 126%; and (e) the ratio is 8.0 which corresponds to an increase of 200%. The increase in total intraindividual variation is defined as $I = \dfrac{\sqrt{\sigma_B^2 + \sigma_A^2} - \sigma_B}{\sigma_B}$ where σ_B^2 is the biological variance and σ_A^2 is the analytical variance. If we denote σ_A^2/σ_B^2 by x, I can be expressed as a function of x, i.e., $I(x) = \sqrt{x + 1} - 1$.

allowable analytical variance be equal to $\frac{1}{4}$ the biological variance. As noted above, when the ratio of the analytical variance over the biological variance equals 0.25 the increase of the intraindividual variation is 11.8% (Figure 14). If we want to use a subject as his own referent and we assume a homeostatic model for the biological time-series, then the biological variance is equal to the physiological day-to-day intraindividual variance. For illustrative purposes we will examine the increase in intraindividual variations for various magnitudes of analytical variations and determine these values for the enzyme activities as examples. The physiological day-to-day variations for AP, GGT, LDH, AST, ALT, and CPK are found in Table 3. We will assume four different degrees of analytical variation: 2%, 5%, 12%, and 30%, all in terms of coefficient of variation. Figure 15 presents the calculated increase in variation. The dashed line in the figure represents the maximum allowable increase according to Cotlove *et al.* For

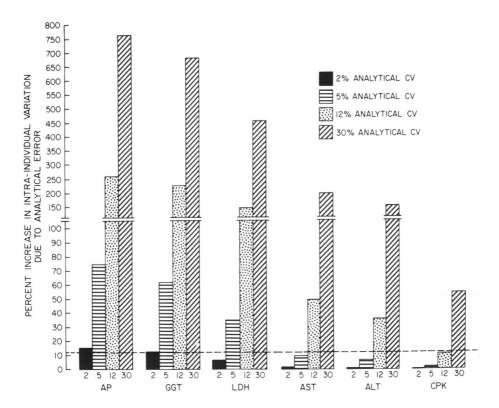

Figure 15. 100 × Increase in intraindividual variation due to analytical error for alkaline phosphatase (AP), gamma-glutamyl transpeptidase (GGT), lactate dehydrogenase (LDH), aspartate aminotransferase (AST), alanine aminotransferase (ALT), and creatine phosphokinase (CPK) as a function of analytical coefficient of variation.

AP and GGT the physiological intraindividual variation is very small: thus, an analytical CV of 12% would cause more than a 200% increase in the intraindividual variation. For creatine kinase and for the aminotransferases, a 12% analytical CV would result in less than a 50% increase in the intraindividual variation. Using Cotlove's guidelines for the allowable analytical variation, we must recommend maximum analytical CV as follows: AP, 2%; GGT, 2%; LDH, 3%; AST, 5%; ALT, 6%; and CPK, 13%.

7. Conclusion

We have observed that for many analytes commonly measured in blood or serum the value of the physiological day-to-day variation is much smaller than the value of the interindividual variation of mean values as measured in groups of healthy subjects. This observation supports the use of subject-specific reference intervals rather than group-specific reference intervals as being more sensitive in detecting a deviation from "normalcy" in the healthy subject. The use of subject-specific reference intervals presupposes the choice of an appropriate biological model, compatible analytical and procedural variations in the establishment of the interval, and maintaining consistency in future analyses. We have assumed a homeostatic model in reviewing the literature on the physiological day-to-day variation of various analytes in the healthy subject. Finally, we have examined the contribution of the analytical variation to the intraindividual day-to-day variation.

8. References

1. E. K. Harris, Effects of intra- and inter-individual variation on the appropriate use of normal ranges, *Clin. Chem., 20,* 1535 (1974).
2. B. E. Statland, P. Winkel, and L. M. Killingsworth, Factors contributing to intraindividual variation of serum constituents: 6. Physiologic day-to-day variation of concentration values of ten specific proteins in the sera of healthy subjects, *Clin. Chem., 22,* 1635 (1976).
3. B. E. Statland, P. Winkel, and R. E. Cross, Physiologic variation of enzyme activities in serum, presented at Seventh International Symposium on Clinical Enzymology, Venice, Italy, April 6 (1976).
4. B. E. Statland, P. Winkel, and H. Bokelund, Factors contributing to intra-individual variation of serum constituents: 2. Effects of exercise and diet on variation of serum constituents in healthy subjects. *Clin. Chem., 19,* 1380 (1973).
5. S. King, B. E. Statland, and J. Savory, The effects of a short burst of exercise on activity values of enzymes in sera of healthy subjects, *Clin. Chim. Acta., 72,* 211 (1976).

6. D. Freer and B. E. Statland, The effects of moderate ethanol consumption on the activity values of enzymes in serum of healthy subjects: 1. Short-term effects, *Clin. Chem.*, in press (1977).

7. D. Freer and B. E. Statland, The effects of moderate ethanol consumption on the activity values of serum of healthy subjects: 2. Intermediate-term effects, *Clin. Chem.* in press (1977).

8. B. E. Statland, H. Bokelund, and P. Winkel, Factors contributing to intra-individual variation of serum constituents: 4. Effects of posture and tourniquet application on variation of serum constituents in healthy subjects, *Clin. Chem.*, 20, 1513, 1974.

9. D. S. Young, L. C. Pestaner, and V. Gibberman, Effects of drugs on clinical laboratory tests, *Clin. Chem.*, 21, ID, 1975.

10. B. E. Statland and P. Winkel, Effects of non-analytical factors on the intra-individual variation of analytes in the blood of healthy subjects: Consideration of preparation of the subject and time of venipuncture, *CRC Reviews in Clinical Laboratory Science*, in press (1977).

11. E. K. Harris, Some theory of reference values. I. Stratified normal ranges and a method for following an individual's clinical laboratory values, *Clin. Chem.*, 21, 1457 (1975).

12. E. K. Harris, Some theory of reference values: II. Comparison of some statistical models of intra-individual variation in blood constituents, *Clin. Chem.*, 22, 1343 (1976).

13. B. E. Statland, P. Winkel, and H. Bokelund, Variations of serum iron concentration in young healthy men: Within-day and day-to-day changes, *Clin. Biochem.*, 9, 26 (1974).

14. P. Winkel, P. Gaede, and J. Lyngbye, Method for monitoring plasma progesterone concentrations in pregnancy, *Clin. Chem.*, 22, 422 (1976).

15. P. Winkel, H. Bokelund, and B. E. Statland, Factors contributing to intra-individual variation of serum constituents: 5. Short-term day-to-day and within-hour variation of serum constituents in healthy subjects, *Clin. Chem.*, 20, 1520 (1974).

16. P. Winkel, B. E. Statland, and H. Bokelund, The effects of time of venipuncture on variation of serum constituents: Consideration of within-day and day-to-day changes in a group of healthy young men, *Am. J. Clin. Path.*, 64, 433 (1975).

17. G. Z. Williams, D. S. Young, M. R. Stein, and E. Cotlove, Biological and analytic components of variation in long-term studies of serum constituents in normal subjects: 1. Objectives, subject selection laboratory procedures, and estimation of analytic deviation, *Clin. Chem.*, 16, 1016 (1970).

18. D. S. Young, E. K. Harris, and E. Cotlove, Biological and analytic components of variation in long-term studies of serum constituents in normal subjects: 4. Results of a study design to eliminate long-term analytic deviations, *Clin. Chem.*, 17, 403 (1971).

19. W. G. Glenn and I. L. Shannon, Normal human serum parameters for simulated altitude and aerospace flights. III. Estimation of change in serum potassium, sodium, and chloride. *Aerosp. Med.*, 37, 1008 (1966).

20. K. O. Pedersen, On the cause and degree of intra-individual serum calcium variability, *Scand. J. Clin. Lab. Invest.*, 30, 191 (1972).

21. H. A. Frank and M. H. Carr, "Normal" serum electrolytes with a note on seasonal and menstrual variation, *J. Lab. Clin. Med.*, 49, 246 (1957).

22. J. G. Henrotte and P. S. Krishnaraj, Climatic variation of plasma potassium in men, *Nature*, 195, 184 (1962).

23. M. McLaughlin, P. R. Raggatt, A. Fairney, D. J. Brown, E. Lester, and M. R. Wills, Seasonal variations in serum 25-hydroxycholecalciferol in healthy people, *Lancet*, 1(857), 536 (1974).

24. T. H. Milby, J. E. Mitchell, and T. S. Freeman, Seasonal neonatal hyperbilirubine-mia, *Pediatrics*, *43*, 601 (1969).

25. E. J. W. Bowie, W. N. Tauxe, W. E. Sjoberg, and M. Y. Yamaguchi, Daily variation in the concentration of iron in serum, *Am. J. Clin. Pathol.*, *40*, 491 (1963).

26. K. Hoyer, Physiologic variations in the iron content of human blood serum, *Acta. Med. Scand.*, *119*, 562 (1944).

27. B. E. Statland and P. Winkel, The relationship of the day-to-day variation of serum iron concentration values to the iron binding capacity values in a group of healthy young women, *Am. J. Clin. Pathol.*, *67*, 84 (1977).

28. L. Garby, W. Schneider, O. Sundquist, and J. C. Vuille, A ferro-erythrokinetic model and its properties, *Acta. Physiol. Scand.*, *49*, Suppl 216, 1 (1963).

29. J. C. Vuille, Computer simulation of ferrokinetic models, *Acta. Physiol. Scand.*, *65*, Suppl 253, 1 (1965).

30. C. Munkner, Fasting concentrations of non-esterified fatty acids in diabetic and non-diabetic plasma and diurnal variations in normal subjects, *Scand. J. Clin. Lab. Invest.*, *11*, 388 (1959).

31. J. Hammond, P. Wentz, B. E. Statland, J. C. Phillips, and P. Winkel, Daily variation of lipids and hormones in sera of healthy subjects, *Clin. Chim. Acta.*, *73*, 347 (1976).

32. G. R. Warnick, J. J. Albers, Physiological and analytical variation in cholesterol and triglycerides, *Lipids*, *11*, 203 (1975).

33. L. E. Hollister, W. G. Beckman, and M. Baker, Comparative variability of serum cholesterol and serum triglycerides, *Am. J. Med. Sci.*, *248*, 329 (1964).

34. W. Leonhardt *et al.*, Long-term and short-term observations of the intra-individual variations of lipids in serum, Second European Congress of Clinical Chemistry, Prague, October, 1976.

35. J. H. Fuller, S. L. Grainer, R. J. Jarrett, and H. Keen, Possible seasonal variation of plasma lipids in a healthy population, *Clin. Chem. Acta.*, *52*, 305 (1974).

36. C. B. Thomas, H. W. D. Holljes, and F. F. Eisenberg, Observations on seasonal variations in total serum cholesterol level among healthy young prisoners, *Ann. Intern. Med.*, *54*, 413 (1961).

37. J. W. Winkelman, D. C. Cannon, V. J. Pilleggi, and A. H. Reed, Estimation of norms from a controlled sample survey. II. Influence of body habitus, oral contraceptives, and other factors on values for the normal range derived from the SMA 12/60 screening group of tests, *Clin. Chem.*, *19*, 488 (1973).

38. F. Wroblewski and J. S. LaDue, Lactic dehydrogenase activity in blood, *Proc. Soc. Exp. Biol. Med.*, *90*, 210 (1955).

39. B. E. Statland, P. Winkel, S. C. Harris, M. J. Burdsall, and A. M. Saunders, A study of variation of concentration values of leukocyte types: 1. Biological components of variation in healthy subjects. *Advances in Automated Analysis*, Technicon International Congress, Mediad incorporated, Tarrytown, N.Y., December, 1976.

40. B. E. Statland, P. Winkel, S. C. Harris, M. J. Burdsall, and A. M. Saunders, Evaluation of biological sources of variation of leukocyte counts and other hemato-logical quantities using very precise automated analyzers, *Am. J. Clin. Pathol.*, in press (1977).

41. B. J. Bain and J. M. England, Variations in leukocyte count during menstrual cycle, *Br. Med. J.*, *2*, 473 (1975).

42. C. L. Pathak and B. S. Kahali, Cyclic variations in the eosinophil count during the phases of the menstrual cycle, *J. Clin. Endocrinol.*, *17*, 862 (1957).

43. H. Pepper and S. Lindsay, Levels of eosinophils, platelets, leukocytes and 17-hydroxycorticosteroids during normal menstrual cycle, *Proc. Soc. Exper. Biol. Med.*, *104*, 145 (1960).

44. H. Rorsman, Normal variation in the count of circulating basophil leukocytes in man, *Acta Allergol. 17*, 49 (1962).
45. H. Bokelund, P. Winkel, and B. E. Statland, Factors contributing to intra-individual variation of serum constituents: 3. Use of randomized duplicate serum specimens to evaluate sources of analytical errors, *Clin. Chem., 20*, 1507 (1974).
46. E. Cotlove, E. K. Harris, and G. Z. Williams, Biological and analytical components of variation in long-term studies of serum constituents in normal subjects: 3. Physiological and medical implications, *Clin. Chem., 16*, 1028 (1970).
47. R. Dybkaer, (chairman), Commission on Quantities and Units, Section on Clinical Chemistry, IUPAC, and Expert Panel on Quantities and Units, Committee on Standards, IFCC: Quantities and units in clinical chemistry. Recommendation 1973, *Pure and Appl. Chem., 37*, 517–572, (1974).
48. R. Dybkaer, (chairman), Commission on Quantities and Units, Section on Clinical Chemistry, IUPAC, and Expert Panel on Quantities and Units, Committee on Standards, IFCC: List of quantities in clinical chemistry, *Pure and Appl. Chem., 37*, 547–572, (1974).

9. Note Added in Proof

The Commission on Quantities and Units in Clinical Chemistry (International Union of Pure and Applied Chemistry) and the Expert Panel of Quantities and Units (International Federation of Clinical Chemistry) have recommended a careful distinction of the terms "quantity," "numerical value," and "unit." A quantity is a number × unit. Thus, if one desires to follow the recommendations of the joint IUPAC–IFCC group, one should label the axis in the figures referring to numerical values as quantity/unit, e.g., glucose/(mmol/liter). We suggest following this principle; unfortunately, many of the figures used in this review were reprinted from previous publications where the nomenclature was not as rigidly controlled. We regret this; however, we trust the reader will understand the meaning intended by all figures. We have attempted to correct all ambiguities in the text proper to be consistent with the recommendations concerning nomenclature. In addition, the joint committee has recommended certain base units for "substance concentration" and for "number concentration." Again we found it difficult to adhere in all cases in that the information and figures used in this review were based on previous work where the authors may not have always used the SI (Standardized International) Unit. The reader is referred to references 47 and 48 for a clear definition of the terms "quantity," "number," and "units" as recommended by the panel and also for a list of recommended base units for clinical chemistry quantities.

Index